Advanced Textbooks in Control and Signal Processing

Series Editors

Professor Michael J. Grimble, Professor of Industrial Systems and Director
Professor Michael A. Johnson, Professor Emeritus of Control Systems and Deputy Director

Industrial Control Centre, Department of Electronic and Electrical Engineering,
University of Strathclyde, Graham Hills Building, 50 George Street, Glasgow G1 1QE, UK

W.H. Kwon and S. Han

Receding Horizon Control

Model Predictive Control for State Models

With 51 Figures

Wook Hyun Kwon, PhD
Soohee Han, PhD
School of Electrical Engineering and Computer Science
Seoul National University
San 56-1 Shinlim-dong Gwanak-ku
Seoul 151-742
Korea

British Library Cataloguing in Publication Data
Kwon, W. H. (Wook Hyun)
 Receding horizon control: model predictive control for
 state models. - (Advanced textbooks in control and signal
 processing)
 1. Predictive control
 I. Title II. Han, S.
 629.8
ISBN-10: 1846280249

Library of Congress Control Number: 2005927907

Advanced Textbooks in Control and Signal Processing series ISSN 1439-2232
ISBN-10: 1-84628-024-9   e-ISBN: 1-84628-017-6   Printed on acid-free paper
ISBN-13: 978-1-84628-024-5

Typesetting: Camera ready by authors
Production: LE-TEX Jelonek, Schmidt & Vöckler GbR, Leipzig, Germany
Printed in Germany

9 8 7 6 5 4 3 2 1

Springer Science+Business Media
springeronline.com

Dedicated to Allan and Myrna Pearson

Series Editors' Foreword

The topics of control engineering and signal processing continue to flourish and develop. In common with general scientific investigation, new ideas, concepts and interpretations emerge quite spontaneously and these are then discussed, used, discarded or subsumed into the prevailing subject paradigm. Sometimes these innovative concepts coalesce into a new sub-discipline within the broad subject tapestry of control and signal processing. This preliminary battle between old and new usually takes place at conferences, through the Internet and in the journals of the discipline. After a little more maturity has been acquired by the new concepts then archival publication as a scientific or engineering monograph may occur.

A new concept in control and signal processing is known to have arrived when sufficient material has evolved for the topic to be taught as a specialised tutorial workshop or as a course to undergraduate, graduate or industrial engineers. *Advanced Textbooks in Control and Signal Processing* are designed as a vehicle for the systematic presentation of course material for both popular and innovative topics in the discipline. It is hoped that prospective authors will welcome the opportunity to publish a structured and systematic presentation of some of the newer emerging control and signal processing technologies in the textbook series.

In the 1970s many new control ideas were emerging. The optimal control paradigm initiated by the seminal papers of Kalman was growing into maturity For example, in 1971 the *IEEE Transactions on Automatic Control* published the famous special issue on LQG optimal control edited by Athans. The groundswell of ideas included finite-time optimal control solutions, and the concept of the separation principle partitioning solutions into control and estimation; these were influential concepts for later theorists. At the same time, the rapidly advancing power of digital control was being exploited by industry for the control of ever increasing numbers of process systems. Some control schemes of the 1970s were driven by the practical and commercial needs of industrial processes to operate within process regions bounded by constraints. Dynamic matrix control and model algorithmic control were typical software products that used models, cost functions and prediction to handle constrained industrial control problems.

The practical and theoretical experience of the 1970s finally produced generalised predictive control, model predictive control and the clear use of a receding horizon control principle as key methods for industrial control in the 1980s and 1990s. Today, model predictive control and the receding horizon principle is the subject of a small but influential set of theoretical and practical textbooks.

Wook Hyun Kwon's work on the receding horizon principle and model predictive control is grounded in the optimal control research of the late 1970s and has followed and contributed to the development of the field ever since. In this textbook, Professor Kwon and his colleague Soohee Han present their distinctive version of this important control methodology. The book can be used for a course study or for self study. Each chapter has a detailed problems section. MATLAB® codes for various worked examples are given in an appendix so that the reader can try the various control and filter methods. Starred sections alert the reader and the lecturer to material that can be omitted if study time is constrained.

The opening chapter of the book succinctly reviews the concepts of receding horizon control and filters, the output feedback receding control problem and the use of prediction. A neat section on the advantages and disadvantages of the receding horizon control paradigm gives a useful insight into the reasons for the success of these methods. The most difficult task is then to construct a book framework for presenting material that has been evolving and developing since 1970s. The route planned by the authors is direct and effective. First, there is optimal control (Chapter 2). This covers the finite-time and infinite time horizon varieties for general systems, LQ, H-infinity, LQG and H_2 optimal control. The chapter interleaves a section on Kalman and H-infinity filters as needed by the appropriate optimal control solutions. Chapter 3 introduces the receding horizon version of these optimal control formulations. A key feature of Chapter 3 is that state feedback is the active control mechanism. However, before moving to the challenging topic of output feedback receding horizon control, a chapter on receding horizon filtering is presented. The filters described in this Chapter 4 are used in the key results on output feedback receding horizon control. An important staging point of the textbook is Chapter 5 which presents detailed solutions to the important practical problem of output feedback receding horizon control. The exhaustive material of this chapter also includes a global optimization approach to the solution of the output feedback receding horizon control problem where a separation principle between control and estimation is proven. This is a highly satisfying outcome to the route taken so far by the authors. The final two chapters of the book then tackle the addition of constraints to the receding horizon control problem (Chapter 6) and receding horizon control with nonlinear dynamical process models (Chapter 7).

Overall, the textbook gives a well-structured presentation to receding horizon control methods using state-space models. The comprehensiveness of the presentation enables flexibility in selecting sections and topics to support an advanced course. The postgraduate student and the academic researcher will find many topics of interest for further research and contemplation in this fine addition to the series of advanced course textbooks in control and signal processing.

M.J. Grimble and M.A. Johnson
Industrial Control Centre
Glasgow, Scotland, U.K.
March 2005

Preface

This book introduces some essentials of receding horizon control (RHC) that have been emerged as a successful feedback strategy in many industry fields, including process industries in particular. RHC is sometimes called receding-horizon predictive control (RHPC) and is better known as model predictive control (MPC) for state-space models.

RHC is based on the conventional optimal control that is obtained by minimization or mini-maximization of some performance criterion either for a fixed finite horizon or for an infinite horizon. RHC introduces a new concept of *the receding horizon* that is different from a fixed finite horizon and an infinite horizon. The basic concept of RHC is as follows. At the current time, optimal controls, either open loop or closed loop, are obtained on a fixed finite horizon from the current time. Among the optimal controls on the fixed finite horizon, only the first one is implemented as the current control law. The procedure is then repeated at the next time with a fixed finite horizon from the next time. Owing to this unique characteristic, there are several advantages for wide acceptance of RHC, such as closed-loop structure, guaranteed stability, good tracking performance, input/output (I/O) constraint handling, simple computation, and easy extension to nonlinear systems.

Historically, generalized predictive control (GPC) and MPC has been investigated and implemented for industrial applications independently. Originally, RHC dealt with state-space models, while GPC and MPC dealt with I/O models. These three controls are equivalent to one another when the problem formulation is the same. Since the recent problem formulations for the above three predictive controls are based on state-space models, RHC based on state-space models in this book will be useful in understanding the global picture of the predictive controls.

Discrete-time systems are discussed in this book, since they are useful for modern computer applications and they are easier to convey basic concepts compared with continuous systems. Most results in this book for discrete-time systems can be obtained for continuous-time systems. This book starts from simpler systems of linear systems without constraints and then moves to linear systems with constraints and to nonlinear systems for better understanding, although we can begin with nonlinear systems for generality. Both minimization and mini-maximization optimal problems are dealt with for completeness. Also, both state feedback and output feedback controls are dealt with if possible. For output controls, a new type of receding horizon finite impulse response (FIR) filter that utilizes measurement data on the receding filter horizon is introduced, which may be similar, in concept, to RHC. This filter is used as a state estimator and also as a part of the output feedback RHC.

In this book, optimal solutions of RHCs, the stability and the performance of the closed-loop system, and robustness with respect to disturbances are dealt with. Robustness with respect to parameter uncertainties is not covered in this book.

This book is organized as follows. After a brief introduction to concepts and advantages of RHCs in Chapter 1, conventional optimal controls such as dynamic programming and the minimum principle for nonlinear systems and LQ, LQG, and H_∞ controls for linear systems are introduced in Chapter 2, since RHCs are based on these conventional optimal controls. In Chapter 3, state feedback RHCs are investigated in depth. In Chapter 4, state estimators such as receding horizon FIR filters are introduced as alternative filters to conventional filters, which have some connections with RHCs. In Chapter 5, output feedback RHCs will be introduced with a finite memory structure and an unbiased condition. In Chapter 6, RHCs are given for linear systems with input and state constraints. In Chapter 7, RHCs for nonlinear systems are explained. In Chapters 6 and 7, introductory topics are covered in this book. Some fundamental theories necessary for RHC are given in the appendices. Sections denoted by an asterisk can be skipped when a course cannot cover all the materials of the book.

In each chapter, we introduce references that help challenging readers obtain a detailed knowledge of related areas. We tried to include the important literature, but this may not be complete. If we have missed citing some important references, we sincerely apologize for that.

The first author appreciates the constant support of Allan and Myrna Pearson and Thomas Kailath since he was at Brown University as a graduate student and at Stanford University as a visiting professor respectively. We would like to express our appreciation to Ki back Kim, Young Sam Lee, and Young Il Lee who finished their Ph.Ds at our laboratory and provided some input to this book. Also, Choon Ki Ahn, Zhong Hua Quan, Bo Kyu Kwon, Jung Hun Park, and some other graduate students were of great help to us in developing this book.

Seoul, Korea   *Wook Hyun Kwon*
January 2005   *Soohee Han*

Contents

1

Introduction

1.1 Control Systems

In this section, we will briefly discuss important topics for control systems, such as models, control objectives, control structure, and performance critera.

Models

The basic variables of dynamical control systems are input, state, and output variables that consist of controlled outputs and measured outputs, as in Figure 1.1. The input variable is the control variable and the measured output is used for feedback. Usually, output variables are a subset of whole state variables. In dynamical control systems, there can be several undesirable elements, such as disturbances, noises, nonlinear elements, time-delay, uncertainties of dynamics and its parameters, constraints in input and state variables, etc. All or some parts of undesirable elements exist in each system, depending on the system characteristics. A model can be represented as a stochastic system with noises or a deterministic system with disturbances. A model can be a linear or nonlinear system. Usually, dynamic models are described by state-space systems, but sometimes with input and output models.

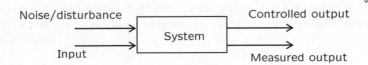

Fig. 1.1. Control systems

Control Objectives

There can be several objectives for control systems. The control is designed such that the controlled output tracks the reference signal under the above undesirable elements. Actually, the closed-loop stability and the tracking performance even under the undesirable elements are known to be important control objectives.

In order to achieve control objectives easily, the real system is separated into a nominal system without model uncertainties and an additional uncertain system representing the above undesirable elements. The control can be designed first for the nominal system and then for the uncertain system. In this case, the control objectives can be divided into simpler intermediate control objectives, such as

- Nominal stability: closed-loop stability for nominal systems.
- Nominal performance: tracking performance for nominal systems.
- Robust stability: closed-loop stability for uncertain systems (robust stability).
- Robust performance: tracking performance for uncertain systems (robust tracking performance).

The first three are considered to be the most important.

Control Structure

If all the states are measured we can use state feedback controls such as in Figure 1.2 (a). However, if only the outputs can be measured, then we have to use output feedback controls such as in Figure 1.2 (b).

The state feedback controls are easier to design than the output feedback controls, since state variables contain all the system information. Static feedback controls are simpler in structure than dynamic feedback controls, but may exist in limited cases. They are often used for state feedback controls. Dynamic feedback controls are easier to design than static feedback controls, but the dimension of the overall systems increases. They are often used for output feedback controls. Finite memory feedback controls, or simply finite memory controls, which are linear combinations of finite measured inputs and outputs, can be another option, which will be explained extensively in this book later. The feedback control is required to be linear or allowed to be nonlinear.

Performance Criterion

There are several approaches for control designs to meet control objectives. Optimal control has been one of the widely used methods. An optimal control is obtained by minimizing or maximizing a certain performance criterion. It is also obtained by mini-maximizing or maxi-minimizing a certain performance

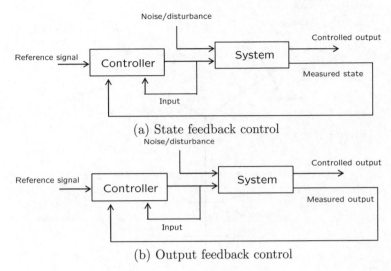

Fig. 1.2. Feedback controls

criterion. Optimal controls are given on the finite horizon and also on the infinite horizon. For linear systems, popular optimal controls based on minimizing are the LQ controls for state feedback controls and the LQG controls for output feedback controls. Popular optimal controls based on mini-maximizing are H_∞ controls.

The optimal controls are often given in open-loop controls for nonlinear systems. However, optimal controls for linear systems often lead to feedback controls. Therefore, special care should be taken to obtain a closed-loop control for nonlinear systems.

Even if optimal controls are obtained by satisfying some performance criteria, they need to meet the above-mentioned control objectives: closed-loop stability, tracking performance, robust stability, and robust tracking performance. Closed-loop optimal controls on the infinite horizon tend to meet the tracking performance and the closed-loop stability under some conditions. However, it is not so easy to achieve robust stability with respect to model uncertainties.

Models, control structure, and performance criteria can be summarized visually in Figure 1.3.

1.2 Concept of Receding Horizon Controls

In conventional optimal controls, either finite horizons or infinite horizons are dealt with. Often, feedback control systems must run for a sufficiently long period, as in electrical power generation plants and chemical processes. In these ongoing processes, finite horizon optimal controls cannot be adopted, but

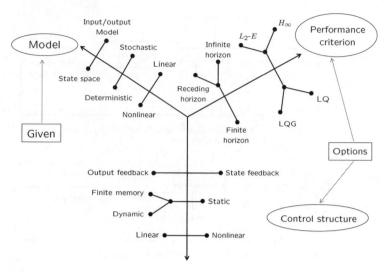

Fig. 1.3. Components of control systems

infinite horizon optimal controls must be used. In addition, we can introduce a new type of control, RHC that is based on optimal control.

The basic concept of RHC is as follows. At the current time, the optimal control is obtained, either closed-loop type, or open-loop type, on a finite fixed horizon from the current time k, say $[k, k + N]$. Among the optimal controls on the entire fixed horizon $[k, k + N]$, only the first one is adopted as the current control law. The procedure is then repeated at the next time, say $[k+1, k+1+N]$. The term "receding horizon" is introduced, since the horizon recedes as time proceeds. There is another type of control, i.e. intervalwise RHC, that will be explained later.

The concept of RHC can be easily explained by using a company's investment planning to maximize the profit. The investment planning should be continued for the years to come as in feedback control systems. There could be three policies for a company's investment planning:

(1) One-time long-term planning

Investment planning can be carried over a fairly long period, which is closer to infinity, as in Figure 1.4. This policy corresponds to the infinite horizon optimal control obtained over $[k, \infty]$.

(2) Periodic short-term planning

Instead of the one-time long-term planning, we can repeat short-term investment planning, say investment planning every 5-years, which is given in Figure 1.5.

(3) Annual short-term planning

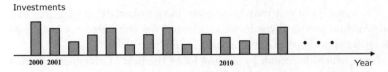

Fig. 1.4. One-time long-term planning

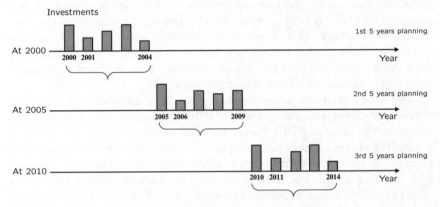

Fig. 1.5. Periodic short-term planning

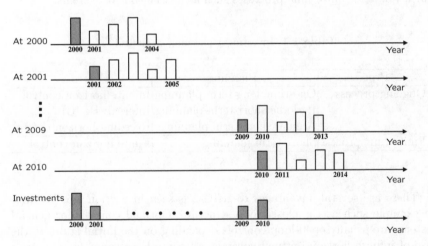

Fig. 1.6. Annual short-term planning

For a new policy, it may be good to have a short-term planning every year and the first year's investment is selected for the current year's investment policy. This concept is depicted in Figure 1.6.

Now, which investment planning looks the best? Obviously we must determine the definition of the "best". The meaning of the "best" is very subjective and can be different depending on the individual. Among investment plannings, any one can be the best policy, depending on the perception of each person. The above question, about which investment planning is the best, was asked of students in a class without explaining many details. An average seven or eight out of ten persons selected the third policy, i.e. annual short-term planning, as the best investment planning. This indicates that annual short-term planning can have significant meanings.

The above examples are somewhat vague in a mathematical sense, but are adequate to explain the concepts of RHC. An annual short-term planning is exactly the same as RHC. Therefore, RHC may have significant meanings that will be clearer in the coming chapters. The periodic short-term investment planning in Figure 1.5 corresponds to intervalwise RHC. The term "intervalwise receding horizon" is introduced since the horizon recedes intervalwise or periodically as time proceeds. Different types of investment planning are compared with the different types of optimal control in Table 1.1. It is noted that the short-term investment planning corresponds to finite horizon control, which works for finite time processes such as missile control systems.

Table 1.1. Investment planning vs control

Process	Planning	Control
Ongoing process	One-time long-term planning	Infinite horizon control
	Periodic short-term planning	Intervalwise RHC
	Annual short-term planning	Receding horizon control
Finite time process	Short-term planning	Finite horizon control

There are several advantages to RHCs, as seen in Section 1.5. We take an example such as the closed-loop structure. Optimal controls for general systems are usually open-loop controls depending on the initial state. In the case of infinite horizon optimal controls, all controls are open-loop controls except the initial time. In the case of intervalwise RHCs, only the first control on each horizon is a closed-loop control and others are open-loop controls. In the case of the RHCs, we always have closed-loop controls due to the repeated computation and the implementation of only the first control.

1.3 Receding Horizon Filters and Output Feedback Receding Horizon Controls

RHC is usually represented by a state feedback control if states are available. However, full states may not be available, since measurement of all states may be expensive or impossible. From measured inputs and outputs, we can construct or estimate all states. This is often called a filter for stochastic systems or a state observer for deterministic systems. Often, it is called a filter for both systems. The well-known Luenberger observer for deterministic state-space signal models and the Kalman filter for stochastic state-space signal models are infinite impulse response (IIR) type filters. This means that the state observer utilizes all the measured data up to the current time k from the initial time k_0.

Instead, we can utilize the measured data on the recent finite time $[k - N_f, k]$ and obtain an estimated state by a linear combination of the measured inputs and outputs over the receding finite horizon with some weighting gains to be chosen so that the error between the real state and the estimated one is minimized. N_f is called the filter horizon size and is a design parameter. We will call this filter as the receding horizon filter. This is an FIR-type filter. This concept is depicted in Figure 1.7. It is noted that in the signal

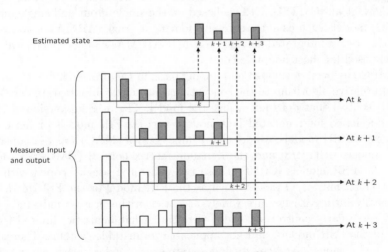

Fig. 1.7. Receding horizon filter

processing area the FIR filter has been widely used for unmodelled signals due to its many good properties, such as guaranteed stability, linear phase (zero error), robustness to temporary parameter changes and round-off error, etc. We can also expect such good properties for the receding horizon filters.

An output feedback RHC can be made by blind combination of a state feedback RHC with a receding horizon filter, just like a combination of an LQ regulator and a Kalman filter. However, we can obtain an output feedback RHC by an optimal approach, not by a blind combination. The output feedback RHC is obtained as a linear combination of the measured inputs and outputs over the receding finite horizon with some weighting gains to be chosen so that the given performance criterion is optimized. This control will be called a receding horizon finite memory control (FMC). The above approach is comparable to the well-known LQG problem, where the control is assumed to be a function of all the measured inputs and outputs from the initial time. The receding horizon FMC are believed to have properties that are as good as the receding horizon FIR filters and state feedback RHCs.

1.4 Predictive Controls

An RHC is one type of predictive control. There are two other well-known predictive controls, generalized predictive control (GPC) and model predictive control (MPC). Originally, these three control strategies had been investigated independently.

GPC was developed in the self-tuning and adaptive control area. Some control strategies that achieve minimum variance were adopted in the self-tuning control [ÅW73] [CG79]. The general frame work for GPC was suggested by Clark et al. [CMT87]. GPC is based on the single input and single output (SISO) models such as auto regressive moving average (ARMA) or controlled auto regressive integrated moving average (CARIMA) models which have been widely used for most adaptive controls.

MPC has been developed on a model basis in the process industry area as an alternative algorithm to the conventional proportional integrate derivative (PID) control that does not utilize the model. The original version of MPC was developed for truncated I/O models, such as FIR models or finite step response (FSR) models. Model algorithmic control (MAC) was developed for FIR models [RRTP78] and the dynamic matrix control (DMC) was developed for FSR models [CR80]. These two control strategies coped with I/O constraints. Since I/O models such as the FIR model or the FSR model are physically intuitive, they are widely accepted in the process industry. However, these early control strategies were somewhat heuristic, limited to the FIR or the FSR models, and not applicable to unstable systems. Thereafter, lots of extensions have been made for state-space models, as shown in survey papers listed in Section 1.7.

RHC has been developed in academia as an alternative control to the celebrated LQ controls. RHC is based on the state-space framework. The stabilizing property of RHC has been shown for case of both continuous and discrete systems using the terminal equality constraint [KP77a] [KP77c]. Thereafter, it has been extended to tracking controls, output feedback controls, and nonlinear controls [KG88] [MM90]. The state and input constraints were not considered in the early developments, but dealt in later works, as seen in the above-mentioned survey papers.

The term "predictive" appears in GPC since the minimum variance is given in predicted values on the finite future time. The term "predictive" appears in MPC since the performance is given in predictive values on the finite future time that can be computed by using the model. The performance for RHC is the same as one for MPC. Thus, the term "predictive" can be incorporated in RHC as receding horizon predictive control (RHPC).

Since I/O models on which GPC and the early MPC are based can be represented in state-space frameworks, GPC and the early MPC can be obtained from predictive controls based on state-space models. In this book, the predictive controls based on the state space model will be dealt with in terms of RHC instead of MPC although MPC based on the state-space model is the same as RHC.

1.5 Advantages of Receding Horizon Controls

RHC has made a significant impact on industrial control engineering and is being increasingly applied in process controls. In addition to industrial applications, RHC has been considered to be a successful control theory in academia. In order to exploit the reasons for such a preference, we may summarize several advantages of RHC over other existing controls.

- Applicability to a broad class of systems. The optimization problem over the finite horizon, on which RHC is based, can be applied to a broad class of systems, including nonlinear systems and time-delayed systems. Analytical or numerical solutions often exist for such systems.
- Systematic approach to obtain a closed loop control. While optimal controls for linear systems with input and output constraints or nonlinear systems are usually open-loop controls, RHCs always provide closed-loop controls due to the repeated computation and the implementation of only the first control.
- Constraint handling capability. For linear systems with the input and state constraints that are common in industrial problems, RHC can be easily and efficiently computed by using mathematical programming, such as quadratic programming (QP) and semidefinite programming (SDP). Even for nonlinear systems, RHC can handle input and state constraints numerically in many case due to the optimization over the finite horizon.
- Guaranteed stability. For linear and nonlinear systems with input and state constraints, RHC guarantees the stability under weak conditions. Optimal control on the infinite horizon, i.e. the steady-state optimal control, can also be an alternative. However, it has guaranteed stability only if it is obtained in a closed form that is difficult to find out.

- Good tracking performance. RHC presents good tracking performance by utilizing the future reference signal for a finite horizon that can be known in many cases. In infinite horizon tracking control, all future reference signals are needed for the tracking performance. However, they are not always available in real applications and the computation over the infinite horizon is almost impossible. In PID control, which has been most widely used in the industrial applications, only the current reference signal is used even when the future reference signals are available on a finite horizon. This PID control might be too short-sighted for the tracking performance and thus has a lower performance than RHC, which makes the best of all future reference signals.

- Adaptation to changing parameters. RHC can be an appropriate strategy for known time-varying systems. RHC needs only finite future system parameters for the computation of the current control, while infinite horizon optimal control needs all future system parameters. However, all future system parameters are not always available in real problems and the computation of the optimal control over the infinite horizon is very difficult and requires infinite memories for future controls.

  Since RHC is computed repeatedly, it can adapt to future system parameters changes that can be known later, not at the current time, whereas the infinite horizon optimal controls cannot adapt, since they are computed once in the first instance.

- Good properties for linear systems. It is well-known that steady-state optimal controls such as LQ, LQG and H_∞ controls have good properties, such as guaranteed stability under weak conditions and a certain robustness. RHC also possesses these good properties. Additionally, there are more design parameters, such as final weighting matrices and a horizon size, that can be tuned for a better performance.

- Easier computation compared with steady-state optimal controls. Since computation is carried over a finite horizon, the solution can be obtained in an easy batch form for a linear system. For linear systems with input and state constraints, RHC is easy to compute by using mathematical programming, such as QP and SDP, while an optimal control on the infinite horizon is hard to compute. For nonlinear systems with input and state constraints, RHC is relatively easier to compute numerically than the steady-state optimal control because of the finite horizon.

- Broad industrial applications. Owing to the above advantages, there exist broad industrial applications for RHC, particularly in industrial processes. This is because industrial processes have limitations on control inputs and require states to stay in specified regions, which can be efficiently handled by RHC. Actually, the most profitable operation is often obtained when a process works around a constraint. For this reason, how to handle the constraint is very important. Conventional controls behave conservatively, i.e. far from the optimal operation, in order to satisfy constraints since the constraint cannot be dealt with in the design phase. Since the dynamics of the system are relative slow, it is possible to make the calculation of the RHC each time within a sampling time.

There are some disadvantages of RHC.

- Longer computation time compared with conventional nonoptimal controls. The absolute computation time of RHC may be longer compared with conventional nonoptimal controls, particularly for nonlinear systems, although the computation time of RHC at each time can be smaller than the corresponding infinite horizon optimal control. Therefore, RHC may not be fast enough to be used as a real-time control for certain processes. However, this problem may be overcome by the high speed of digital processors, together with improvements in optimization algorithms.
- Difficulty in the design of robust controls for parameter uncertainties. System properties, such as robust stability and robust performance due to parameter uncertainties, are usually intractable in optimization problems on which RHCs are based. The repeated computation for RHC makes it more difficult to analyze the robustness. However, the robustness with respect to external disturbances can be dealt with somewhat easily, as seen in minimax RHCs.

1.6 About This Book

In this book, RHCs are extensively presented for linear systems, constrained linear systems, and nonlinear systems. RHC can be of both a state feedback type and an output feedback type. They are derived with different criteria, such as minimization, mini-maximization, and sometimes the mixed of them. FIR filters are introduced for state observers and utilized for output feedback receding RHC.

In this book, optimal solutions of RHC, the stability and the performance of the closed-loop system, and robustness with respect to disturbances are dealt with. Robustness with respect to parameter uncertainties are not covered in this book.

In Chapter 2, existing optimal controls for nonlinear systems are reviewed for minimum and minimax criteria. LQ controls and H_∞ controls are also reviewed with state and output feedback types. Solutions via the linear matrix inequality (LMI) and SDP are introduced for the further use.

In Chapter 3, state feedback LQ and H_∞ RHCs are discussed. In particular, state feedback LQ RHCs are extensively investigated and used for the subsequent derivations of other types of control. Monotonicity of the optimal cost and closed-loop stability are introduced in detail.

In Chapter 4, as state observers, FIR filters are introduced to utilize only recent finite measurement data. Various FIR filters are introduced for minimum, minimax and mixed criteria. Some filters of IIR type are also introduced as dual filters to RHC.

In Chapter 5, output feedback controls, called FMC, are investigated. They are obtained in an optimal manner, rather than by blindly combining filters and state feedback controls together. The globally optimal FMC with a quadratic cost is shown to be separated into the optimal FIR filter in Chapter 4 and RHC in Chapter 3.

In Chapter 6, linear systems with state and input constraints are discussed, which are common in industrial processes, particularly in chemical processes. Feasibility and stability are discussed. While the constrained RHC is difficult to obtain analytically, it is computed easily via SDP.

In Chapter 7, RHC are extended to nonlinear systems. It is explained that the receding horizon concept can be applied easily to nonlinear systems and that stability can be dealt with similarly. The control Lyapunov function is introduced as a cost monotonicity condition for stability. Nonlinear RHCs are obtained based on a terminal equality constraint, a free terminal cost, and a terminal invariance set.

Some fundamental theories necessary to investigate the RHC are given in appendices, such as matrix equality, matrix calculus, systems theory, random variable, LMI and SDP. A survey on applications of RHCs is also listed in Appendix E.

MATLAB® programs of a few examples are listed in Appendix F and you can obtain program files of several examples at *http://cisl.snu.ac.kr/rhc* or the Springer website.

Sections denoted with an asterisk contain H_2 control problems and topics related to the general system matrix A. They can be skipped when a course cannot cover all the materials of the book. Most proofs are provided in order to make this book more self-contained. However, some proofs are left out in Chapter 2 and in appendices due to the lack of space and broad scope of this book.

Notation

This book covers wide topics, including state feedbacks, state estimations and output feedbacks, minimizations, mini-maximizations, stochastic systems, deterministic systems, constrained systems, nonlinear systems, etc. Therefore, notation may be complex. We will introduce global variables and constants that represent the same meaning throughout the chapters, However, they may be used as local variables in very limited cases.

We keep the widely used and familiar notation for matrices related to states and inputs, and introduce new notation using subindices for matrices related to the additional external inputs, such as noises and disturbances. For example, we use $x_{i+1} = Ax_i + Bu_i + B_w w_i$ instead of $x_{i+1} = Ax_i + B_2 u_i + B_1 w_i$ and $y_i = Cx_i + C_w w_i$ instead of $y_i = C_1 x_i + C_2 w_i$. These notations can make it easy to recognize the matrices with their related variables. Additionally, we easily obtain the existing results without external inputs by just removing them, i.e. setting to zero, from relatively complex results with external inputs.

Just setting to $B_w = 0$ and $C_w = 0$ yields the results based on the simpler model $x_{i+1} = Ax_i + Bu_i$, $y_i = Cx_i$. The global variables and constants are as follows:

- System

$$x_{i+1} = Ax_i + Bu_i + B_w w_i$$
$$y_i = Cx_i + D_u u_i + D_w w_i \text{ (or } D_v v_i)$$
$$z_i = C_z x_i + D_{zu} u_i + D_{zw} w_i$$

  x_i : state
  u_i : input
  y_i : measured output
  z_i : controlled output
  w_i : noise or disturbance
  v_i : noise
  Note that, for stochastic systems, D_v is set to I and the notation G is often used instead of B_w.

- Time indices
  i, j : time sequence index
  k : current time index

- Time variables for controls and filters
  $u_{k+j|k}$, u_{k+j} : control ahead of j steps from the current time k as a reference
  $x_{k+j|k}$, x_{k+j} : state ahead of j steps from the current time k as a reference
  $\hat{x}_{k|l}$: estimated state at time k based on the observed data up to time l
  Note that $u_{k+j|k}$ and $x_{k+j|k}$ appear in RHC and $\hat{x}_{k|l}$ in filters. We use the same notation, since there will be no confusion in contexts.

- Dimension
  n : dimension of x_i
  m : dimension of u_i
  l : dimension of w_i
  p : dimension of y_i
  q : dimension of z_i

- Feedback gain
  H : feedback gain for control
  F : feedback gain for filter
  Γ : feedback gain for disturbance

- Controllability and observability
  G_o : observability Grammian
  G_c : controllability Grammian
  n_o : observability index
  n_c : controllability index

- Weighting matrices

  Q : weighting matrix of state

  R : weighting matrix of input

  Q_f : weighting matrix of final state

- Matrix decomposition

$$Q = Q^{\frac{1}{2}}Q^{\frac{1}{2}} = C^T C$$
$$R = R^{\frac{1}{2}}R^{\frac{1}{2}}$$

- Covariance

  Q_w : covariance of system noise

  R_v : covariance of measurement noises

- Solution to Riccati equation.

  K_i, K_{i,i_f} : Solution to Riccati equation in LQ control

  P_i, P_{i,i_f} : Solution to Riccati equation in Kalman filter

  M_i, M_{i,i_f} : Solution to Riccati equation in H_∞ control

  S_i, S_{i,i_f} : Solution to Riccati equation in H_∞ filter

- Horizon size for performance

  N_c : control horizon size or control horizon

  N_f : filter horizon size or filter horizon

  N_p : prediction horizon size or prediction horizon. This is often called performance horizon or cost horizon.

  Note that N_c and N_f are simply written as N when the meaning is clear.

- System and performance criteria for nonlinear systems

  f : system function

  g : cost function of intermediate state

  h : cost function of terminal state

- Performance criteria :

  We have several notations depending on interested variables.

  J

  $J(x_i, i)$

  $J(x_i, i, i_f)$

  $J(x_i, i, u_i)$

  $J(x_i, i, u_{i+.})$

  $J(x_i, i, u_i, w_i)$

  where $u_{i+.} = \{u_{i+j}, \ j \geq 0\}$ and the second argument i is usually removed.

  Note that a performance criterion is often called a cost function.

- Ellipsoid

  $$\mathcal{E}_{P,\alpha} = \{x | x^T P x \leq \alpha\}$$

- Others

  $G(z)$: transfer function

  p_i : costate in optimization method

  I : identity matrix

1.7 References

There are several survey papers on the existing results of predictive control [RRTP78, GPM89, RMM94, Kwo94, May95, LC97, May97, CA98, ML99, MRRS00, KHA04]. There are useful proceedings of conferences or workshops on MPC [Cla94, KGC97, AZ99, Gri02]. In [Kwo94], a very wide ranging list of references up to 1994 is provided. [AZ99, Gri02] would be helpful for those who are interested in nonlinear MPC. In [MRRS00], the stability and the optimality for constrained and nonlinear predictive controls for state-space models are well summarized and categorized. Recent predictive controls for nonlinear systems are surveyed in [KHA04]. In particular, comparisons among industrial MPCs are well presented in [QB97], [QB00], [QB03] from a practical point of view.

There are several books on predictive controls [BGW90, Soe92, Mos95, MSR96, AZ00, KC00, Mac02, Ros03, HKH02, CB04]. The GPC for unconstrained linear systems and its monotonicity conditions for guaranteeing stability are given in [BGW90]. The book [Soe92] provides a comprehensive exposition on GPC and its relationship with MPC. The book [Mos95] covers predictive controls and adaptive predictive controls for unconstrained linear systems with a common unifying framework. The book [AZ00] covers nonlinear RHC theory, computational aspects of on-line optimization and application issues. In the book [Mac02], industrial case studies are illustrated and several commercial predictive control products are introduced. Implementation issues for predictive controls are dealt with in the book [CB04].

2

Optimal Controls on Finite and Infinite Horizons: A Review

2.1 Introduction

In this chapter, important results on optimal controls are reviewed.

Optimal controls depend on the performance criterion that should reflect the designer's concept of good performance. Two important performance criteria are considered for optimal controls. One is for minimization and the other for minimaximization.

Both nonlinear and linear optimal controls are reviewed. First, the general results for nonlinear systems are introduced, particularly with the dynamic programming and a minimum principle. Then, the optimal controls for linear systems are obtained as a special case. Actually, linear quadratic and H_∞ optimal controls are introduced for both state feedback and output feedback controls. Tracking controls are also introduced for future use.

Optimal controls are discussed for free and fixed terminal states. The former may or may not have a terminal cost. In particular, a nonzero terminal cost for the free terminal state is called a free terminal cost in the subsequent chapters. In addition, a fixed terminal state is posed as a terminal equality constraint in the subsequent chapters. The optimal controls for the fixed terminal and nonzero reference case will be derived in this chapter. They are important for RHC. However, they are not common in the literature.

Linear optimal controls are transformed to SDP using LMIs for easier computation of the control laws. This numerical method can be useful for obtaining optimal controls in constrained systems, which will be discussed later.

Most results given in this chapter lay the foundation for the subsequent chapters on receding horizon controls.

Proofs are generally given in order to make our presentation in this book more self-contained, though they appear in the existing literature. H_2 filters and H_2 controls are important, but not used for subsequent chapters; thus, they are summarized without proof.

The organization of this chapter is as follows. In Section 2.2, optimal controls for general systems such as dynamic programming and the minimum principle are dealt with for both minimum and minimax criteria. In Section 2.3, linear optimal controls, such as the LQ control based on the minimum criterion and H_∞ control based on the minimax criterion, are introduced. In Section 2.4, the Kalman filter on the minimum criterion and the H_∞ filter on the minimax criterion are discussed. In Section 2.5, LQG control on the minimum criterion and the output feedback H_∞ control on the minimax criterion are introduced for output feedback optimal controls. In Section 2.6, the infinite horizon LQ and H_∞ control are represented in LMI forms. In Section 2.7, H_2 controls are introduced as a general approach for LQ control.

2.2 Optimal Control for General Systems

In this section, we consider optimal controls for general systems. Two approaches will be taken. The first approach is based on the minimization and the second approach is based on the minimaximization.

2.2.1 Optimal Control Based on Minimum Criterion

Consider the following discrete-time system:

$$x_{i+1} = f(x_i, u_i, i), \ x_{i_0} = x_0 \tag{2.1}$$

where $x_i \in \Re^n$ and $u_i \in \Re^m$ are the state and the input respectively, and may be required to belong to the given sets, i.e. $x_i \in \mathcal{X} \in \Re^n$ and $u_i \in \mathcal{U} \in \Re^m$.

A performance criterion with the free terminal state is given by

$$J(x_{i_0}, i_0, u) = \sum_{i=i_0}^{i_f-1} g(x_i, u_i, i) \ + \ h(x_{i_f}, i_f) \tag{2.2}$$

i_0 and i_f are the initial and terminal time. $g(\cdot, \cdot, \cdot)$ and $h(\cdot, \cdot)$ are specified scalar functions. We assume that i_f is fixed here for simplicity. Note that x_{i_f} is free for the performance criterion (2.2). However, x_{i_f} can be fixed. A performance criterion with the fixed terminal state is given by

$$J(x_{i_0}, i_0, u) = \sum_{i=i_0}^{i_f-1} g(x_i, u_i, i) \tag{2.3}$$

subject to

$$x_{i_f} = x_{i_f}^r \tag{2.4}$$

where $x_{i_f}^r$ is given.

Here, the optimal control problem is to find an admissible control $u_i \in \mathcal{U}$ for $i \in [i_0, i_f - 1]$ that minimizes the cost function (2.2) or (2.3) with the constraint (2.4).

The Principle of Optimality and Dynamic Programming

If S-a-D is the optimal path from S to D with the cost J^*_{SD}, then a-D is the optimal path from a to D with J^*_{aD}, as can be seen in Figure 2.1. This property is called *the principle of optimality*. Thus, an optimal policy has the property that, whatever the initial state and initial decision are, the remaining decisions must constitute an optimal policy with regard to the state resulting from the first decision.

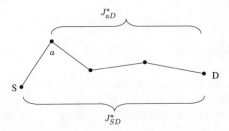

Fig. 2.1. Optimal path from S to D

Now, assume that there are allowable paths S-a-D, S-b-D, S-c-D, and S-d-D and optimal paths from a, b, c, and d to D are J^*_{aD}, J^*_{bD}, J^*_{cD}, and J^*_{dD} respectively, as can be seen in Figure 2.2. Then, the optimal trajectory that starts at S is found by comparing

$$J^*_{SaD} = J_{Sa} + J^*_{aD}$$
$$J^*_{SbD} = J_{Sb} + J^*_{bD}$$
$$J^*_{ScD} = J_{Sc} + J^*_{cD}$$
$$J^*_{SdD} = J_{Sd} + J^*_{dD} \qquad (2.5)$$

The minimum of these costs must be the one associated with the optimal decision at point S. Dynamic programming is a computational technique which extends the above decision-making concept to the sequences of decisions which together define an optimal policy and trajectory.

Assume that the final time i_f is specified. If we consider the performance criterion (2.2) subject to the system (2.1), the performance criterion of dynamic programming can be represented by

$$J(x_i, i, u) = g(x_i, u_i, i) + J^*(x_{i+1}, i+1), \quad i \in [i_0, i_f - 1] \qquad (2.6)$$
$$J^*(x_i, i) = \min_{u_\tau, \tau \in [i, i_f - 1]} J(x_i, i, u)$$
$$= \min_{u_i}\{g(x_i, u_i, i) + J^*(f(x_i, u_i, i), i+1)\} \qquad (2.7)$$

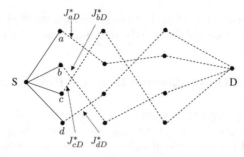

Fig. 2.2. Paths from S through a, b, c, and d to D

where

$$J^*(x_{i_f}, i_f) = h(x_{i_f}, i_f) \tag{2.8}$$

For the fixed terminal state, $J^*(x_{i_f}, i_f) = h(x_{i_f}, i_f)$ is fixed since x_{i_f} and i_f are constants.

It is noted that the dynamic programming method gives a closed-loop control, while the method based on the minimum principle considered next gives an open-loop control for most nonlinear systems.

Pontryagin's Minimum Principle

We assume that the admissible controls are constrained by some boundaries, since in realistic systems control constraints do commonly occur. Physically realizable controls generally have magnitude limitations. For example, the thrust of a rocket engine cannot exceed a certain value and motors provide a limited torque.

By definition, the optimal control u^* makes the performance criterion J a local minimum if

$$J(u) - J(u^*) = \triangle J \geq 0$$

for all admissible controls sufficiently close to u^*. If we let $u = u^* + \delta u$, the increment in J can be expressed as

$$\triangle J(u^*, \delta u) = \delta J(u^*, \delta u) + \text{higher order terms}$$

Hence, the necessary conditions for u^* to be the optimal control are

$$\delta J(u^*, \delta u) \geq 0$$

if u^* lies on the boundary during any portion of the time interval $[i_0,\ i_f]$ and

$$\delta J(u^*, \delta u) = 0$$

if u^* lies within the boundary during the entire time interval $[i_0,\ i_f]$.

We form the following augmented cost functional:

$$J_a = \sum_{i=i_0}^{i_f-1} \left\{ g(x_i, u_i, i) + p_{i+1}^T [f(x_i, u_i, i) - x_{i+1}] \right\} + h(x_{i_f}, i_f)$$

by introducing the Lagrange multipliers $p_{i_0}, p_{i_0+1}, \cdots, p_{i_f}$. For simplicity of the notation, we denote $g(x_i^*, u_i^*, i)$ by g and $f(x_i^*, u_i^*, i)$ by f respectively. Then, the increment of J_a is given by

$$\triangle J_a = \sum_{i=i_0}^{i_f-1} \left\{ g(x_i^* + \delta x_i, u_i^* + \delta u_i, i) + [p_{i+1}^* + \delta p_{i+1}]^T \right.$$

$$\times [f(x_i^* + \delta x_i, u_i^* + \delta u_i, i) - (x_{i+1}^* + \delta x_{i+1})] \right\} + h(x_{i_f}^* + \delta x_{i_f}, i_f)$$

$$- \sum_{i=i_0}^{i_f-1} \left\{ g(x_i^*, u_i^*, i) + p_{i+1}^{*T}[f(x_i^*, u_i^*, i) - x_{i+1}^*] \right\} + h(x_{i_f}^*, i_f)$$

$$= \sum_{i=i_0}^{i_f-1} \left\{ \left[\frac{\partial g}{\partial x_i} \right]^T \delta x_i + \left[\frac{\partial g}{\partial u_i} \right]^T \delta u_i + p_{i+1}^{*T} \left[\frac{\partial f}{\partial x_i} \right]^T \delta x_i \right.$$

$$+ p_{i+1}^{*T} \left[\frac{\partial f}{\partial u_i} \right]^T \delta u_i + \delta p_{i+1}^T f(x_i^*, u_i^*, i)$$

$$\left. - \delta p_{i+1}^T x_{i+1}^* - p_{i+1}^{*T} \delta x_{i+1} \right\} + \left[\frac{\partial h}{\partial x_{i_f}} \right]^T \delta x_{i_f}$$

$$+ \text{ higher order terms} \tag{2.9}$$

To eliminate δx_{i+1}, we use the fact

$$\sum_{i=i_0}^{i_f-1} p_{i+1}^{*T} \delta x_{i+1} = p_{i_f}^{*T} \delta x_{i_f} + \sum_{i=i_0}^{i_f-1} p_i^T \delta x_i$$

Since the initial state x_{i_0} is given, it is apparent that $\delta x_{i_0} = 0$ and p_{i_0} can be chosen arbitrarily. Now, we have

$$\triangle J_a = \sum_{i=i_0}^{i_f-1} \left\{ \left[\frac{\partial g}{\partial x_i} - p_i^* + \frac{\partial f}{\partial x_i} p_{i+1}^* \right]^T \delta x_i + \left[\frac{\partial g}{\partial u_i} + \frac{\partial f}{\partial u_i} p_{i+1}^* \right]^T \delta u_i \right.$$

$$\left. + \delta p_{i+1}^T \left[f(x_i^*, u_i^*, i) - x_{i+1}^* \right] \right\} + \left[\frac{\partial h}{\partial x_{i_f}} - p_{i_f}^* \right]^T \delta x_{i_f}$$

$$+ \text{ higher order terms}$$

Note that variable δx_i for $i = i_0 + 1, \cdots, i_f$ are all arbitrary. Define the function $\mathcal{H}$, called the *Hamiltonian*

$$\mathcal{H}(x_i, u_i, p_{i+1}, i) \triangleq g(x_i, u_i, i) + p_{i+1}^T f(x_i, u_i, i)$$

If the state equations are satisfied, and p_i^* is selected so that the coefficient of δx_i is identically zero, that is,

$$x_{i+1}^* = f(x_i^*, u_i^*, i) \tag{2.10}$$

$$p_i^* = \frac{\partial g}{\partial x_i} + \frac{\partial f}{\partial x_i} p_{i+1}^* \tag{2.11}$$

$$x_{i_0}^* = x_0 \tag{2.12}$$

$$p_{i_f}^* = \frac{\partial h}{\partial x_{i_f}} \tag{2.13}$$

then we have

$$\triangle J_a = \sum_{i=i_0}^{i_f-1} \left\{ \left[\frac{\partial \mathcal{H}}{\partial u}(x_i^*, u_i^*, p_i^*, i) \right]^T \delta u_i \right\} + \text{higher order terms}$$

The first-order approximation to the change in $\mathcal{H}$ caused by a change in u alone is given by

$$\left[\frac{\partial \mathcal{H}}{\partial u}(x_i^*, u_i^*, p_i^*, i) \right]^T \delta u_i \approx \mathcal{H}(x_i^*, u_i^* + \delta u_i, p_{i+1}^*, i) - \mathcal{H}(x_i^*, u_i^*, p_{i+1}^*, i)$$

Therefore,

$$\triangle J(u^*, \delta u) = \sum_{i=i_0}^{i_f-1} \left[\mathcal{H}(x_i^*, u_i^* + \delta u_i, p_{i+1}^*, i) - \mathcal{H}(x_i^*, u_i^*, p_{i+1}^*, i) \right]$$

$$+\text{higher order terms} \tag{2.14}$$

If $u^* + \delta u$ is in a sufficiently small neighborhood of u^*, then the higher order terms are small, and the summation (2.14) dominates the expression for $\triangle J_a$. Thus, for u^* to be an optimal control, it is necessary that

$$\sum_{i=i_0}^{i_f-1} \left[\mathcal{H}(x_i^*, u_i^* + \delta u_i, p_{i+1}^*, i) - \mathcal{H}(x_i^*, u_i^*, p_{i+1}^*, i) \right] \geq 0 \tag{2.15}$$

for all admissible δu. We assert that in order for (2.15) to be satisfied for all admissible δu in the specified neighborhood, it is necessary that

$$\mathcal{H}(x_i^*, u_i^* + \delta u_i, p_{i+1}^*, i) - \mathcal{H}(x_i^*, u_i^*, p_{i+1}^*, i) \geq 0 \tag{2.16}$$

for all admissible δu_i and for all $i = i_0, \cdots, i_f$. In order to prove the inequality (2.15), consider the control

$$\begin{aligned} u_i &= u_i^*, \quad i \notin [i_1, i_2] \\ u_i &= u_i^* + \delta u_i, \quad i \in [i_1, i_2] \end{aligned} \tag{2.17}$$

where $[i_1, i_2]$ is a nonzero time interval, i.e. $i_1 < i_2$ and δu_i is an admissible control variation that satisfies $u^* + \delta u \in \mathcal{U}$.

Suppose that inequality (2.16) is not satisfied in the interval $[i_1, i_2]$ for the control described in (2.17). So, we have

$$\mathcal{H}(x_i^*, u_i, p_{i+1}^*, i) < \mathcal{H}(x_i^*, u_i^*, p_{i+1}^*, i)$$

in the interval $[i_1, i_2]$ and the following inequality is obtained:

$$\sum_{i=i_0}^{i_f-1} \left[\mathcal{H}(x_i^*, u_i, p_{i+1}^*, i) - \mathcal{H}(x_i^*, u_i^*, p_{i+1}^*, i) \right]$$

$$= \sum_{i=i_1}^{i_2} \left[\mathcal{H}(x_i^*, u_i, p_{i+1}^*, i) - \mathcal{H}(x_i^*, u_i^*, p_{i+1}^*, i) \right] < 0$$

Since the interval $[i_1, i_2]$ can be anywhere in the interval $[i_0, i_f]$, it is clear that if

$$\mathcal{H}(x_i^*, u_i, p_{i+1}^*, i) < \mathcal{H}(x_i^*, u_i^*, p_{i+1}^*, i)$$

for any $i \in [i_0, i_f]$, then it is always possible to construct an admissible control, as in (2.17), which makes $\triangle J_a < 0$, thus contradicting the optimality of the control u_i^*. Therefore, a necessary condition for u_i^* to minimize the functional J_a is

$$\mathcal{H}(x_i^*, u_i^*, p_{i+1}^*, i) \leq \mathcal{H}(x_i^*, u_i, p_{i+1}^*, i) \tag{2.18}$$

for all $i \in [i_0, i_f]$ and for all admissible controls. The inequality (2.18) indicates that *an optimal control must minimize the Hamiltonian*. Note that we have established a necessary, but not, in general, sufficient, condition for optimality. An optimal control must satisfy the inequality (2.18). However, there may be controls that satisfy the minimum principle that are not optimal.

We now summarize the principle results. In terms of the Hamiltonian, the necessary conditions for u_i^* to be an optimal control are

$$x_{i+1}^* = \frac{\partial \mathcal{H}}{\partial p_{i+1}}(x_i^*, u_i^*, p_{i+1}^*, i) \tag{2.19}$$

$$p_i^* = \frac{\partial \mathcal{H}}{\partial x}(x_i^*, u_i^*, p_{i+1}^*, i) \tag{2.20}$$

$$\mathcal{H}(x_i^*, u_i^*, p_{i+1}^*, i) \leq \mathcal{H}(x_i^*, u_i, p_{i+1}^*, i) \tag{2.21}$$

for all admissible u_i and $i \in [i_0, i_f - 1]$, and two boundary conditions

$$x_{i_0} = x_0, \quad p_{i_f}^* = \frac{\partial h}{\partial x_{i_f}}(x_{i_f}^*, i_f)$$

The above result is called Pontryagin's minimum principle. The minimum principle, although derived for controls in the given set $\mathcal{U}$, can also be applied to problems in which the admissible controls are not bounded. In this case, for u_i^* to minimize the Hamiltonian it is necessary (but not sufficient) that

$$\frac{\partial \mathcal{H}}{\partial u_i}(x_i^*, u_i^*, p_{i+1}^*, i) = 0, \quad i \in [i_0, i_f - 1] \tag{2.22}$$

If (2.22) is satisfied and the matrix

$$\frac{\partial^2 \mathcal{H}}{\partial u_i^2}(x_i^*, u_i^*, p_{i+1}^*, i)$$

is positive definite, then this is sufficient to guarantee that u_i^* makes J_a a local minimum. If the Hamiltonian can be expressed in the form

$$\mathcal{H}(x_i, u_i, p_{i+1}, i) = c_0(x_i, p_{i+1}, i) + \left[c_1(x_i, p_{i+1}, i)\right]^T u_i + \frac{1}{2}u_i^T R u_i$$

where $c_0(\cdot, \cdot, \cdot)$ and $c_1(\cdot, \cdot, \cdot)$ are a scalar and an $m \times 1$ vector function respectively, that do not have any term containing u_i, then (2.22) and $\partial^2 \mathcal{H}/\partial u_i^2 > 0$ are necessary and sufficient for $\mathcal{H}(x_i^*, u_i^*, p_{i+1}^*, i)$ to be a global minimum.

For a fixed terminal state, δx_{i_f} in the last term of (2.9) is equal to zero. Thus, (2.13) is not necessary, which is replaced with $x_{i_f} = x_{i_f}^r$.

2.2.2 Optimal Control Based on Minimax Criterion

Consider the following discrete-time system:

$$x_{i+1} = f(x_i, u_i, w_i, i), \qquad x_{i_0} = x_0 \tag{2.23}$$

with a performance criterion

$$J(x_{i_0}, i_0, u, w) = \sum_{i=i_0}^{i_f - 1} [g(x_i, u_i, w_i, i)] + h(x_{i_f}, i_f) \tag{2.24}$$

where $x_i \in \Re^n$ is the state, $u_i \in \Re^m$ is the input and $w_i \in \Re^l$ is the disturbance. The input and the disturbance are required to belong to the given sets, i.e. $u_i \in \mathcal{U}$ and $w_i \in \mathcal{W}$. Here, the fixed terminal state is not dealt with because the minimax problem in this case does not make sense.

The minimax criterion we are dealing with is related to a difference game. We want to minimize the performance criterion, while disturbances try to maximize one. A pair policies $(u, w) \in \mathcal{U} \times \mathcal{W}$ is said to constitute a saddle-point solution if, for all $(u, w) \in \mathcal{U} \times \mathcal{W}$,

$$J(x_{i_0}, i_0, u^*, w) \leq J(x_{i_0}, i_0, u^*, w^*) \leq J(x_{i_0}, i_0, u, w^*) \tag{2.25}$$

We may think that u^* is the best control, while w^* is the worst disturbance. The existence of these u^* and w^* is guaranteed by specific conditions.

The control u^* makes the performance criterion (2.24) a local minimum if

$$J(x_{i_0}, i_0, u, w) - J(x_{i_0}, i_0, u^*, w) = \triangle J \geq 0$$

for all admissible controls. If we let $u = u^* + \delta u$, the increment in J can be expressed as

$$\triangle J(u^*, \delta u, w) = \delta J(u^*, \delta u, w) + \text{higher order terms}$$

Hence, the necessary conditions for u^* to be the optimal control are

$$\delta J(u^*, \delta u, w) \geq 0$$

if u^* lies on the boundary during any portion of the time interval $[i_0, \ i_f]$ and

$$\delta J(u^*, \delta u, w) = 0$$

if u^* lies within the boundary during the entire time interval $[i_0, \ i_f]$.

Meanwhile, the disturbance w^* makes the performance criterion (2.24) a local maximum if

$$J(u, w) - J(u, w^*) = \triangle J \leq 0$$

for all admissible disturbances. Taking steps similar to the case of u^*, we obtain the necessary condition

$$\delta J(u, w^*, \delta w) \leq 0$$

if w^* lies on the boundary during any portion of the time interval $[i_0, \ i_f]$ and

$$\delta J(u, w^*, \delta w) = 0$$

if w^* lies within the boundary during the entire time interval $[i_0, \ i_f]$.

We now summarize the principle results. In terms of the Hamiltonian, the necessary conditions for u_i^* to be an optimal control are

$$x_{i+1}^* = \frac{\partial \mathcal{H}}{\partial p_{i+1}}(x_i^*, u_i^*, w_i^*, p_{i+1}^*, i)$$

$$p_i^* = \frac{\partial \mathcal{H}}{\partial x_i}(x_i^*, u_i^*, w_i^*, p_{i+1}^*, i)$$

$$\mathcal{H}(x_i^*, u_i^*, w_i, p_{i+1}^*, i) \leq \mathcal{H}(x_i^*, u_i^*, w_i^*, p_{i+1}^*, i) \leq \mathcal{H}(x_i^*, u_i, w_i^*, p_{i+1}^*, i)$$

for all admissible u_i and w_i on the $i \in [i_0, i_f - 1]$, and two boundary conditions

$$x_{i_0} = x_0, \quad p_{i_f}^* = \frac{\partial h}{\partial x_{i_f}}(x_{i_f}^*, i_f)$$

Now, a dynamic programming for minimaxization criterion is explained. Let there exist a function $J^*(x_i, i)$, $i \in [i_0, i_f - 1]$ such that

$$J^*(x_i, i) = \min_{u \in \mathcal{U}} \max_{w \in \mathcal{W}} \left[g(x_i, u_i, w_i, i) + J^*(f(x_i, u_i, w_i, i), i+1) \right]$$

$$= \max_{w \in \mathcal{W}} \min_{u \in \mathcal{U}} \left[g(x_i, u_i, w_i, i) + J^*(f(x_i, u_i, w_i, i), i+1) \right]$$

$$J^*(x_{i_f}, i_f) = h(x_{i_f}, i_f) \tag{2.26}$$

Then a pair of u and w that is generated by (2.26) provides a saddle point with the corresponding value given by $J^*(x_{i_0}, i_0)$.

2.3 Linear Optimal Control with State Feedback

2.3.1 Linear Quadratic Controls Based on Minimum Criterion

In this section, an LQ control in a tracking form for discrete time-invariant systems is introduced in a state-feedback form. We consider the following discrete time-invariant system:

$$x_{i+1} = Ax_i + Bu_i$$
$$z_i = C_z x_i \tag{2.27}$$

There are two methods which are used to obtain the control of minimizing the chosen cost function. One is dynamic programming and the other is the minimum principle of Pontryagin. The minimum principle of Pontryagin and dynamic programming were briefly introduced in the previous section. In the method of dynamic programming, an optimal control is obtained by employing the intuitively appealing concept called the principle of optimality. Here, we use the minimum principle of Pontryagin in order to obtain an optimal finite horizon LQ tracking control (LQTC).

We can divide the terminal states into two cases. The first case is a free terminal state and the second case is a fixed terminal state. In the following, we will derive two kinds of LQ controls in a tracking form.

1) Free Terminal State

The following quadratic performance criterion is considered:

$$J(z^r, u.) = \sum_{i=i_0}^{i_f - 1} [(z_i - z_i^r)^T \bar{Q}(z_i - z_i^r) + u_i^T R u_i]$$
$$+ [z_{i_f} - z_{i_f}^r]^T \bar{Q}_f [z_{i_f} - z_{i_f}^r] \tag{2.28}$$

Here, $x_i \in \Re^n$, $u_i \in \Re^m$, $z_i \in \Re^p$, z_i^r, $\bar{Q} > 0$, $R > 0$, $\bar{Q}_f > 0$ are the state, the input, the controlled output, the command signal or the reference signal, the state weighting matrix, the input weighting matrix, and the terminal weighting matrix respectively. Here, $z_{i_0}^r$, $z_{i_0+1}^r$, $\cdots$, $z_{i_f}^r$ are command signals which are assumed to be available over the future horizon $[i_0, i_f]$.

For tracking problems, with $\bar{Q} > 0$ and $\bar{Q}_f > 0$ in (2.28), there is a tendency that $z_i \to z_i^r$. In order to derive the optimal tracking control which minimizes the performance criterion (2.28), it is convenient to express the performance criterion (2.28) with the state x_i instead of z_i. It is well known that for a given $p \times n$ ($p \le n$) full rank matrix C_z there always exist some $n \times p$ matrices L such that $C_z L = I_{p \times p}$. Let

$$x_i^r = L z_i^r \tag{2.29}$$

The performance criterion (2.28) is then rewritten as

$$
J(x^r, u) = \sum_{i=i_0}^{i_f-1} [(x_i - x_i^r)^T C_z^T \bar{Q} C_z (x_i - x_i^r) + u_i^T R u_i]
$$
$$
+ [x_{i_f} - x_{i_f}^r]^T C_z^T \bar{Q}_f C_z [x_{i_f} - x_{i_f}^r] \tag{2.30}
$$

The performance criterion (2.30) can be written as

$$
J(x^r, u) = \sum_{i=i_0}^{i_f-1} \left[(x_i - x_i^r)^T Q (x_i - x_i^r) + u_i^T R u_i \right]
$$
$$
+ (x_{i_f} - x_{i_f}^r)^T Q_f (x_{i_f} - x_{i_f}^r) \tag{2.31}
$$

where

$$Q = C_z^T \bar{Q} C_z \quad \text{and} \quad Q_f = C_z^T \bar{Q}_f C_z \tag{2.32}$$

Q and Q_f in (2.31) can be independent design parameters ignoring the relation (2.32). That is, the matrices Q and Q_f can be positive definite, if necessary, though Q and Q_f in (2.32) are semidefinite when C_z is not of full rank. In this book, Q and Q_f in (2.31) are independent design parameters. However, whenever necessary, we will make some connections to (2.32).

We first form a Hamiltonian:

$$\mathcal{H}_i = [(x_i - x_i^r)^T Q (x_i - x_i^r) + u_i^T R u_i] + p_{i+1}^T [A x_i + B u_i] \tag{2.33}$$

where $i \in [i_0, i_f - 1]$. According to (2.20) and (2.2.1), we have

$$p_i = \frac{\partial \mathcal{H}_i}{\partial x_i} = 2Q(x_i - x_i^r) + A^T p_{i+1} \tag{2.34}$$

$$p_{i_f} = \frac{\partial h(x_{i_f}, i_f)}{\partial x_{i_f}} = 2Q_f(x_{i_f} - x_{i_f}^r) \tag{2.35}$$

where $h(x_{i_f}, i_f) = (x_{i_f} - x_{i_f}^r)^T Q_f (x_{i_f} - x_{i_f}^r)$.

A necessary condition for u_i to minimize $\mathcal{H}_i$ is $\frac{\partial \mathcal{H}_i}{\partial u_i} = 0$. Thus, we have

$$\frac{\partial \mathcal{H}_i}{\partial u_i} = 2Ru_i + B^T p_{i+1} = 0 \qquad (2.36)$$

Since the matrix $\frac{\partial^2 \mathcal{H}_i}{\partial u_i^2} = 2R$ is positive definite and $\mathcal{H}_i$ is a quadratic form in u, the solution of (2.36) is an optimal control to minimize $\mathcal{H}_i$. The optimal solution u_i^* is

$$u_i^* = -\frac{1}{2} R^{-1} B^T p_{i+1} \qquad (2.37)$$

If we assume that

$$p_i = 2K_{i,i_f} x_i + 2g_{i,i_f} \qquad (2.38)$$

the solution to the optimal control problem can be reduced to finding the matrices K_{i,i_f} and g_{i,i_f}. From (2.35), the boundary conditions are given by

$$K_{i_f,i_f} = Q_f \qquad (2.39)$$
$$g_{i_f,i_f} = -Q_f x_{i_f}^r \qquad (2.40)$$

Substituting (2.27) into (2.38) and replacing u_i with (2.37), we have

$$\begin{aligned}
p_{i+1} &= 2K_{i+1,i_f} x_{i+1} + 2g_{i+1,i_f} \\
&= 2K_{i+1,i_f}(Ax_i + Bu_i) + 2g_{i+1,i_f} \\
&= 2K_{i+1,i_f}\left(Ax_i - \frac{1}{2} BR^{-1} B^T p_{i+1}\right) + 2g_{i+1,i_f} \qquad (2.41)
\end{aligned}$$

Solving for p_{i+1} in (2.41) yields the following equation:

$$p_{i+1} = [I + K_{i+1,i_f} BR^{-1} B^T]^{-1} [2K_{i+1,i_f} Ax_i + 2g_{i+1,i_f}] \qquad (2.42)$$

Substituting for p_{i+1} from (2.37), we can write

$$u_i^* = -R^{-1} B^T [I + K_{i+1,i_f} BR^{-1} B^T]^{-1} [K_{i+1,i_f} Ax_i + g_{i+1,i_f}] \qquad (2.43)$$

What remains to do is to find K_{i,i_f} and g_{i,i_f}. If we put the equation (2.42) into the equation (2.34), we have

$$\begin{aligned}
p_i &= 2Q(x_i - x_i^r) + A^T [I + K_{i+1,i_f} BR^{-1} B^T]^{-1} [2K_{i+1,i_f} Ax_i + 2g_{i+1,i_f}], \\
&= 2[Q + A^T (I + K_{i+1,i_f} BR^{-1} B^T)^{-1} K_{i+1,i_f} A]x_i \\
&\quad + 2[-Qx_i^r + A^T (I + K_{i+1,i_f} BR^{-1} B^T)^{-1} g_{i+1,i_f}] \qquad (2.44)
\end{aligned}$$

The assumption (2.38) holds by choosing K_{i,i_f} and g_{i,i_f} as

$$\begin{aligned}
K_{i,i_f} &= A^T[I + K_{i+1,i_f}BR^{-1}B^T]^{-1}K_{i+1,i_f}A + Q \\
&= A^T K_{i+1,i_f}A - A^T K_{i+1,i_f}B(R + B^T K_{i+1,i_f}B)^{-1}B^T K_{i+1,i_f}A \\
&\quad + Q \tag{2.45} \\
g_{i,i_f} &= A^T[I + K_{i+1,i_f}BR^{-1}B^T]^{-1}g_{i+1,i_f} - Qx_i^r \tag{2.46}
\end{aligned}$$

where the second equality comes from

$$[I + K_{i+1,i_f}BR^{-1}B^T]^{-1} = I - K_{i+1,i_f}B(R + B^T K_{i+1,i_f}B)^{-1}B^T \tag{2.47}$$

using the matrix inversion lemma (A.2) in Appendix A. The optimal control derived until now is summarized in the following theorem.

Theorem 2.1. *In the system (2.27), the LQTC for the free terminal state is given as (2.43) for the performance criterion (2.31). K_{i,i_f} and g_{i,i_f} in (2.43) are obtained from Riccati Equation (2.45) and (2.46) with boundary condition (2.39) and (2.40).*

Depending on Q_f, K_{i,i_f} may be nonsingular (positive definite) or singular (positive semidefinite). This property will be important for stability and the inversion of the matrix K_{i,i_f} in coming sections.

For a zero reference signal, g_{i,i_f} becomes zero so that we have

$$u_i^* = -R^{-1}B^T[I + K_{i+1,i_f}BR^{-1}B^T]^{-1}K_{i+1,i_f}Ax_i \tag{2.48}$$

The performance criterion (2.31) associated with the optimal control (2.43) is given in the following theorem.

Theorem 2.2. *The optimal cost $J^*(x_i)$ with the reference value can be given*

$$J^*(x_i) = x_i^T K_{i,i_f}x_i + 2x_i^T g_{i,i_f} + w_{i,i_f} \tag{2.49}$$

where

$$w_{i,i_f} = w_{i+1,i_f} + x_i^{rT}Qx_i^r - g_{i+1,i_f}^T B(B^T K_{i+1,i_f}B + R)^{-1}B^T g_{i+1,i_f} \tag{2.50}$$

with boundary condition $w_{i_f,i_f} = x_{i_f}^{rT}Q_f x_{i_f}^r$.

Proof. A long and tedious calculation is required to obtain the optimal cost using the result of Theorem 2.1. Thus, we derive the optimal cost using dynamic programming, where the optimal control and the optimal cost are obtained simultaneously.

Let $J^*(x_{i+1})$ denote the optimal cost associated with the initial state x_{i+1} and the interval $[i+1, i_f]$. Suppose that the optimal cost $J^*(x_{i+1})$ is given as

$$J^*(x_{i+1}) = x_{i+1}^T K_{i+1,i_f}x_{i+1} + 2x_{i+1}^T g_{i+1,i_f} + w_{i+1,i_f} \tag{2.51}$$

where w_{i+1,i_f} will be determined later. We wish to calculate the optimal cost $J^*(x_i)$ from (2.51).

By applying the principle of optimality, $J^*(x_i)$ can be represented as follows:

$$J^*(x_i) = \min_{u_i} \left[(x_i - x_i^r)^T Q(x_i - x_i^r) + u_i^T R u_i + J^*(x_{i+1}) \right] \quad (2.52)$$

(2.52) can be evaluated backward by starting with the condition $J^*(x_{i_f}) = (x_{i_f} - x_{i_f}^r)^T Q_f(x_{i_f} - x_{i_f}^r)$.

Substituting (2.27) and (2.51) into (2.52), we have

$$\begin{aligned} J^*(x_i) = \min_{u_i} &\Big[(x_i - x_i^r)^T Q(x_i - x_i^r) + u_i^T R u_i + x_{i+1}^T K_{i+1,i_f} x_{i+1} \\ &+ 2x_{i+1}^T g_{i+1,i_f} + w_{i+1,i_f} \Big] \end{aligned} \quad (2.53)$$

$$\begin{aligned} = \min_{u_i} &\Big[(x_i - x_i^r)^T Q(x_i - x_i^r) + u_i^T R u_i \\ &+ (Ax_i + Bu_i)^T K_{i+1,i_f}(Ax_i + Bu_i) + 2(Ax_i + Bu_i)^T g_{i+1,i_f} \\ &+ w_{i+1,i_f} \Big] \end{aligned} \quad (2.54)$$

Note that $J^*(x_i)$ has a quadratic equation with respect to u_i and x_i. For a given x_i, the control u_i is chosen to be optimal according to (2.54). Taking derivatives of (2.54) with respect to u_i to obtain

$$\frac{\partial J^*(x_i)}{\partial u_i} = 2Ru_i + 2B^T K_{i+1,i_f} Bu_i + 2B^T K_{i+1,i_f} Ax_i + 2B^T g_{i+1,i_f} = 0$$

we have the following optimal control u_i:

$$u_i = -(R + B^T K_{i+1,i_f} B)^{-1}[B^T K_{i+1,i_f} Ax_i + B^T g_{i+1,i_f}] \quad (2.55)$$
$$= -L_{1,i} x_i + L_{2,i} g_{i+1,i_f} \quad (2.56)$$

where

$$L_{1,i} \overset{\triangle}{=} [R + B^T K_{i+1,i_f} B]^{-1} B^T K_{i+1,i_f} A \quad (2.57)$$
$$L_{2,i} \overset{\triangle}{=} -[R + B^T K_{i+1,i_f} B]^{-1} B^T \quad (2.58)$$

It is noted that the optimal control u_i in (2.56) is the same as (2.43) derived from the minimum principle. How to obtain the recursive equations of K_{i+1,i_f} and g_{i+1,i_f} is discussed later.

From definitions (2.57) and (2.58), we have the following relations:

$$\begin{aligned} A^T K_{i+1,i_f} B[R + B^T K_{i+1,i_f} B]^{-1} B^T K_{i+1,i_f} A &= A^T K_{i+1,i_f} B L_{1,i} \\ &= L_{1,i}^T B^T K_{i+1,i_f} A \\ &= L_{1,i}^T [R + B^T K_{i+1,i_f} B] L_{1,i} \end{aligned}$$
$$\quad (2.59)$$

where the most left side is equivalent to the second term of Riccati Equation (2.45) and these relations are useful for representing the Riccati equation in terms of closed-loop system $A - BL_{1,i}$.

Substituting (2.56) into (2.54) yields

$$J^*(x_i) = x_i^T \left[Q + L_{1,i}^T RL_{1,i} + (A - BL_{1,i})^T K_{i+1,i_f}(A - BL_{1,i}) \right] x_i$$

$$+ 2x_i^T \left[-L_{1,i}^T RL_{2,i} g_{i+1,i_f} + (A - BL_{1,i})^T K_{i+1,i_f} BL_{2,i} g_{i+1,i_f} \right.$$

$$+ (A - BL_{1,i})^T g_{i+1,i_f} - Qx_i^r \Big] + g_{i+1,i_f}^T L_{2,i}^T RL_{2,i} g_{i+1,i_f}$$

$$+ g_{i+1,i_f}^T L_{2,i}^T B^T K_{i+1,i_f} BL_{2,i} g_{i+1,i_f} + 2g_{i+1,i_f}^T L_{2,i}^T B^T g_{i+1,i_f} + w_{i+1,i_f}$$

$$+ x_i^{r T} Qx_i^r \tag{2.60}$$

where the terms are arranged according to the order of x_i. The quadratic terms with respect to x_i in (2.60) can be reduced to $x_i^T K_{i,i_f} x_i$ from Riccati Equation given by

$$K_{i,i_f} = [A - BL_{1,i}]^T K_{i+1,i_f}[A - BL_{1,i}] + L_{1,i}^T RL_{1,i} + Q \tag{2.61}$$

which is the same as (2.45) according to the relation (2.59).

The first-order coefficients with respect to x_i in (2.60) can be written as

$$- L_{1,i}^T RL_{2,i} g_{i+1,i_f} + (A - BL_{1,i})^T K_{i+1,i_f} BL_{2,i} g_{i+1,i_f}$$

$$+ (A - BL_{1,i})^T g_{i+1,i_f} - Qx_i^r$$

$$= -L_{1,i}^T RL_{2,i} g_{i+1,i_f} + A^T K_{i+1,i_f} BL_{2,i} g_{i+1,i_f} - L_{1,i}^T B^T K_{i+1,i_f} BL_{2,i} g_{i+1,i_f}$$

$$+ A^T g_{i+1,i_f} - L_{1,i}^T B^T g_{i+1,i_f} - Qx_i^r$$

$$= -A^T [K_{i+1,i_f}^T B(R + B^T K_{i+1,i_f} B)^{-1} B^T - I] g_{i+1,i_f} - Qx_i^r$$

$$= A^T [I + K_{i+1,i_f}^T BR^{-1} B]^{-1} g_{i+1,i_f} - Qx_i^r$$

which can be reduced to g_{i,i_f} if it is generated from (2.46).

The terms without x_i in (2.60) can be written as

$$g_{i+1,i_f}^T L_{2,i}^T RL_{2,i} g_{i+1,i_f} + g_{i+1,i_f}^T L_{2,i}^T B^T K_{i+1,i_f} BL_{2,i} g_{i+1,i_f}$$

$$+ 2g_{i+1,i_f}^T L_{2,i}^T B^T g_{i+1,i_f} + w_{i+1,i_f} + x_i^{r T} Qx_i^r$$

$$= g_{i+1,i_f}^T B[R + B^T K_{i+1,i_f} B]^{-1} B^T g_{i+1,i_f}$$

$$- 2g_{i+1,i_f}^T B[R + B^T K_{i+1,i_f} B]^{-1} B^T g_{i+1,i_f} + w_{i+1,i_f} + x_i^{r T} Qx_i^r$$

$$= -g_{i+1,i_f}^T B[R + B^T K_{i+1,i_f} B]^{-1} B^T g_{i+1,i_f} + w_{i+1,i_f} + x_i^{r T} Qx_i^r$$

which can be reduced to w_{i,i_f} if it is defined as (2.50). If g_{i,i_f} and w_{i,i_f} are chosen as (2.46) and (2.50), then $J^*(x_i)$ is in a form such as (2.51), i.e.

$$J^*(x_i) = x_i^T K_{i,i_f} x_i + 2x_i^T g_{i,i_f} + w_{i,i_f} \qquad (2.62)$$

Now, we have only to find the boundary value of g_{i,i_f} and w_{i,i_f}. $J^*(x_{i_f})$ should be equal to the performance criterion for the final state. Thus, w_{i_f,i_f} and g_{i_f,i_f} should be chosen as $w_{i_f,i_f} = x_{i_f}^{rT} Q_f x_{i_f}^r$, and $g_{i_f,i_f} = -Q_f x_{i_f}^r$ so that we have

$$\begin{aligned} J^*(x_{i_f}) &= x_{i_f}^T K_{i_f,i_f} x_{i_f} + 2x_{i_f}^T g_{i_f,i_f} + w_{i_f,i_f} \\ &= x_{i_f}^T Q_f x_{i_f} - 2x_{i_f}^T Q_f x_{i_f}^r + x_{i_f}^{rT} Q_f x_{i_f}^r \\ &= (x_{i_f} - x_{i_f}^r)^T Q_f (x_{i_f} - x_{i_f}^r) \end{aligned}$$

This completes the proof.   ∎

The result of Theorem 2.2 will be utilized only for zero reference signals in subsequent sections.

For positive definite Q_f and nonsingular matrix A, we can have another form of the above control (2.43). Let $\hat{P}_{i,i_f} = K_{i,i_f}^{-1}$ if the inverse of K_{i,i_f} exists. Then (2.45) can be represented by

$$\hat{P}_{i,i_f}^{-1} = A^T [I + \hat{P}_{i+1,i_f}^{-1} BR^{-1}B^T]^{-1} \hat{P}_{i+1,i_f}^{-1} A + Q \qquad (2.63)$$

$$\hat{P}_{i,i_f} = \left\{ A^T [\hat{P}_{i+1,i_f} + BR^{-1}B^T]^{-1} A + Q \right\}^{-1} \qquad (2.64)$$

Let $P_{i,i_f} = \hat{P}_{i,i_f} + BR^{-1}B^T$. Then

$$\begin{aligned} P_{i,i_f} &= (A^T P_{i+1,i_f}^{-1} A + Q)^{-1} + BR^{-1}B^T \\ &= A^{-1}(P_{i+1,i_f}^{-1} + A^{-T}QA^{-1})^{-1}A^{-T} + BR^{-1}B^T \\ &= A^{-1}[I + P_{i+1,i_f} A^{-T}QA^{-1}]^{-1} P_{i+1,i_f} A + BR^{-1}B^T \quad (2.65) \end{aligned}$$

$$\begin{aligned} g_{i,i_f} &= A^T [I + \hat{P}_{i+1,i_f}^{-1} BR^{-1}B^T]^{-1} g_{i+1,i_f} - Qx_i^r \\ &= A^T [\hat{P}_{i+1,i_f} + BR^{-1}B^T]^{-1} \hat{P}_{i+1,i_f} g_{i+1,i_f} - Qx_i^r \\ &= A^T P_{i+1,i_f}^{-1} (P_{i+1,i_f} - BR^{-1}B^T) g_{i+1,i_f} - Qx_i^r \quad (2.66) \end{aligned}$$

with the boundary condition

$$P_{i_f,i_f} = Q_f^{-1} + BR^{-1}B^T \qquad (2.67)$$

Using the following relation:

$$\begin{aligned} -[I &+ R^{-1}B^T \hat{P}_{i+1,i_f}^{-1} B]^{-1} R^{-1}B^T \hat{P}_{i+1,i_f}^{-1} Ax_i \\ &= -R^{-1}B^T \hat{P}_{i+1,i_f}^{-1} [I + BR^{-1}B^T \hat{P}_{i+1,i_f}^{-1}]^{-1} Ax_i \\ &= -R^{-1}B^T [\hat{P}_{i+1,i_f} + BR^{-1}B^T]^{-1} Ax_i \\ &= -R^{-1}B^T P_{i+1,i_f}^{-1} Ax_i \end{aligned}$$

and

$$-[I + R^{-1}B^T \hat{P}_{i+1,i_f}^{-1}B]^{-1}R^{-1}Bg_{i+1,i_f}$$
$$= R^{-1}B[I + \hat{P}_{i+1,i_f}^{-1}BR^{-1}B^T]^{-1}g_{i+1,i_f}$$
$$= R^{-1}BP_{i+1,i_f}^{-1}(P_{i+1,i_f} - BR^{-1}B^T)g_{i+1,i_f}$$

we can represent the control in another form:

$$u_i^* = -R^{-1}B^T P_{i+1,i_f}^{-1}[Ax_i + (P_{i+1,i_f} - BR^{-1}B^T)g_{i+1,i_f}] \qquad (2.68)$$

where P_{i+1,i_f} and g_{i+1,i_f} are obtained from (2.65) and (2.66) with boundary conditions (2.67) and (2.40) respectively.

2) Fixed Terminal State

Here, the following performance criterion is considered:

$$J(x^r, u) = \sum_{i=i_0}^{i_f-1}\left[(x_i - x_i^r)^T Q(x_i - x_i^r) + u_i^T Ru_i\right] \qquad (2.69)$$

$$x_{i_f} = x_{i_f}^r \qquad (2.70)$$

For easy understanding, we start off from a simple case.

Case 1: zero state weighting

Now, our terminal objective will be to make x_{i_f} match exactly the desired final reference state $x_{i_f}^r$. Since we are demanding that x_{i_f} be equal to a known desired $x_{i_f}^r$, the final state has no effect on the performance criterion (2.31). It is therefore redundant to include a final state weighting term in a performance criterion. Accordingly, we may as well set $Q_f = 0$.

Before we go to the general problem, we first consider a simple case for the following performance criterion:

$$J_{i_0} = \frac{1}{2}\sum_{i=i_0}^{i_f-1} u_i^T Ru_i \qquad (2.71)$$

where $Q = 0$. Observe that the weighting matrix for the state becomes zero.

As mentioned before, we require the control to drive x_{i_0} exactly to

$$x_{i_f} = x_{i_f}^r \qquad (2.72)$$

using minimum control energy. The terminal condition can be expressed by

$$x_{i_f} = A^{i_f-i_0}x_{i_0} + \sum_{i=i_0}^{i_f-1} A^{i_f-i-1}Bu_i = x_{i_f}^r \qquad (2.73)$$

We try to find the optimal control among ones satisfying (2.73). It can be seen that both the performance criterion and the constraint are expressed in terms of the control, not including the state, which makes the problem tractable. Introducing a Lagrange multiplier λ, we have

$$J_{i_0} = \frac{1}{2} \sum_{i=i_0}^{i_f-1} u_i^T R u_i + \lambda^T (A^{i_f-i_0} x_{i_0} + \sum_{i=i_0}^{i_f-1} A^{i_f-i-1} B u_i - x_{i_f}^r) \quad (2.74)$$

Take the derivative on both sides of Equation (2.74) with respect to u_i to obtain

$$R u_i + B^T (A^T)^{i_f-i-1} \lambda = 0 \quad (2.75)$$

Thus,

$$u_i = -R^{-1} B^T (A^T)^{i_f-i-1} \lambda \quad (2.76)$$

Substituting (2.76) into (2.73) and solving for λ yields

$$\lambda = -G_{i_0,i_f}^{-1} (x_{i_f}^r - A^{i_f-i_0} x_{i_0}) \quad (2.77)$$

where

$$G_{i_0,i_f} = \sum_{i=i_0}^{i_f-1} A^{i_f-i-1} B R^{-1} B^T (A^T)^{i_f-i-1}. \quad (2.78)$$

Actually, G_{i_0,i_f} is a controllability Gramian of the systems (2.27). In the case of controllable systems, G_{i_0,i_f} is guaranteed to be nonsingular if $i_f - i_0$ is more than or equal to the controllability index n_c.

The optimal open-loop control is given by

$$u_i^* = R^{-1} B^T (A^T)^{i_f-i-1} G_{i_0,i_f}^{-1} (x_{i_f}^r - A^{i_f-i_0} x_{i_0}) \quad (2.79)$$

It is noted that the open-loop control is defined for all $i \in [i_0, \ i_f - 1]$. Since i_0 is arbitrary, we can obtain the closed-loop control by replacing with i such as,

$$u_i^* = R^{-1} B^T (A^T)^{i_f-i-1} G_{i,i_f}^{-1} (x_{i_f}^r - A^{i_f-i} x_i) \quad (2.80)$$

It is noted that the closed-loop control can be defined only on i that is less than or equal $i_f - n_c$. After the time $i_f - n_c$, the open-loop control can be used, if necessary. In Figure 2.3, the regions of the closed- and open-loop control are shown respectively.

The above solutions can also be obtained with the formal procedure using the minimum principle, but it is given in a closed form from this procedure. Thus, the control after $i_f - n_c$ cannot be obtained.

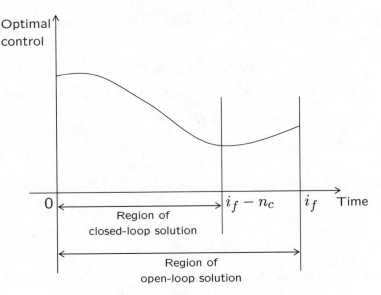

Fig. 2.3. Region of the closed-loop solution

Case 2: nonzero state weighting

We derive the optimal solution for the fixed terminal state from that for the free terminal state by setting $Q_f = \infty I$. We assume that A is nonsingular.

Since $K_{i_f,i_f} = Q_f = \infty I$, the boundary condition of P_{i,i_f} becomes

$$P_{i_f,i_f} = BR^{-1}B^T \tag{2.81}$$

from (2.67). From (2.65), we know that Equation (2.81) is satisfied with another terminal condition:

$$P_{i_f+1,i_f+1} = 0 \tag{2.82}$$

It is noted that P_{i+1,i_f} can be singular on $[i_f - n_c + 2, i_f]$. Therefore, g_{i,i_f} cannot be generated from (2.66) and the control (2.68) does not make sense. However, the control for the zero reference signal can be represented as

$$u_i^* = -R^{-1}B^T P_{i+1,i_f}^{-1} Ax_i \tag{2.83}$$

where g_{i,i_f} is not necessary.

For nonzero reference signals we will take an approach called the sweep method. The state and costate equations are the same as those of the free terminal case:

$$u_i = -R^{-1}B^T p_{i+1} \tag{2.84}$$
$$x_{i+1} = Ax_i - BR^{-1}B^T p_{i+1} \tag{2.85}$$
$$p_i = Q(x_i - x_i^r) + A^T p_{i+1} \tag{2.86}$$

We try to find an optimal control to ensure $x_{i_f} = x_{i_f}^r$. Assume the following relation:

$$p_i = K_i x_i + M_i p_{i_f} + g_i \tag{2.87}$$

where we need to find S_i, M_i, and g_i satisfying the boundary conditions

$$K_{i_f} = 0$$
$$M_{i_f} = I$$
$$g_{i_f} = 0$$

respectively. Combining (2.85) with (2.87) yields the following optimal trajectory:

$$x_{i+1} = (I + BR^{-1}B^T K_{i+1})^{-1}(Ax_i - BR^{-1}B^T M_{i+1}p_{i_f}$$
$$- BR^{-1}B^T g_{i+1}) \tag{2.88}$$

Substituting (2.87) into (2.86) provides

$$K_i x_i + M_i p_{i_f} + g_i = Q(x_i - x_i^r) + A^T[K_{i+1}x_{i+1} + M_{i+1}p_{i_f} + g_{i+1}] \tag{2.89}$$

Substituting x_{i+1} in (2.88) into (2.89) yields

$$[-K_i + A^T K_{i+1}(I + BR^{-1}B^T K_{i+1})^{-1}A + Q]x_i +$$
$$[-M_i - A^T K_{i+1}(I + BR^{-1}B^T K_{i+1})^{-1}BR^{-1}B^T M_{i+1} + A^T M_{i+1}]p_{i_f} +$$
$$[-g_i + A^T g_{i+1} - A^T K_{i+1}(I + BR^{-1}B^T K_{i+1})^{-1}BR^{-1}B^T g_{i+1} - Qx_i^r] = 0$$

Since this equality holds for all trajectories x_i arising from any initial condition x_{i_0}, each term in brackets must vanish. The matrix inversion lemma, therefore, yields the Riccati equation

$$K_i = A^T K_{i+1}(I + BR^{-1}B^T K_{i+1})^{-1}A + Q \tag{2.90}$$

and the auxiliary homogeneous difference equation

$$M_i = A^T M_{i+1} - A^T K_{i+1}(I + BR^{-1}B^T K_{i+1})^{-1}BR^{-1}B^T M_{i+1}$$
$$g_i = A^T g_{i+1} - A^T K_{i+1}(I + BR^{-1}B^T K_{i+1})^{-1}BR^{-1}B^T g_{i+1} - Qx_i^r$$

We assume that $x_{i_f}^r$ is a linear combination of x_i, p_{i_f}, and some specific matrix N_i for all i, i.e.

$$x_{i_f}^r = U_i x_i + S_i p_{i_f} + h_i \tag{2.91}$$

Evaluating for $i = i_f$ yields

$$U_{i_f} = I \tag{2.92}$$
$$S_{i_f} = 0 \tag{2.93}$$
$$h_{i_f} = 0 \tag{2.94}$$

Clearly, then

$$U_i = M_i^T \tag{2.95}$$

The left-hand side of (2.91) is a constant, so take the difference to obtain

$$0 = U_{i+1}x_{i+1} + S_{i+1}p_{i_f} + h_{i+1} - U_i x_i - S_i p_{i_f} - h_i \tag{2.96}$$

Substituting x_{i+1} in (2.88) into (2.96) and rearranging terms, we have

$$[U_{i+1}\{A - B(B^T K_{i+1}B + R)^{-1}B^T K_{i+1}A\} - U_i]x_i$$
$$+ [S_{i+1} - S_i - U_{i+1}B(B^T K_{i+1}B + R)^{-1}B^T M_{i+1}]p_{i_f}$$
$$+ h_{i+1} - h_i - U_{i+1}(I + BR^{-1}B^T K_{i+1})^{-1}BR^{-1}B^T g_{i+1} = 0 \tag{2.97}$$

The first term says that

$$U_i = U_{i+1}\{A - B(B^T K_{i+1}B + R)^{-1}B^T K_{i+1}A\} \tag{2.98}$$

The second and third terms now yield the following recursive equations:

$$S_i = S_{i+1} - M_{i+1}^T B(B^T K_{i+1}B + R)^{-1}B^T M_{i+1} \tag{2.99}$$
$$h_i = h_{i+1} - M_{i+1}^T (I + BR^{-1}B^T K_{i+1})^{-1}BR^{-1}B^T g_{i+1} \tag{2.100}$$

We are now in a position to determine p_{i_f}. From (2.91), we have

$$p_{i_f} = S_{i_0}^{-1}(x_{i_f}^r - M_{i_0}^T x_{i_0} - h_{i_0}) \tag{2.101}$$

We can now finally compute the optimal control

$$u_i = -R^{-1}B^T[K_{i+1}x_{i+1} + M_{i+1}p_{i_f} + g_{i+1}] \tag{2.102}$$

by substituting (2.87) into (2.84).

u_i can be represented in terms of the current state x_i:

$$u_i = -R^{-1}B^T(I + K_{i+1}BR^{-1}B^T)^{-1}[K_{i+1}Ax_i + M_{i+1}p_{i_f} + g_{i+1}],$$
$$= -R^{-1}B^T(I + K_{i+1}BR^{-1}B^T)^{-1}[K_{i+1}Ax_i + M_{i+1}S_{i_0}^{-1}(x_{i_f}^r$$
$$- M_{i_0}^T x_{i_0} - h_{i_0}) + g_{i+1}] \tag{2.103}$$

What we have done so far is summarized in the following theorem.

Theorem 2.3. *The LQTC for the fixed terminal state is given in (2.103). S_i, M_i, P_i, g_i, h_i are as follows:*

$$K_i = A^T K_{i+1}(I + BR^{-1}B^T K_{i+1})^{-1}A + Q$$
$$M_i = A^T M_{i+1} - A^T K_{i+1}(I + BR^{-1}B^T K_{i+1})^{-1}BR^{-1}B^T M_{i+1}$$
$$S_i = S_{i+1} - M_{i+1}^T B(B^T K_{i+1}B + R)^{-1}B^T M_{i+1}$$
$$g_i = A^T g_{i+1} - A^T K_{i+1}(I + BR^{-1}B^T K_{i+1})^{-1}BR^{-1}B^T g_{i+1} - Qx_i^r$$
$$h_i = h_{i+1} - M_{i+1}^T (I + BR^{-1}B^T K_{i+1})^{-1}BR^{-1}B^T g_{i+1}$$

where

$$K_{i_f} = 0, \ M_{i_f} = I, \ S_{i_f} = 0, \ g_{i_f} = 0, \ h_{i_f} = 0$$

∎

It is noted that the control (2.103) is a state feedback control with respect to the current state and an open-loop control with respect to the initial state, which looks somewhat awkward at first glance. However, if the receding horizon scheme is adopted, then we can obtain the state feedback control that requires only the current state, not other past states. That will be covered in the next chapter.

Replacing i_0 with i in (2.103) yields the following closed-loop control:

$$\begin{aligned} u_i = -R^{-1}B^T(I + K_{i+1}BR^{-1}B^T)^{-1}[K_{i+1}Ax_i + M_{i+1}S_i^{-1} \\ \times (x_{i_f}^r - M_i^T x_i - h_i) + g_{i+1}] \end{aligned} \tag{2.104}$$

where S_i is guaranteed to be nonsingular on $i \leq i_f - n_c$.

If Q in (2.103) becomes zero, then (2.103) is reduced to (2.79), which is left as a problem at the end of this chapter.

For the zero reference signal, g_i and h_i in Theorem 2.3 become zero due to $x_i^r = 0$. Thus, we have

$$u_i = -R^{-1}B^T(I + K_{i+1}BR^{-1}B^T)^{-1}[K_{i+1}A - M_{i+1}S_i^{-1}M_i^T]x_i \tag{2.105}$$

in the form of the closed-loop control. As seen above, it is a little complex to obtain the closed-form solution for the fixed terminal state problem with nonzero reference signals.

Example 2.1

The LQTC (2.103) with the fixed terminal state is a new type of a tracking control. It is demonstrated through a numerical example.

Consider the following state space model:

$$x_{k+1} = \begin{bmatrix} 0.013 & 0.811 & 0.123 \\ 0.004 & 0.770 & 0.096 \\ 0.987 & 0.903 & 0.551 \end{bmatrix} x_k + \begin{bmatrix} 0.456 \\ 0.018 \\ 0.821 \end{bmatrix} u_k \tag{2.106}$$

Q and R in the performance criterion (2.69) are set to $100I$ and I respectively. The reference signal and state trajectories can be seen in Figure 2.4 where the fixed terminal condition is met. A batch form solution for the fixed terminal state is given in Section 3.5, and its computation turns out to be the same as that of (2.103). ∎

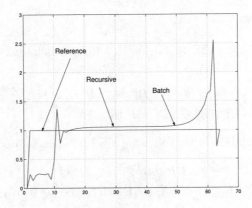

(a)   First component of the state

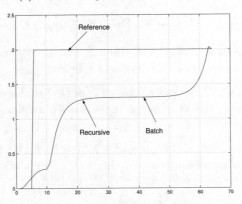

(b)   Second component of the state

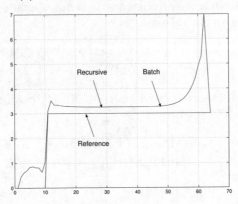

(c)   Third component of the state

Fig. 2.4. State trajectory of Example 2.1

Infinite Horizon Case

If i_f goes to ∞ in (2.43), (2.45), and (2.46), an infinite horizon LQTC is given by

$$
\begin{aligned}
u_i^* &= -[I + R^{-1}B^T K_\infty B]^{-1} R^{-1} B^T [K_\infty A x_i + g_{i+1,\infty}] \\
&= -[R + B^T K_\infty B]^{-1} B^T [K_\infty A x_i + g_\infty]
\end{aligned}
\tag{2.107}
$$

where K_∞ is a solution of the following algebraic Riccati Equation (ARE):

$$
\begin{aligned}
K_\infty &= A^T [I + K_\infty B R^{-1} B^T]^{-1} K_\infty A + Q \tag{2.108} \\
&= A^T K_\infty A - A^T K_\infty B [R + B K_\infty B^T]^{-1} B^T K_\infty A + Q \tag{2.109}
\end{aligned}
$$

and g_∞ is given by

$$
g_\infty = A^T [I + K_\infty B R^{-1} B^T]^{-1} g_\infty - Q \bar{x}^r
\tag{2.110}
$$

with a fixed reference signal $\bar{x}^r$. The stability and an existence of the solution to the Riccati equation are summarized as follows:

Theorem 2.4. *If (A, B) is controllable and $(A, Q^{\frac{1}{2}})$ is observable, the solution to Riccati Equation (2.108) is unique and positive definite, and the stability of u_i (2.107) is guaranteed.*

We can see a proof of Theorem 2.4 in much of the literature, e.g. in [Lew86a]. The conditions on controllability and observability in Theorem 2.4 can be weakened to the reachability and detectability.

Here, we shall present the return difference equality for the infinite horizon LQ control and introduce some robustness in terms of gain and phase margins. From the following simple relation:

$$
\begin{aligned}
K_\infty - A^T K_\infty A &= (z^{-1}I - A)^T K_\infty (zI - A) + (z^{-1}I - A)^T K_\infty A \\
&\quad + A^T K_\infty (zI - A)
\end{aligned}
\tag{2.111}
$$

$K_\infty - A^T K_\infty A$ in (2.111) is replaced with $-A^T K_\infty B (B^T K_\infty B + R)^{-1} B^T K_\infty A - Q$ according to (2.109) to give

$$
\begin{aligned}
(z^{-1}I - A)^T K_\infty (zI - A) + (z^{-1}I - A)^T K_\infty A + A^T K_\infty (zI - A) \\
+ A^T K_\infty B (B^T K_\infty B + R)^{-1} B^T K_\infty A = Q
\end{aligned}
\tag{2.112}
$$

Pre- and post-multiply (2.112) by $B^T(z^{-1}I - A)^{-T}$ and $(zI - A)^{-1}B$ respectively to get

$$
\begin{aligned}
B^T K_\infty B + B^T K_\infty A (zI - A)^{-1} B + B^T (z^{-1}I - A)^{-T} A^T K_\infty B \\
+ B^T (z^{-1}I - A)^{-T} A^T K_\infty B (B^T K_\infty B + R)^{-1} B^T K_\infty A (zI - A)^{-1} B \\
= B^T (z^{-1}I - A)^{-T} Q (zI - A)^{-1} B
\end{aligned}
\tag{2.113}
$$

Adding R to both sides of (2.113) and factorizing it yields the following equation:

$$B^T(z^{-1}I - A)^{-T}Q(zI - A)^{-1}B + R$$
$$= [I + \mathcal{K}_\infty(z^{-1}I - A)^{-1}B]^T(B^TK_\infty B + R)[I + \mathcal{K}_\infty(zI - A)^{-1}B] \quad (2.114)$$

where $\mathcal{K}_\infty = [R + B^TK_\infty B]^{-1}B^TK_\infty A$.

From Equation (2.114), we are in a position to check gain and phase margins for the infinite horizon LQ control. First, let $F(z)$ be $I + \mathcal{K}_\infty(zI - A)^{-1}B$, which is called a return difference matrix. It follows from (2.114) that

$$B^TK_\infty B + R = F^{-T}(z^{-1})[R + B^T(z^{-1}I - A)^{-T}Q(zI - A)^{-1}B]F^{-1}(z)$$

which implies that

$$\bar{\sigma}(B^TK_\infty B + R) \geq \bar{\sigma}^2(F^{-1}(z))$$
$$\times \underline{\sigma}[R + B^T(z^{-1}I - A)^{-T}Q(zI - A)^{-1}B] \quad (2.115)$$

Note that $\bar{\sigma}(M^TSM) \geq \underline{\sigma}(S)\bar{\sigma}^2(M)$ for $S \geq 0$. Recalling the two facts $\bar{\sigma}[F^{-1}(z)] = \underline{\sigma}^{-1}[F(z)]$ and $\bar{\sigma}(I - z^{-1}A) \leq 1 + \bar{\sigma}(A)$ for $|z| = 1$, we have

$$\underline{\sigma}[R + B^T(z^{-1}I - A)^{-T}Q(zI - A)^{-1}B]$$
$$\geq \underline{\sigma}(R)\underline{\sigma}[I + R^{-\frac{1}{2}}B^T(z^{-1}I - A)^{-T}Q(zI - A)^{-1}BR^{-\frac{1}{2}}]$$
$$\geq \underline{\sigma}(R)[1 + \underline{\sigma}^{-1}(R)\underline{\sigma}^2(B)\underline{\sigma}(Q)\bar{\sigma}^{-2}(I - z^{-1}A)\alpha]$$
$$\geq \frac{\underline{\sigma}(R)}{\bar{\sigma}(R)}[\bar{\sigma}(R) + \underline{\sigma}^2(B)\underline{\sigma}(Q)\{1 + \bar{\sigma}(A)\}^{-2}\alpha] \quad (2.116)$$

where α is 1 when $p \leq q$ and 0 otherwise. Recall that the dimensions of inputs u_i and outputs y_i are p and q respectively. Substituting (2.116) into (2.115) and arranging terms yields

$$\underline{\sigma}^2[F(z)] \geq \frac{\underline{\sigma}(R)/\bar{\sigma}(R)}{\bar{\sigma}(R) + \bar{\sigma}^2(B)\bar{\sigma}(K_\infty)}[\bar{\sigma}(R) + \underline{\sigma}^2(B)\underline{\sigma}(Q)\{1 + \bar{\sigma}(A)\}^{-2}\alpha]$$
$$\triangleq R_f^2 \quad (2.117)$$

Let a circle of radius R_f centered at $(-1, 0)$ be $C(-1, R_f)$. The Nyquist plot of the open-loop system of the optimal regulator lies outside $C(-1, R_f)$ for an SISO system, as can be seen in Figure 2.5; the guaranteed gain margins GM of a control are given by

$$(1 + R_f)^{-1} \leq GM \leq (1 - R_f)^{-1} \quad (2.118)$$

and the phase margins PM of the control are given by

$$-2\sin^{-1}(\frac{R_f}{2}) \leq PM \leq 2\sin^{-1}(\frac{R_f}{2}) \quad (2.119)$$

It is noted that margins for discrete systems are smaller than those for continuous systems, i.e. $0.5 \leq GM < \infty$, and $-\pi/3 \leq PM \leq \pi/3$.

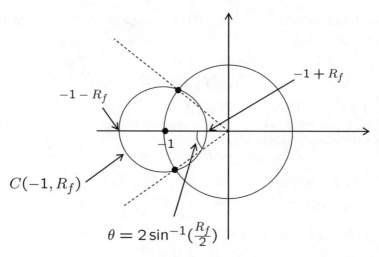

Fig. 2.5. Nyquist plot

2.3.2 H_∞ Control Based on Minimax Criterion

In this subsection we derive an H_∞ tracking control (HTC) for discrete time-invariant systems in a state-feedback form. Consider the following discrete time-invariant system:

$$x_{i+1} = Ax_i + B_w w_i + Bu_i$$
$$\hat{z}_i = \begin{bmatrix} Q^{\frac{1}{2}} x_i \\ R^{\frac{1}{2}} u_i \end{bmatrix} \tag{2.120}$$

where $x_i \in \Re^n$ denotes the state, $w_i \in \Re^l$ the disturbance, $u_i \in \Re^m$ the control input, and $\hat{z}_i \in \Re^{q+n}$ the controlled variable which needs to be regulated. The H_∞ norm of $T_{\hat{z}w}(e^{jw})$ can be represented as

$$\|T_{\hat{z}w}(e^{jw})\|_\infty = \sup_{w_i} \bar{\sigma}(T_{\hat{z}w}(e^{jw})) = \sup_{w_i} \frac{\sum_{i=i_0}^\infty [x_i^T Q x_i + u_i^T R u_i]}{\sum_{i=i_0}^\infty w_i^T w_i}$$

$$= \sup_{\|w_i\|_2 = 1} \sum_{i=i_0}^\infty [x_i^T Q x_i + u_i^T R u_i] = \gamma^{*2} \tag{2.121}$$

where $T_{\hat{z}w}(e^{jw})$ is a transfer function from w_i to $\hat{z}_i$ and $\bar{\sigma}(\cdot)$ is the maximum singular value. $\hat{z}_i$ in (2.120) is chosen to make a quadratic cost function as (2.121).

The H_∞ norm of the systems is equal to the induced L_2 norm. The H_∞ control is obtained so that the H_∞ norm is minimized with respect to u_i.

However, it is hard to achieve an optimal H_∞ control. Instead of the above performance criterion, we can introduce a suboptimal control such that

$\|T_{\hat{z}w}(e^{jw})\|_\infty < \gamma^2$ for some positive gamma γ^2 ($> \gamma^{*2}$). For $\|T_{\hat{z}w}(e^{jw})\|_\infty < \gamma^2$ we will obtain a control so that the following inequality is satisfied:

$$\frac{\sum_{i=i_0}^\infty [x_i^T Q x_i + u_i^T R u_i]}{\sum_{i=i_0}^\infty w_i^T w_i} < \gamma^2 \tag{2.122}$$

for all w_i. Observe that the gain from $\|w_i\|_2^2$ to $\|\hat{z}_i\|_2^2$ in (2.122) is always less than γ, so that the maximum gain, i.e. H_∞ norm, is also less than γ^2.

From simple algebraic calculations, we have

$$\sum_{i=i_0}^\infty [x_i^T Q x_i + u_i^T R u_i - \gamma^2 w_i^T w_i] < 0 \tag{2.123}$$

from (2.122). Since the inequality (2.123) should be satisfied for all w_i, the value of the left side of (2.123) should be always negative, i.e.

$$\sup_{w_i}\left\{ \sum_{i=i_0}^\infty [x_i^T Q x_i + u_i^T R u_i - \gamma^2 w_i^T w_i] \right\} < 0 \tag{2.124}$$

In order to check whether the feasible solution to (2.124) exists, we try to find out a control minimizing the left side of the inequality (2.124) and the corresponding optimal cost. If this optimal cost is positive, then we cannot obtain the control satisfying the H_∞ norm. Unlike an LQ control, the fixed terminal state is impossible in H_∞ controls. We focus only on the free terminal state.

1) Free Terminal State

When dealing with the finite horizon case, we usually include a weighting matrix for the terminal state, such as

$$\max_{w_i}\left\{ \sum_{i=i_0}^{i_f-1} \left[x_i^T Q x_i + u_i^T R u_i - \gamma^2 w_i^T w_i \right] + x_{i_f}^T Q_f x_{i_f} \right\} < 0 \tag{2.125}$$

A feasible solution u_i in (2.125) can be obtained from the following difference game problem:

$$\min_u \max_w \ J(u, w) \tag{2.126}$$

where

$$J(u, w) = \sum_{i=i_0}^{i_f-1} \left[x_i^T Q x_i + u_i^T R u_i - \gamma^2 w_i^T w_i \right] + x_{i_f}^T Q_f x_{i_f} \tag{2.127}$$

Note that the initial state is assumed to be zero in H_∞ norm in (2.121). However, in the difference game problem (2.126)–(2.127), the nonzero initial state can be handled.

Until now, the regulation problem has been considered. If a part of the state should be steered according to the given reference signal, we can consider

$$
\begin{aligned}
J(u, w) = \sum_{i=i_0}^{i_f-1} & [\, (z_i - z_i^r)^T \bar{Q}(z_i - z_i^r) + u_i R u_i - \gamma^2 w_i R_w w_i] \\
& + (z_{i_f} - z_{i_f}^r)^T \bar{Q}_f (z_{i_f} - z_{i_f}^r)
\end{aligned}
$$

instead of (2.127). Here, $z_i = C_z x_i$ is expected to approach z_i^r.

It is well known that for a given $p \times n (p \leq n)$ full rank matrix C_z there always exist some $n \times p$ matrices L such that $C_z L = I_{p \times p}$. For example, we can take $L = C_z^T (C_z C_z^T)^{-1}$. Let $x_i^r = L y_i^r$. $J(u, w)$ is rewritten as

$$
\begin{aligned}
J(u, w) = \sum_{i=i_0}^{i_f-1} & \left[(x_i - x_i^r)^T Q(x_i - x_i^r) + u_i^T R u_i - \gamma^2 w_i^T R_w w_i \right] \\
& + (x_{i_f} - x_{i_f}^r)^T Q_f (x_{i_f} - x_{i_f}^r)
\end{aligned} \tag{2.128}
$$

where $Q = C_z^T \bar{Q} C_z$ and $Q_f = C_z^T \bar{Q}_f C_z$ or Q_f and Q are independent design parameters.

For the optimal solution, we first form the following Hamiltonian:

$$
\begin{aligned}
\mathcal{H}_i = & [(x_i - x_i^r)^T Q(x_i - x_i^r) + u_i^T R u_i - \gamma^2 w_i^T R_w w_i] \\
& + p_{i+1}^T (A x_i + B_w w_i + B u_i), \quad i = i_0, \cdots, i_f - 1
\end{aligned}
$$

The necessary conditions for u_i and w_i to be the saddle points are

$$
x_{i+1} = \frac{\partial \mathcal{H}}{\partial p_{i+1}} = A x_i + B_w w_i + B u_i \tag{2.129}
$$

$$
p_i = \frac{\partial \mathcal{H}}{\partial x_i} = 2Q(x_i - x_i^r) + A^T p_{i+1} \tag{2.130}
$$

$$
0 = \frac{\partial \mathcal{H}}{\partial u_i} = 2R u_i + B^T p_{i+1} \tag{2.131}
$$

$$
0 = \frac{\partial \mathcal{H}}{\partial w_i} = -2\gamma^2 R_w w_i + B_w^T p_{i+1} \tag{2.132}
$$

$$
p_{i_f} = \frac{\partial h(x_{i_f})}{\partial x_{i_f}} = 2Q_f(x_{i_f} - x_{i_f}^r) \tag{2.133}
$$

where

$$
h(x_{i_f}) = (x_{i_f} - x_{i_f}^r)^T Q_f (x_{i_f} - x_{i_f}^r) \tag{2.134}
$$

Assume

$$
p_i = 2M_{i,i_f} x_i + 2g_{i,i_f} \tag{2.135}
$$

From (2.129), (2.131), and (2.135), we have

$$\frac{\partial \mathcal{H}_i}{\partial u_i} = 2Ru_i + B^T p_{i+1}$$

$$= 2Ru_i + 2B^T M_{i+1,i_f} x_{i+1} + 2B^T g_{i+1,i_f}$$

$$= 2Ru_i + 2B^T M_{i+1,i_f} [Ax_i + B_w w_i + Bu_i] + 2B^T g_{i+1,i_f}$$

Therefore,

$$\frac{\partial^2 \mathcal{H}_i}{\partial u_i^2} = 2R + 2B^T M_{i+1,i_f} B$$

It is apparent that $\frac{\partial^2 \mathcal{H}_i}{\partial u_i^2} > 0$ for $i = i_0, \cdots, i_f - 1$. Similarly, we have

$$\frac{\partial \mathcal{H}_i}{\partial w_i} = -2\gamma^2 R_w w_i + 2B_w^T M_{i+1,i_f} [Ax_i + B_w w_i + Bu_i] + 2B_w^T g_{i+1,i_f}$$

Therefore,

$$\frac{\partial^2 \mathcal{H}_i}{\partial w_i^2} = -2\gamma^2 R_w + 2B_w^T M_{i+1,i_f} B_w$$

From these, the difference game problem (2.126) for the performance criterion (2.128) has a unique solution if and only if

$$R_w - \gamma^{-2} B_w^T M_{i+1,i_f} B_w > 0, \quad i = i_0, \cdots, i_f - 1 \qquad (2.136)$$

We proceed to obtain the optimal solution w_i^* and u_i^*. Eliminating u_i and w_i in (2.129) using (2.131) and (2.132) yields

$$x_{i+1} = Ax_i + \frac{1}{2}(-BR^{-1}B^T + \gamma^{-2} B_w R_w^{-1} B_w^T) p_{i+1} \qquad (2.137)$$

From (2.135) and (2.137) we obtain

$$p_{i+1} = 2M_{i+1,i_f} x_{i+1} + 2g_{i+1,i_f}$$

$$= 2M_{i+1,i_f} Ax_i + M_{i+1,i_f}(-BR^{-1}B^T + \gamma^{-2} B_w R_w^{-1} B_w^T) p_{i+1} + 2g_{i+1,i_f}$$

Therefore,

$$p_{i+1} = 2[I + M_{i+1,i_f}(BR^{-1}B^T - \gamma^{-2} B_w R_w^{-1} B_w^T)]^{-1}(M_{i+1,i_f} Ax_i + g_{i+1,i_f})$$

Let

$$\Lambda_{i+1,i_f} = I + M_{i+1,i_f}(BR^{-1}B^T - \gamma^{-2} B_w R_w^{-1} B_w^T) \qquad (2.138)$$

Then p_{i+1} is rewritten as

$$p_{i+1} = 2\Lambda_{i+1,i_f}^{-1}[M_{i+1,i_f} Ax_i + g_{i+1,i_f}] \qquad (2.139)$$

If we substitute (2.139) into (2.130), then we obtain

$$p_i = 2Q(x_i - x_i^r) + 2A^T \Lambda_{i+1,i_f}^{-1}[M_{i+1,i_f} Ax_i + g_{i+1,i_f}]$$

$$= 2[A^T \Lambda_{i+1,i_f}^{-1} M_{i+1,i_f} A + Q]x_i + 2A^T \Lambda_{i+1,i_f}^{-1} g_{i+1,i_f} - 2Qx_i^r$$

Therefore, from (2.133) and the assumption (2.135), we have

$$M_{i,i_f} = A^T \Lambda_{i+1,i_f}^{-1} M_{i+1,i_f} A + Q \qquad (2.140)$$

$$M_{i_f,i_f} = Q_f \qquad (2.141)$$

and

$$g_{i,i_f} = A^T \Lambda_{i+1,i_f}^{-1} g_{i+1,i_f} - Qx_i^r \qquad (2.142)$$

$$g_{i_f,i_f} = -Q_f x_{i_f}^r \qquad (2.143)$$

for $i = i_0, \cdots, i_f - 1$. From (2.136), we have

$$I - \gamma^{-2} R_w^{-\frac{1}{2}} B_w^T M_{i+1,i_f} B_w R_w^{-\frac{1}{2}} > 0 \qquad (2.144)$$

$$I - \gamma^{-2} M_{i+1,i_f}^{\frac{1}{2}} B_w R_w^{-1} B_w^T M_{i+1,i_f}^{\frac{1}{2}} > 0 \qquad (2.145)$$

where the second inequality comes from the fact that $I - SS^T > 0$ implies $I - S^T S > 0$. $\Lambda_{i+1,i_f}^{-1} M_{i+1,i_f}$ in the right side of (2.140) can be written as

$$\Lambda_{i+1,i_f}^{-1} M_{i+1,i_f}$$

$$= \left[I + M_{i+1,i_f} (BR^{-1}B^T - \gamma^{-2} B_w R_w^{-1} B_w^T) \right]^{-1} M_{i+1,i_f}$$

$$= M_{i+1,i_f}^{\frac{1}{2}} \left[I + M_{i+1,i_f}^{\frac{1}{2}} (BR^{-1}B^T - \gamma^{-2} B_w R_w^{-1} B_w^T) M_{i+1,i_f}^{\frac{1}{2}} \right]^{-1} M_{i+1,i_f}^{\frac{1}{2}}$$

$$\geq 0$$

where the last inequality holds because of (2.145). Therefore, M_{i,i_f} generated by (2.140) is always nonnegative definite.

From (2.131) and (2.132), the H_∞ controls are given by

$$u_i^* = -R^{-1}B^T \Lambda_{i+1,i_f}^{-1} [M_{i+1,i_f} Ax_i + g_{i+1,i_f}] \qquad (2.146)$$

$$w_i^* = \gamma^{-2} R_w^{-1} B_w^T \Lambda_{i+1,i_f}^{-1} [M_{i+1,i_f} Ax_i + g_{i+1,i_f}] \qquad (2.147)$$

It is noted that u_i^* is represented by

$$u_i^* = H_i x_i + v_i$$

where

$$H_i = -R^{-1}B^T \Lambda_{i+1,i_f}^{-1} M_{i+1,i_f} A$$

$$v_i = -R^{-1}B^T \Lambda_{i+1,i_f}^{-1} g_{i+1,i_f}$$

Here, H_i is the feedback gain matrix and v_i can be viewed as a command signal.

The optimal cost can be represented as

$$J^* = x_i^T M_{i,i_f} x_i + 2x_i^T g_{i,i_f} + h_{i,i_f} \qquad (2.148)$$

where M_{i,i_f} and g_{i,i_f} are defined as (2.140) and (2.142) respectively. The derivation for h_{i,i_f} is left as an exercise.

The saddle-point value of the difference game with (2.128) for a zero reference signal is given as

$$J^*(x_i, i, i_f) = x_i^T M_{i,i_f} x_i \qquad (2.149)$$

Since M_{i,i_f} is nonnegative definite, the saddle-point value (2.149) is nonnegative.

To conclude, the solution of the HTC problem can be reduced to finding M_{i,i_f} and g_{i,i_f} for $i = i_0, \cdots, i_f - 1$. The Riccati solution M_{i,i_f} is a symmetric matrix, which can be found by solving (2.140) backward in time using the boundary condition (2.141). In a similar manner, g_{i,i_f} can be found by solving (2.142) backward in time using the boundary condition (2.143).

For a regulation problem, i.e. $x_i^r = 0$, the control (2.146) and the disturbance (2.147) can also be represented as

$$\begin{bmatrix} u_i^* \\ w_i^* \end{bmatrix} = -R_{c,i}^{-1} \begin{bmatrix} B^T \\ B_w^T \end{bmatrix} M_{i+1,i_f} A x_i \qquad (2.150)$$

$$M_{i,i_f} = A^T M_{i+1,i_f} A - A^T M_{i+1,i_f} \begin{bmatrix} B & B_w \end{bmatrix} R_{c,i}^{-1} \begin{bmatrix} B^T \\ B_w^T \end{bmatrix} M_{i+1,i_f} A$$

where

$$R_{c,i} = \begin{bmatrix} B^T \\ B_w^T \end{bmatrix} M_{i+1,i_f} \begin{bmatrix} B & B_w \end{bmatrix} + \begin{bmatrix} R & 0 \\ 0 & -\gamma^2 R_w \end{bmatrix}$$

It is observed that optimal solutions u^* and w^* in (2.150) look like an LQ solution.

For a positive definite Q_f and a nonsingular matrix A, we can have another form of the control (2.146) and the disturbance (2.147). Let

$$\Pi = BR^{-1}B - \gamma^{-2} B_w R_w^{-1} B_w^T \qquad (2.151)$$

It is noted that M_{i,i_f} is obtained from K_{i,i_f} of the LQ control by replacing $BR^{-1}B^T$ by Π. If M_{i,i_f} is nonsingular at $i \leq i_f$, then there exists the following quantity:

$$P_{i,i_f} = M_{i,i_f}^{-1} + \Pi$$

In terms of P_{i,i_f}, (2.146) and (2.147) are represented as

$$u_i^* = -R^{-1}B^T P_{i+1,i_f}^{-1}[Ax_i + (P_{i+1,i_f} - \Pi)g_{i+1,i_f}] \qquad (2.152)$$

$$w_i^* = \gamma^{-2} R_w^{-1} B_w^T P_{i+1,i_f}^{-1}[Ax_i + (P_{i+1,i_f} - \Pi)g_{i+1,i_f}] \qquad (2.153)$$

where

$$P_{i,i_f} = A^{-1}P_{i+1,i_f}[I + A^{-1}QA^{-1}P_{i+1,i_f}]^{-1}A^{-1} + \Pi \tag{2.154}$$

and

$$g_{i,i_f} = -A^T P_{i+1,i_f}^{-1}(P_{i+1,i_f} - \Pi)g_{i+1,i_f} - Qx_i^r \tag{2.155}$$

with

$$P_{i_f,i_f} = M_{i_f,i_f}^{-1} + \Pi = Q_f^{-1} + \Pi > 0, \quad g_{i_f,i_f} = -Q_f x_{i_f}^r \tag{2.156}$$

Here, Q_f must be nonsingular.

Note that P_{i,i_f}^{-1} is well defined only if M_{i,i_f} satisfies the condition

$$R_w - \gamma^{-2}B_w^T M_{i+1,i_f} B_w > 0 \tag{2.157}$$

which is required for the existence of the saddle-point.

The terminal weighting matrix Q_f cannot be arbitrarily large, since M_{i,i_f} generated from the large $M_{i_f,i_f} = Q_f$ is also large and thus the inequality condition (2.136) may not be satisfied. That is why the terminal equality constraint for case of the RH H_∞ control does not make sense.

Infinite Horizon Case

From the finite horizon H_∞ control of a form (2.150), we now turn to the infinite horizon H_∞ control, which is summarized in the following theorem.

Theorem 2.5. *Suppose that (A, B) is stabilizable and $(A, Q^{\frac{1}{2}})$ is observable. For the infinite horizon performance criterion*

$$\inf_{u_i} \sup_{w_i} \sum_{i=i_0}^{\infty} \left[x_i^T Q x_i + u_i^T R u_i - \gamma^2 w_i^T R_w w_i \right] \tag{2.158}$$

the H_∞ control and the worst-case disturbance are given by

$$\begin{bmatrix} u_i^* \\ w_i^* \end{bmatrix} = -R_{c,\infty}^{-1} \begin{bmatrix} B^T \\ B_w^T \end{bmatrix} M_\infty A x_i$$

$$M_\infty = A^T M_\infty A - A^T M_\infty \begin{bmatrix} B & B_w \end{bmatrix} R_{c,\infty}^{-1} \begin{bmatrix} B^T \\ B_w^T \end{bmatrix} M_\infty A \tag{2.159}$$

where

$$R_{c,\infty} = \begin{bmatrix} B^T \\ B_w^T \end{bmatrix} M_\infty \begin{bmatrix} B & B_w \end{bmatrix} + \begin{bmatrix} R & 0 \\ 0 & -\gamma^2 R_w \end{bmatrix}$$

if and only if the following conditions are satisfied:

(1) there exists a solution M_∞ satisfying (2.159);
(2) the matrix

$$A - \begin{bmatrix} B & B_w \end{bmatrix} R_{c,\infty}^{-1} \begin{bmatrix} B^T \\ B_w^T \end{bmatrix} M_\infty A \tag{2.160}$$

is stable;
(3) the numbers of the positive and negative eigenvalues of $\begin{bmatrix} R & 0 \\ 0 & -\gamma^2 R_w \end{bmatrix}$
are the same as those of $R_{c,\infty}$;
(4) $M_\infty \geq 0$.

We can see a proof of Theorem 2.5 in much of the literature including textbooks listed at the end of this chapter. In particular, its proof is made on the Krein space in [HSK99].

2.4 Optimal Filters

2.4.1 Kalman Filter on Minimum Criterion

Here, we consider the following stochastic model:

$$x_{i+1} = Ax_i + Bu_i + Gw_i \tag{2.161}$$
$$y_i = Cx_i + v_i \tag{2.162}$$

At the initial time i_0, the state x_{i_0} is a Gaussian random variable with a mean $\bar{x}_{i_0}$ and a covariance P_{i_0}. The system noise $w_i \in \Re^p$ and the measurement noise $v_i \in \Re^q$ are zero-mean white Gaussian and mutually uncorrelated. The covariances of w_i and v_i are denoted by Q_w and R_v respectively, which are assumed to be positive definite matrices. We assume that these noises are uncorrelated with the initial state x_{i_0}.

In practice, the state may not be available, so it should be estimated from measured outputs and known inputs. Thus, a state estimator, called a filter, is needed. This filter can be used for an output feedback control. Now, we will seek a derivation of a filter which estimates the state x_i from measured data and known inputs so that the error between the real state and the estimated state is minimized. When the filter is designed, the input signal is assumed to be known, and thus it is straightforward to handle the input signal.

A filter, called the Kalman filter, is derived for the following performance criterion:

$$E[(x_i - \hat{x}_{i|i})^T (x_i - \hat{x}_{i|i}) | Y_i] \tag{2.163}$$

where $\hat{x}_{i|j}$ is denoted by the estimated value at time i based on the measurement up to j and $Y_i = [y_{i_0}, \cdots, y_i]^T$. Note that $\hat{x}_{i|i}$ is a function of Y_i. $\hat{x}_{i+1|i}$

and $\hat{x}_{i|i}$ are often called a predictive estimated value and a filtered estimated value respectively.

From Appendix C.1 we have the optimal filter

$$\hat{x}_{i|i} = E[x_i|Y_i] \tag{2.164}$$

We first obtain a probability density function of x_i given Y_i and then find out the mean of it.

By the definition of the conditional probability, we have

$$p(x_i|Y_i) = \frac{p(x_i, Y_i)}{p(Y_i)} = \frac{p(x_i, y_i, Y_{i-1})}{p(y_i, Y_{i-1})} \tag{2.165}$$

The numerator in (2.165) can be represented in terms of the conditional expectation as follows:

$$\begin{aligned} p(x_i, y_i, Y_{i-1}) &= p(y_i|x_i, Y_{i-1})p(x_i, Y_{i-1}) \\ &= p(y_i|x_i, Y_{i-1})p(x_i|Y_{i-1})p(Y_{i-1}) \\ &= p(y_i|x_i)p(x_i|Y_{i-1})p(Y_{i-1}) \end{aligned} \tag{2.166}$$

where the last equality comes from the fact that Y_{i-1} is redundant information if x_i is given. Substituting (2.166) into (2.165) yields

$$\begin{aligned} p(x_i|Y_i) &= \frac{p(y_i|x_i)p(x_i|Y_{i-1})p(Y_{i-1})}{p(y_i, Y_{i-1})} = \frac{p(y_i|x_i)p(x_i|Y_{i-1})p(Y_{i-1})}{p(y_i|Y_{i-1})p(Y_{i-1})} \\ &= \frac{p(y_i|x_i)p(x_i|Y_{i-1})}{p(y_i|Y_{i-1})} \end{aligned} \tag{2.167}$$

For the given Y_i, the denominator $p(y_i|Y_{i-1})$ is fixed. Two conditional probability densities in the numerator of Equation (2.167) can be evaluated from the statistical information. For the given x_i, y_i follows the normal distribution, i.e. $y_i \sim \mathcal{N}(Cx_i, R_v)$. The conditional probability $p(x_i|Y_{i-1})$ is also normal. Since $E[x_i|Y_{i-1}] = \hat{x}_{i|i-1}$ and $E[(x_i - \hat{x}_{i|i-1})(x_i - \hat{x}_{i|i-1})^T|Y_{i-1}] = P_{i|i-1}$, $p(x_i|Y_{i-1})$ is a normal probability function, i.e. $\mathcal{N}(\hat{x}_{i|i-1}, P_{i|i-1})$. Therefore, we have

$$p(y_i|x_i) = \frac{1}{\sqrt{(2\pi)^m|R_v|}} \exp\left\{-\frac{1}{2}[y_i - Cx_i]^T R_v^{-1}[y_i - Cx_i]\right\}$$

$$p(x_i|Y_{i-1}) = \frac{1}{\sqrt{(2\pi)^n|P_{i|i-1}|}} \exp\left\{-\frac{1}{2}[x_i - \hat{x}_{i|i-1}]^T P_{i|i-1}^{-1}[x_i - \hat{x}_{i|i-1}]\right\}$$

from which, using (2.167), we find that

$$\begin{aligned} p(x_i|Y_i) = C \exp\{-\frac{1}{2}[y_i - Cx_i]R_v^{-1}[y_i - Cx_i]\} \times \\ \exp\{-\frac{1}{2}[x_i - \hat{x}_{i|i-1}]P_{i|i-1}^{-1}[x_i - \hat{x}_{i|i-1}]\} \end{aligned} \tag{2.168}$$

where $\mathcal{C}$ is the constant involved in the denominator of (2.167).

We are now in a position to find out the mean of $p(x_i|Y_i)$. Since the Gaussian probability density function has a peak value at the average, we will find x_i that sets the derivative of (2.168) to zero. Thus, we can obtain the following equation:

$$-2C^T R_v^{-1}(y_i - Cx_i) + 2P_{i|i-1}^{-1}(x_i - \hat{x}_{i|i-1}) = 0 \qquad (2.169)$$

Denoting the solution x_i to (2.169) by $\hat{x}_{i|i}$ and arranging terms give

$$\begin{aligned}
\hat{x}_{i|i} &= (I + P_{i|i-1}C^T R_v^{-1}C)^{-1}\hat{x}_{i|i-1} \\
&\quad + (I + P_{i|i-1}C^T R_v^{-1}C)^{-1}P_{i|i-1}C^T R_v^{-1}y_i \qquad (2.170) \\
&= [I - P_{i|i-1}C^T(CP_{i|i-1}C^T + R_v)^{-1}C]\hat{x}_{i|i-1} \\
&\quad + P_{i|i-1}C^T(R_v + CP_{i|i-1}C^T)^{-1}y_i \qquad (2.171) \\
&= \hat{x}_{i|i-1} + K_i(y_i - C\hat{x}_{i|i-1}) \qquad (2.172)
\end{aligned}$$

where

$$K_i \triangleq P_{i|i-1}C^T(R_v + CP_{i|i-1}C^T)^{-1} \qquad (2.173)$$

$\hat{x}_{i+1|i}$ can be easily found from the fact that

$$\begin{aligned}
\hat{x}_{i+1|i} &= E[x_{i+1}|Y_i] = AE[x_i|Y_i] + GE[w_i|Y_i] + Bu_i \\
&= A\hat{x}_{i|i} + Bu_i \qquad (2.174) \\
&= A\hat{x}_{i|i-1} + AK_i(y_i - C\hat{x}_{i|i-1}) + Bu_i \qquad (2.175)
\end{aligned}$$

$P_{i+1|i}$ can be obtained recursively from the error dynamic equations.

Subtracting x_i from both sides of (2.172) yields the following error equation:

$$\tilde{x}_{i|i} = [I - K_iC]\tilde{x}_{i|i-1} - K_iv_i \qquad (2.176)$$

where $\tilde{x}_{i|i} \triangleq \hat{x}_{i|i} - x_i$ and $\tilde{x}_{i|i-1} = \hat{x}_{i|i-1} - x_i$. From (2.175) and (2.161), an additional error equation is obtained as

$$\tilde{x}_{i+1|i} = A\tilde{x}_{i|i} - Gw_i \qquad (2.177)$$

From (2.176) and (2.177), $P_{i|i}$ and $P_{i+1|i}$ are represented as

$$\begin{aligned}
P_{i|i} &= (I - K_iC)P_{i|i-1}(I - K_iC)^T + K_iR_vK_i = (I - K_iC)P_{i|i-1} \\
P_{i+1|i} &= AP_{i|i}A^T + GQ_wG^T \\
&= AP_{i|i-1}A^T + GQ_wG^T \\
&\quad - AP_{i|i-1}C^T(R_v + CP_{i|i-1}C^T)^{-1}CP_{i|i-1}A^T \qquad (2.178)
\end{aligned}$$

The initial values $\hat{x}_{i_0|i_0-1}$ and $P_{i_0|i_0-1}$ are given by $E[x_{i_0}]$ and $E[(\hat{x}_{i_0} - x_{i_0})(\hat{x}_{i_0} - x_{i_0})^T]$, which are *a priori* knowledge.

The Kalman filter can be represented as follows:

$$\hat{x}_{i+1|i} = A\hat{x}_{i|i-1} + AP_iC^T(CP_iC^T + R_v)^{-1}(y_i - C\hat{x}_{i|i-1}) \quad (2.179)$$

where

$$
\begin{aligned}
P_{i+1} &= AP_iA^T - AP_iC^T(R_v + CP_iC^T)^{-1}CP_iA^T + GQ_wG^T \\
&= A[I + P_iC^TR_v^{-1}C]^{-1}P_iA^T + GQ_wG^T \quad (2.180)
\end{aligned}
$$

with the given initial condition P_{i_0}. Note that P_i in (2.179) is used instead of $P_{i|i-1}$.

Throughout this book, we use the predicted form $\hat{x}_{i|i-1}$ instead of filtered form $\hat{x}_{i|i}$. For simple notation, $\hat{x}_{i|i-1}$ will be denoted by $\hat{x}_i$ if necessary.

If the index i in (2.180) goes to ∞, then the infinite horizon or steady-state Kalman filter is given by

$$\hat{x}_{i+1|i} = A\hat{x}_{i|i-1} + AP_\infty C^T(CP_\infty C^T + R_v)^{-1}(y_i - C\hat{x}_{i|i-1}) \quad (2.181)$$

where

$$
\begin{aligned}
P_\infty &= AP_\infty A^T - AP_\infty C^T(R_v + CP_\infty C^T)^{-1}CP_\infty A^T + GQ_wG^T \\
&= A[I + P_\infty C^TR_v^{-1}C]^{-1}P_\infty A^T + GQ_wG^T \quad (2.182)
\end{aligned}
$$

As in LQ control, the following theorem gives the result on the condition for the existence of P_∞ and the stability for the infinite horizon Kalman filter.

Theorem 2.6. *If (A, G) is controllable and (A, C) is observable, then there is a unique positive definite solution P_∞ to the ARE (2.182). Additionally, the steady-state Kalman filter is asymptotically stable.*

We can see a proof of Theorem 2.6 in much of the literature including textbooks listed at the end of this chapter. In Theorem 2.6, the conditions on controllability and observability can be weakened to the reachability and detectability.

2.4.2 H_∞ Filter on Minimax Criterion

Here, an H_∞ filter is introduced. Consider the following systems:

$$
\begin{aligned}
x_{i+1} &= Ax_i + B_ww_i + Bu_i \\
y_i &= Cx_i + D_ww_i \\
z_i &= C_zx
\end{aligned}
\quad (2.183)
$$

where $x_i \in \Re^n$ denotes states, $w_i \in \Re^l$ disturbance, $u_i \in \Re^m$ inputs, $y_i \in \Re^p$ measured outputs, and $z_i \in \Re^q$ estimated values. $B_wD_w^T = 0$ and $D_wD_w^T = $

I are assumed for simple calculation. In the estimation problem, the input control has no effect on the design of the estimator, so that B in (2.183) is set to zero and added later .

Our objective is to find a linear estimator $\hat{x}_i = T(y_{i_0}, y_{i_0+1}, \cdots, y_{i-1})$ so that $e_i = z_i - \hat{z}_i$ satisfies the following performance criterion:

$$\sup_{w_i \neq 0} \frac{\sum_{i=i_0}^{i_f} e_i^T e_i}{\sum_{i=i_0}^{i_f} w_i^T w_i} < \gamma^2 \tag{2.184}$$

From the system (2.183), we obtain the following state-space realization that has inputs $[w_i^T \ \hat{z}_i^T]^T$ and outputs $[e_i^T \ y_i^T]$ as

$$\begin{bmatrix} x_{i+1} \\ e_i \\ y_i \end{bmatrix} = \begin{bmatrix} A & B_w & 0 \\ \hline C_z & 0 & -I \\ C & D_w & 0 \end{bmatrix} \begin{bmatrix} x_i \\ w_i \\ \hat{z}_i \end{bmatrix} \tag{2.185}$$

under which we try to find the filter represented by

$$\hat{z}_i = T(y_{i_0}, y_{i_0+1}, \cdots, y_{i-1}) \tag{2.186}$$

The adjoint system of (2.185) can be represented as

$$\begin{bmatrix} \tilde{x}_i \\ \tilde{w}_i \\ \tilde{\hat{z}}_i \end{bmatrix} = \begin{bmatrix} A & B_w & 0 \\ \hline C_z & 0 & -I \\ C & D_w & 0 \end{bmatrix}^T \begin{bmatrix} \tilde{x}_{i+1} \\ \tilde{e}_i \\ \tilde{y}_i \end{bmatrix} = \begin{bmatrix} A^T & C_z^T & C^T \\ \hline B_w^T & 0 & D_w^T \\ 0 & -I & 0 \end{bmatrix} \begin{bmatrix} \tilde{x}_{i+1} \\ \tilde{e}_i \\ \tilde{y}_i \end{bmatrix} \tag{2.187}$$

where $\tilde{x}_{i_f+1} = 0$ and $i = i_f, i_f - 1, \cdots, i_0$. Observe that the input and the output are switched. Additionally, time indices are arranged in a backward way. The estimator that we try to find out is changed as follows:

$$\tilde{y}_i = \tilde{T}(\tilde{\hat{z}}_{i_f}, \tilde{\hat{z}}_{i_f-1}, \cdots, \tilde{\hat{z}}_{i+1}) \tag{2.188}$$

where $\tilde{T}(\cdot)$ is the adjoint system of $T(\cdot)$. Now we are in a position to apply the H_∞ control theory to the above H_∞ filter problem.

The state feedback H_∞ control is obtained from the following adjoint system:

$$\tilde{x}_i = A^T \tilde{x}_{i+1} + C^T \tilde{y}_i + \tilde{e}_i \tag{2.189}$$

$$\tilde{w}_i = B_w^T \tilde{x}_{i+1} + D_w^T \tilde{y}_i \tag{2.190}$$

$$\tilde{\hat{z}}_i = -\tilde{e}_i \tag{2.191}$$

$$\tilde{y}_i = \tilde{T}(\tilde{\hat{z}}_{i_f}, \tilde{\hat{z}}_{i_f-1}, \cdots, \tilde{\hat{z}}_{i+1}) \tag{2.192}$$

From the above system, the $\tilde{y}_i$ and $\tilde{w}_i$ are considered as an input and controlled output respectively. It is noted that time indices are reversed, i.e. we goes from the future to the past.

The controller $\tilde{T}(\cdot)$ can be selected to bound the cost:

$$\max_{\|\tilde{e}_i\|_{2,[0,i_f]}\neq 0} \frac{\sum_{i=i_0}^{i_f} \tilde{w}_i^T \tilde{w}_i}{\sum_{i=i_0}^{i_f} \tilde{e}_i^T \tilde{e}_i} < \gamma^2 \tag{2.193}$$

According to (2.146) and the correspondence

$$Q \longleftarrow B_w B_w^T, \ R \longleftarrow I, \ R_w \longleftarrow I, \ A^T \longleftarrow A, \ B \longleftarrow C^T, \ B_w \longleftarrow C_z$$

the resulting controller and the worst case $\tilde{e}_i$ are given:

$$\tilde{y}_i = -L_{i,i_0}^T \tilde{x}_{i+1}, \ \tilde{e}_i = -N_{i,i_0}^T \tilde{x}_{i+1} \tag{2.194}$$

where

$$L_{i,i_0}^T = C\Gamma_{i,i_0}^{-1} S_{i,i_0} A^T, \ N_{i,i_0}^T = -C_z \Gamma_{i,i_0}^{-1} S_{i,i_0} A^T \tag{2.195}$$

$$S_{i+1,i_0} = A S_{i,i_0} \Gamma_{i,i_0}^{-1} A^T + B_w B_w^T \tag{2.196}$$

$$\Gamma_{i,i_0} = I + (C^T C - \gamma^{-2} C_z^T C_z) S_{i,i_0} \tag{2.197}$$

with $S_{i_0,i_0} = 0$. In Figure 2.6, controls using Riccati solutions are represented in forward and backward ways. The state-space model for the controller is given as

$$\tilde{x}_i = A^T \tilde{x}_{i+1} - C^T L_{i,i_0}^T \tilde{x}_{i+1} + C_z^T \tilde{e}_i \tag{2.198}$$

$$\tilde{y}_i = -L_{i,i_f}^T \tilde{x}_{i+1}, \tag{2.199}$$

which can be represented as

$$\begin{bmatrix} \tilde{x}_i \\ \tilde{y}_i \end{bmatrix} = \left[\begin{array}{c|c} A^T - C^T L_{i,i_0}^T & C_z^T \\ \hline -L_{i,i_0}^T & 0 \end{array} \right] \begin{bmatrix} \tilde{x}_{i+1} \\ \tilde{e}_i \end{bmatrix}$$

$$= \left[\begin{array}{c|c} A^T - C^T L_{i,i_0}^T & -C_z^T \\ \hline -L_{i,i_0}^T & 0 \end{array} \right] \begin{bmatrix} \tilde{x}_{i+1} \\ \tilde{\tilde{z}}_i \end{bmatrix} \tag{2.200}$$

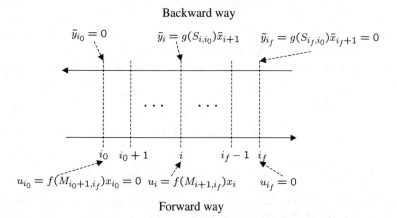

Fig. 2.6. Computation directions for H_∞ filter

which is a state-space realization for the control (2.188). The adjoint system of (2.200) is as follows:

$$\begin{bmatrix} \hat{\eta}_{i+1} \\ \hat{z}_i \end{bmatrix} = \begin{bmatrix} A^T - C^T L_{i,i_0}^T & -C_z^T \\ \hline -L_{i,i_0}^T & 0 \end{bmatrix}^T \begin{bmatrix} \hat{\eta}_i \\ y_i \end{bmatrix} = \begin{bmatrix} A - L_{i,i_0} C & -L_{i,i_0} \\ \hline -C_z & 0 \end{bmatrix} \begin{bmatrix} \hat{\eta}_i \\ y_i \end{bmatrix}$$

which is a state-space realization for the filter (2.186). Rearranging terms, replacing $-\hat{\eta}_i$ with $\hat{x}_i$, and adding the input into the estimator equation yields the H_∞ filter

$$\hat{z}_i = C_z \hat{x}_i, \quad \hat{x}_{i+1} = A\hat{x}_i + Bu_i + L_{i,i_0}(y_i - C\hat{x}_i) \tag{2.201}$$

The H_∞ filter can also be represented as follows:

$$\hat{z}_i = C_z \hat{x}_i, \quad \hat{x}_{i+1} = A\hat{x}_i + AS_{i,i_0} \begin{bmatrix} C^T & C_z^T \end{bmatrix} R_{f,i}^{-1} \begin{bmatrix} y - C\hat{x}_i \\ 0 \end{bmatrix} \tag{2.202}$$

$$S_{i+1,i_0} = AS_{i,i_0} A^T - AS_{i,i_0} \begin{bmatrix} C^T & C_z^T \end{bmatrix} R_{f,i}^{-1} \begin{bmatrix} C \\ C_z \end{bmatrix} S_{i,i_0} A^T + B_w B_w^T \tag{2.203}$$

where

$$R_{f,i} = \begin{bmatrix} I_p & 0 \\ 0 & -\gamma^2 I_q \end{bmatrix} + \begin{bmatrix} C \\ C_z \end{bmatrix} S_{i,i_0} \begin{bmatrix} C^T & C_z^T \end{bmatrix}$$

It is observed that the H_∞ filter of the form (2.202) and (2.203) looks like the Kalman filter.

From the finite horizon H_∞ filter of the form (2.202) and (2.203), we now turn to the infinite horizon H_∞ filter. If the index i goes to ∞, the infinite horizon H_∞ filter is given by

$$\hat{z}_i = C_z \hat{x}_i, \quad \hat{x}_{i+1} = A\hat{x}_i + AS_\infty \begin{bmatrix} C^T & C_z^T \end{bmatrix} R_{f,\infty}^{-1} \begin{bmatrix} y - C\hat{x}_i \\ 0 \end{bmatrix} \tag{2.204}$$

$$S_\infty = AS_\infty A^T - AS_\infty \begin{bmatrix} C^T & C_z^T \end{bmatrix} R_{f,\infty}^{-1} \begin{bmatrix} C \\ C_z \end{bmatrix} S_\infty A^T + B_w B_w^T \tag{2.205}$$

where

$$R_{f,\infty} = \begin{bmatrix} I_p & 0 \\ 0 & -\gamma^2 I_q \end{bmatrix} + \begin{bmatrix} C \\ C_z \end{bmatrix} S_\infty \begin{bmatrix} C^T & C_z^T \end{bmatrix} \tag{2.206}$$

As in the infinite horizon H_∞ control, the following theorem gives the result on the condition for the existence of S_∞ and stability for the infinite horizon H_∞ filter.

Theorem 2.7. *Suppose that (A, B) is stabilizable and $(A, Q^{\frac{1}{2}})$ is observable. For the following infinite horizon performance criterion:*

$$\max_{w_i \neq 0} \frac{\sum_{i=i_0}^{\infty} e_i^T e_i}{\sum_{i=i_0}^{\infty} w_i^T w_i} < \gamma^2 \tag{2.207}$$

the H_∞ filter (2.204) exists if and only if the following things are satisfied:

(1) there exists a solution S_∞ satisfying (2.205);
(2) the matrix

$$A - AS_\infty \begin{bmatrix} C^T & C_z^T \end{bmatrix} R_{f,\infty}^{-1} \begin{bmatrix} C \\ C_z \end{bmatrix} \qquad (2.208)$$

is stable;
(3) the numbers of the positive and negative eigenvalues of $R_{f,\infty}$ in (2.206)
are the same as those of the matrix $\begin{bmatrix} I_p & 0 \\ 0 & -\gamma^2 I_q \end{bmatrix}$;
(4) $S_\infty \geq 0$.

We can see a proof of Theorem 2.7 in a number of references. The H_∞ filter in Theorem 2.7 is obtained mostly by using the duality from the H_∞ control, e.g. in [Bur98]. The Krein space instead of the Hilbert space is used to derive H_∞ filters in [HSK99].

2.4.3 Kalman Filters on Minimax Criterion

We assume that Q_w and R_v are unknown, but are bounded above as follows:

$$Q_w \leq Q_o, \; R_v \leq R_o \qquad (2.209)$$

The Kalman filter can be derived for the minimax performance criterion given by

$$\min_{L_i} \max_{Q_w \leq Q_o, R_v \leq R_o} E[(x_i - \hat{x}_{i|i-1})(x_i - \hat{x}_{i|i-1})^T]$$

From the error dynamics

$$\tilde{x}_{i+1|i} = [A - L_i C]\tilde{x}_{i|i-1} + \begin{bmatrix} G & -L_i \end{bmatrix} \begin{bmatrix} w_i \\ v_i \end{bmatrix}$$

where $\tilde{x}_{i|i-1} = x_i - \hat{x}_{i|i-1}$, the following equality between the covariance matrices at time $i+1$ and i is satisfied:

$$P_{i+1} = [A - L_i C]P_i[A - L_i C]^T + \begin{bmatrix} G & L_i \end{bmatrix} \begin{bmatrix} Q_w & 0 \\ 0 & R_v \end{bmatrix} \begin{bmatrix} G^T \\ L_i^T \end{bmatrix} \qquad (2.210)$$

As can be seen in (2.210), P_i is monotonic with respect to Q_w and R_v, so that taking Q_w and R_v as Q_o and R_o we have

$$P_{i+1} = [A - L_i C]P_i[A - L_i C]^T + \begin{bmatrix} G & L_i \end{bmatrix} \begin{bmatrix} Q_o & 0 \\ 0 & R_o \end{bmatrix} \begin{bmatrix} G^T \\ L_i^T \end{bmatrix} \qquad (2.211)$$

It is well known that the right-hand side of (2.211) is minimized for the solution to the following Riccati equation:

$$P_{i+1} = -AP_i C^T (R_o + CP_i C^T)^{-1} CP_i A^T + AP_i A^T + GQ_o G^T$$

where L_i is chosen as

$$L_i = AP_i C^T (R_o + CP_i C^T)^{-1} \qquad (2.212)$$

It is noted that (2.212) is the same as the Kalman gain with Q_o and R_o.

2.5 Output Feedback Optimal Control

Before moving to an output feedback control, we show that a quadratic performance criterion for deterministic systems with no disturbances can be represented in a square form.

Lemma 2.8. *A quadratic performance criterion can be represented in a perfect square expression for any control,*

$$\sum_{i=i_0}^{i_f-1} \left[x_i^T Q x_i + u_i^T R u_i \right] + x_{i_f}^T Q_f x_{i_f}$$

$$= \sum_{i=i_0}^{i_f-1} \{-\mathcal{K}_i x_i + u_i\}^T [R + B^T K_i B]\{-\mathcal{K}_i x_i + u_i\} + x_{i_0}^T K_{i_0} x_{i_0} \quad (2.213)$$

where $\mathcal{K}_i$ is defined in

$$\mathcal{K}_i \triangleq (B^T K_{i+1} B + R_v)^{-1} B^T K_{i+1} A \quad (2.214)$$

and K_i is the solution to Riccati Equation (2.45).

Proof. Now note the simple identity as

$$\sum_{i=i_0}^{i_f-1} [x_i K_i x_i^T - x_{i+1} K_{i+1} x_{i+1}^T] = x_{i_0}^T K_{i_0} x_{i_0} - x_{i_f}^T K_{i_f} x_{i_f}$$

Then, the second term of the right-hand side can be represented as

$$x_{i_f}^T K_{i_f} x_{i_f} = x_{i_0}^T K_{i_0} x_{i_0} - \sum_{i=i_0}^{i_f-1} [x_i K_i x_i^T - x_{i+1} K_{i+1} x_{i+1}^T]$$

Observe that the quadratic form for x_{i_f} is written in terms of the x_i on $[i_0 \quad i_f - 1]$. Substituting the above equation into the terminal performance criterion yields

$$\sum_{i=i_0}^{i_f-1} \left[x_i^T Q x_i + u_i^T R u_i \right] + x_{i_f}^T Q_f x_{i_f}$$

$$= \sum_{i=i_0}^{i_f-1} \left[x_i^T Q x_i + u_i^T R u_i \right] + x_{i_0}^T K_{i_0} x_{i_0} - \sum_{i=i_0}^{i_f-1} [x_i K_i x_i^T - x_{i+1} K_{i+1} x_{i+1}^T] (2.215)$$

If x_{i+1} is replaced with $Ax_i + Bu_i$, then we have

$$\sum_{i=i_0}^{i_f-1} \left[x_i^T (Q - K_i + A^T K_i A) x_i + u_i^T B^T K_i A x_j \right.$$

$$\left. + x_i^T A^T K_{i+1} B u_i + u_i^T (R + B^T K_i B) u_i \right] + x_{i_0}^T K_{i_0} x_{i_0} \quad (2.216)$$

If K_i satisfies (2.45), then the square completion is achieved as (2.213), This completes the proof. ∎

2.5.1 Linear Quadratic Gaussian Control on Minimum Criterion

Now, we introduce an output feedback LQG control. A quadratic performance criterion is given by

$$J = \sum_{i=i_0}^{i_f-1} E\left[x_i^T Q x_i + u_i^T R u_i \middle| y_{i-1}, y_{i-2}, \cdots, y_{i_0}\right]$$
$$+ E\left[x_{i_f}^T Q_f x_{i_f} \middle| y_{i_f-1}, y_{i_f-2}, \cdots, y_{i_0}\right] \tag{2.217}$$

subject to $u_i = f(y_{i-1}, \cdots, y_{i_0})$. Here, the objective is to find a controller u_i that minimizes (2.217). From now on we will not include the condition part inside the expectation for simplicity. Before obtaining the LQG control, as in the deterministic case (2.213), it is shown that the performance criterion (2.217) can be represented in a square form.

Lemma 2.9. *A quadratic performance criterion can be represented in a perfect square expression for any control,*

$$E\left[\sum_{i=i_0}^{i_f-1}\left[x_i^T Q x_i + u_i^T R u_i\right] + x_{i_f}^T Q_f x_{i_f}\right]$$
$$= E\left[\sum_{i=i_0}^{i_f-1}\{-\mathcal{K}_i x_i + u_i\}^T[R + B^T K_i B]\{-\mathcal{K}_i x_i + u_i\}\right]$$
$$+ \text{tr}\left[\sum_{i=i_0}^{i_f-1} K_i G Q_w G^T\right] + E\left[x_{i_0}^T K_{i_0} x_{i_0}\right] \tag{2.218}$$

where $\mathcal{K}_i$ is defined in

$$\mathcal{K}_i \triangleq (B^T K_{i+1} B + R)^{-1} B^T K_{i+1} A \tag{2.219}$$

and K_i is the solution to Riccati Equation (2.45).

Proof. The relation (2.215) holds even for stochastic systems (2.161)-(2.162). Taking an expectation on (2.215), we have

$$E\left[\sum_{i=i_0}^{i_f-1}\left[x_i^T Q x_i + u_i^T R u_i\right] + x_{i_f}^T Q_f x_{i_f}\right]$$
$$= E\left[\sum_{i=i_0}^{i_f-1}\left[x_i^T Q x_i + u_i^T R u_i\right] + x_{i_0}^T K_{i_0} x_{i_0}\right.$$
$$\left. - \sum_{i=i_0}^{i_f-1}\left[x_i K_i x_i^T - x_{i+1} K_{i+1} x_{i+1}^T\right]\right] \tag{2.220}$$

Replacing x_{i+1} with $Ax_i + Bu_i + Gw_i$ yields

$$E\left[\sum_{i=i_0}^{i_f-1}\left[x_i^T(Q-K_i+A^TK_iA)x_i+u_i^TB^TK_iAx_j\right.\right.$$

$$\left.+x_i^TA^TK_{i+1}Bu_i+u_i^T(R+B^TK_iB)u_i\right]\right]+E\left[x_{i_0}^TK_{i_0}x_{i_0}\right]$$

$$+\text{tr}\left[\sum_{i=i_0}^{i_f-1}K_iGQ_wG^T\right] \tag{2.221}$$

If K_i satisfies (2.140), then the square completion is achieved as (2.218). This completes the proof. ∎

Using Lemma 2.9, we are now in a position to represent the performance criterion (2.217) in terms of the estimated state. Only the first term in (2.221) is dependent on u_i. So, we consider only this term. Let $\hat{x}_{i|i-1}$ be denoted by $\hat{x}_{i|i-1} = E[x_i|y_{i-1}, y_{i-2}, \cdots, y_{i_0}]$. According to (C.2) in Appendix C, we can change the first term in (5.69) to

$$E\sum_{i=i_0}^{i_f-1}(\mathcal{K}_ix_i+u_i)^T\hat{R}_i(\mathcal{K}_ix_i+u_i)=\sum_{i=i_0}^{i_f-1}(\mathcal{K}_i\hat{x}_i+u_i)^T\hat{R}_i(\mathcal{K}_i\hat{x}_i+u_i)$$

$$+\text{tr}\sum_{i=i_0+1}^{i_f}\hat{R}_i^{\frac{1}{2}}\mathcal{K}_i\tilde{P}_i\mathcal{K}_i^T\hat{R}_i^{\frac{1}{2}} \tag{2.222}$$

where

$$\hat{R}_i \triangleq R+B^TK_iB \tag{2.223}$$

and $\tilde{P}_i$ is the variance between $\hat{x}_{i|i-1}$ and x_i. Note that $\mathcal{K}_i\hat{x}_{i|i-1}+u_i = E[\mathcal{K}_ix_i+u_i \mid y_{i-1}, y_{i-2}, \cdots, y_{i_0}]$ and $\text{tr}(\hat{R}_i\mathcal{K}_i\tilde{P}_i\mathcal{K}_i^T) = \text{tr}(\hat{R}_i^{\frac{1}{2}}\mathcal{K}_i\tilde{P}_i\mathcal{K}_i^T\hat{R}_i^{\frac{1}{2}})$.

We try to find the optimal filter gain L_i making the following filter minimizing P_i:

$$\hat{x}_{i+1|i} = A\hat{x}_{i|i-1}+L_i(y_i-C\hat{x}_{i|i-1})+Bu_i \tag{2.224}$$

Subtracting (2.161) from (2.224), we have

$$\tilde{x}_{i+1|i} = \hat{x}_{i+1|i}-x_{i+1} = (A-L_iC)\tilde{x}_{i|i-1}+L_iv_i-Gw_i \tag{2.225}$$

which leads to the following equation:

$$\tilde{P}_{i+1} = (A-L_iC)\tilde{P}_i(A-L_iC)^T+L_iR_vL_i^T+GQ_wG^T \tag{2.226}$$

where $\tilde{P}_i$ is the covariance of $\tilde{x}_{i|i-1}$. As can be seen in (2.226), $\tilde{P}_i$ is independent of u_i, so that $\tilde{P}_i$ and u_i can be determined independently. $\tilde{P}_{i+1}$ in (2.226) can be written as

$$\tilde{P}_{i+1} = \left[L_i(C\tilde{P}_iC^T + R_v) - A\tilde{P}_i \right](C\tilde{P}_iC^T + R)^{-1} \left[L_i(C\tilde{P}_iC^T + R_v) - A\tilde{P}_i \right]^T$$
$$+ GQ_wG^T + A\tilde{P}_i(C\tilde{P}_iC^T + R_v)^{-1}\tilde{P}_iA^T$$
$$\geq GQ_wG^T + A\tilde{P}_i(C\tilde{P}_iC^T + R_v)^{-1}\tilde{P}_iA^T \qquad (2.227)$$

where the equality holds if $L_i = AP_iC^T(R_v + CP_iC^T)^{-1}$.

It can be seen in (2.227) that the covariance P_i generated by the Kalman filter is optimal in view that the covariance $\tilde{P}_i$ of any linear estimator is larger than P_i of the Kalman filter, i.e. $P_i \leq \tilde{P}_i$. This implies that $\text{tr}(P_i) \leq \text{tr}(\tilde{P}_i)$, leading to $\text{tr}(\hat{R}_i^{\frac{1}{2}}K_iP_iK_i^T\hat{R}_i^{\frac{1}{2}}) \leq \text{tr}(\hat{R}_i^{\frac{1}{2}}K_i\tilde{P}_iK_i^T\hat{R}_i^{\frac{1}{2}})$. Thus, the $\hat{x}_{i|i-1}$ minimizing (2.222) is given by the Kalman filter as follows:

$$\hat{x}_{i+1|i} = A\hat{x}_{i|i-1} + [AP_iC^T(R_v + CP_iC^T)^{-1}](y_i - C\hat{x}_{i|i-1}) + Bu_i \quad (2.228)$$
$$P_{i+1} = GQ_wG^T + AP_i(CP_iC^T + R_v)^{-1}P_iA^T \qquad (2.229)$$

with the initial state mean $\hat{x}_{i_0}$ and the initial covariance P_{i_0}. Thus, the following LQG control minimizes the performance criterion:

$$u_i^* = -(B^TK_{i+1}B + R)^{-1}B^TK_{i+1}A\hat{x}_{i|i-1} \qquad (2.230)$$

Infinite Horizon Linear Quadratic Gaussian Control

We now turn to the infinite horizon LQG control. It is noted that, as the horizon N gets larger, (2.217) also becomes larger and finally blows up. So, the performance criterion (2.217) cannot be applied as it is to the infinite horizon case. In a steady state for the infinite horizon case, we may write

$$\min_{u_i} J = \min_{u_i} E[x_i^TQx_i + u_i^TRu_i] \qquad (2.231)$$

$$= \min_{u_i} \frac{1}{2\pi} \int_0^{2\pi} \text{tr}(T(e^{j\omega})T^*(e^{j\omega})) \, d\omega \qquad (2.232)$$

where $T(e^{j\omega})$ is the transfer function from w_i and v_i to u_i and x_i. The infinite horizon LQG control is summarized in the following theorem.

Theorem 2.10. *Suppose that (A, B) and (A, G) are controllable and $(A, Q^{\frac{1}{2}})$ and (A, C) are observable. For the infinite horizon performance criterion (2.232), the infinite horizon LQG control is given by*

$$u_i^* = -(B^TK_\infty B + R)^{-1}B^TK_\infty A\hat{x}_{i|i-1} \qquad (2.233)$$

where

$$K_\infty = A^TK_\infty A - A^TK_\infty B[R + BK_\infty B^T]^{-1}B^TK_\infty A + Q$$
$$\hat{x}_{i+1|i} = A\hat{x}_{i|i-1} + [AP_\infty C^T(R_v + CP_\infty C^T)^{-1}](y_i - C\hat{x}_{i|i-1}) + Bu_i$$
$$P_\infty = GQ_wG^T + AP_\infty(CP_\infty C^T + R_v)^{-1}P_\infty A^T$$

We can see a proof of Theorem 2.10 in many references, e.g. in [Bur98, Lew86b]. The conditions on controllability and observability in Theorem 2.10 can be weakened to the reachability and detectability.

2.5.2 Output Feedback H_∞ Control on Minimax Criterion

Now, we derive the output feedback H_∞ control. The result of the previous H_∞ filter will be used to obtain the output feedback H_∞ control. First, the performance criterion is transformed in perfect square forms with respect to the optimal control and disturbance.

Lemma 2.11. *H_∞ performance criterion can be represented in a perfect square expression for arbitrary control.*

$$\sum_{i=i_0}^{i_f-1}\left[x_i^T Q x_i + u_i^T R u_i - \gamma^2 w_i^T R_w w_i\right] + x_{i_f}^T Q_f x_{i_f}$$

$$= \sum_{i=i_0}^{i_f-1}\left[(u_i - u_i^*)^T \mathcal{V}_i(u_i - u_i^*) - \gamma^2(w_i - w_i^*)^T \mathcal{W}_i(w_i - w_i^*)\right]$$

$$+ x_{i_0}^T M_{i_0} x_{i_0} \tag{2.234}$$

where w_i^* and u_i^* are given as

$$w_i^* = (\gamma^2 R_w - B_w^T M_{i+1} B_w)^{-1} B_w^T M_{i+1}(A x_i + B u_i) \tag{2.235}$$

$$u_i^* = -R^{-1} B^T M_{i+1}[I + (BR^{-1}B^T - \gamma^{-2}B_w R_w^{-1} B_w^T)M_{i+1}]^{-1} A x_i \tag{2.236}$$

$$\mathcal{V}_i = R + B^T M_{i+1}(I - \gamma^{-2}B_w R_w^{-1} B_w^T M_{i+1})^{-1} B$$

$$\mathcal{W}_i = \gamma^2 R_w - B_w^T M_{i+1} B_w$$

and M_i shortened for M_{i,i_f} is given in (2.140).

Proof. Recalling the simple identity as

$$\sum_{i=i_0}^{i_f-1}[x_i M_i x_i^T - x_{i+1} M_{i+1} x_{i+1}^T] = x_{i_0}^T M_{i_0} x_{i_0} - x_{i_f}^T M_{i_f} x_{i_f}$$

we have

$$x_{i_f}^T M_{i_f} x_{i_f} = x_{i_f}^T Q_f x_{i_f} = x_{i_0}^T M_{i_0} x_{i_0} - \sum_{i=i_0}^{i_f-1}[x_i^T M_i x_i - x_{i+1}^T M_{i+1} x_{i+1}] \tag{2.237}$$

By substituting (2.237) into the final cost $x_{i_f}^T Q_f x_{i_f}$, the H_∞ performance criterion of the left-hand side in (2.234) can be changed as

$$\sum_{i=i_0}^{i_f-1} \left[x_i^T Q x_i + u_i^T R u_i - \gamma^2 w_i^T R_w w_i \right] + x_{i_f}^T Q_f x_{i_f}$$

$$= \sum_{i=i_0}^{i_f-1} \left[x_i^T Q x_i + u_i^T R u_i - \gamma^2 w_i^T R_w w_i \right] + x_{i_0}^T M_{i_0} x_{i_0}$$

$$- \sum_{i=i_0}^{i_f-1} [x_i^T M_i x_i - x_{i+1}^T M_{i+1} x_{i+1}]$$

$$= \sum_{i=i_0}^{i_f-1} \left[x_i^T Q x_i + u_i^T R u_i - \gamma^2 w_i^T R_w w_i - x_i^T M_i x_i \right.$$

$$\left. + x_{i+1}^T M_{i+1} x_{i+1} \right] + x_{i_0}^T M_{i_0} x_{i_0} \qquad (2.238)$$

Now, we try to make terms inside the summation represented in a perfect square form. First, time variables are all changed to i. Next, we complete the square with respect to w_i and u_i respectively.

Terms in the summation (2.238) can be arranged as follows:

$$x_{i+1}^T M_{i+1} x_{i+1} - x_i^T M_i x_i + x_i^T Q x_i + u_i^T R u_i - \gamma^2 w_i^T R_w w_i$$

$$+ x_i^T Q x_i + u_i^T R u_i - \gamma^2 w_i^T R_w w_i$$

$$= A_{o,i}^T M_{i+1} A_{o,i} + w_i^T B_w^T M_{i+1} A_{o,i} + A_{o,i}^T M_{i+1} B_w w_i$$

$$- w_i^T (\gamma^2 R_w - B_w^T M_{i+1} B_w) w_i + x_i^T (-M_i + Q) x_i + u_i^T R u_i \qquad (2.239)$$

where $A_{o,i} = A x_i + B u_i$. Terms including w_i in (2.239) can be arranged as

$$w_i^T B_w^T M_{i+1} A_{o,i} + A_{o,i}^T M_{i+1} B_w w_i - w_i^T (\gamma^2 R_w - B_w^T M_{i+1} B_w) w_i$$

$$= -W_i^T (\gamma^2 R_w - B_w^T M_{i+1} B_w)^{-1} W_i$$

$$+ A_{o,i}^T M_{i+1} B_w (\gamma^2 R_w - B_w^T M_{i+1} B_w)^{-1} B_w^T M_{i+1} A_{o,i} \qquad (2.240)$$

where $W_i = (\gamma^2 R_w - B_w^T M_{i+1} B_w) w_i - B_w^T M_{i+1} A_{o,i}$. After completing the square with respect to disturbance w_i, we try to do that for the control u_i. Substituting (2.240) into (2.239) yields

$$-W_i^T (\gamma^2 R_w - B_w^T M_{i+1} B_w)^{-1} W_i$$

$$+ A_{o,i}^T M_{i+1} B_w (\gamma^2 R_w - B_w^T M_{i+1} B_w)^{-1} B_w^T M_{i+1} A_{o,i}$$

$$+ A_{o,i}^T M_{i+1} A_{o,i} + x_i^T (-M_i + Q) x_i + u_i^T R u_i$$

$$= -W_i^T (\gamma^2 R_w - B_w^T M_{i+1} B_w)^{-1} W_i + A_{o,i}^T M_{i+1} (I - \gamma^{-2} B_w R_w^{-1} B_w^T M_{i+1})^{-1}$$

$$\times A_{o,i} + x_i^T (-M_i + Q) x_i + u_i^T R u_i \qquad (2.241)$$

where the last equality comes from

$$P(I - RQ^{-1} R^T P)^{-1} = P + PR(Q - R^T PR)^{-1} R^T P$$

for some matrix P, R, and Q.

The second, the third, and the fourth terms in the right-hand side of (2.241) can be factorized as

$$(Ax_i + Bu_i)^T M_{i+1} \mathcal{Z}_i^{-1} (Ax_i + Bu_i) + x_i^T(-M_i + Q)x_i + u_i^T Ru_i$$
$$= u_i^T[R + B^T M_{i+1} \mathcal{Z}_i^{-1} B]u_i + u_i^T B^T M_{i+1} \mathcal{Z}_i^{-1} Ax_i + x_i^T A^T M_{i+1} \mathcal{Z}_i^{-1} Bu_i$$
$$+ x_i^T[A^T M_{i+1} \mathcal{Z}_i^{-1} A - M_i + Q]x_i = U_i^T[R + B^T M_{i+1} \mathcal{Z}_i^{-1} B]^{-1} U_i \quad (2.242)$$

where

$$U_i = [R + B^T M_{i+1} \mathcal{Z}_i^{-1} B]u_i + B^T M_{i+1} \mathcal{Z}_i^{-1} Ax_i$$
$$\mathcal{Z}_i = I - \gamma^{-2} B_w R_w^{-1} B_w^T M_{i+1}$$

and the second equality comes from the Riccati equation represented by

$$M_i = A^T M_{i+1}(I + (BR^{-1}B^T - \gamma^{-2}B_w R_w^{-1} B_w^T)M_{i+1})^{-1}A + Q$$
$$= A^T M_{i+1}(\mathcal{Z}_i + BR^{-1}B^T M_{i+1})^{-1}A + Q$$
$$= A^T M_{i+1}\left[\mathcal{Z}_i^{-1} - \mathcal{Z}_i^{-1}BR^{-1}B^T(M_{i+1}\mathcal{Z}_i^{-1}BR^{-1}B^T + I)^{-1}M_{i+1}\mathcal{Z}_i^{-1}\right]A$$
$$+ Q$$

or

$$A^T M_{i+1} \mathcal{Z}_i^{-1} A - M_i + Q$$
$$= A^T M_{i+1} \mathcal{Z}_i^{-1} BR^{-1}B^T(M_{i+1}\mathcal{Z}_i^{-1}BR^{-1}B^T + I)^{-1}M_{i+1}\mathcal{Z}_i^{-1}A$$
$$= A^T M_{i+1} \mathcal{Z}_i^{-1} B(B^T M_{i+1}\mathcal{Z}_i^{-1}B + R)^{-1}B^T M_{i+1}\mathcal{Z}_i^{-1}A$$

This completes the proof. ∎

Note that by substituting (5.154) into (5.153), w_i^* can be represented as

$$w_i^* = \gamma^{-2}R_w^{-1}B_w^T M_{i+1}[I + (BR^{-1}B^T - \gamma^{-2}B_w R_w^{-1} B_w^T)M_{i+1}]^{-1}Ax_i$$

which is of very similar form to u_i^*.

For the zero initial state, the inequality

$$\sum_{i=i_0}^{i_f-1}\left[x_i^T Qx_i + u_i^T Ru_i - \gamma^2 w_i^T R_w w_i\right] + x_{i_f}^T Q_f x_{i_f} < 0 \quad (2.243)$$

guarantees the bound on the following ∞ norm:

$$\sup_{\|w_i\|_{2,[0,i_f]} \neq 0} \frac{\sum_{i=i_0}^{i_f-1}(u_i - u_i^*)^T \mathcal{V}_i(u_i - u_i^*)}{\sum_{i=i_0}^{i_f-1}(w_i - w_i^*)^T \mathcal{W}_i(w_i - w_i^*)} < \gamma^2 \quad (2.244)$$

where u_i^* and w_i^* are defined in (5.154) and (5.153) respectively. Here, $R_w = I$, $D_w B_w^T = 0$, and $D_w D_w^T = I$ are assumed for simple calculation.

According to (2.244), we should design u_i so that the H_∞ norm between the weighted disturbance deviation

$$\triangle w_i \overset{\triangle}{=} \mathcal{W}_i^{\frac{1}{2}} w_i - \mathcal{W}_i^{\frac{1}{2}} (\gamma^2 I - B_w^T M_{i+1} B_w)^{-1} B_w^T M_{i+1} (Ax_i + Bu_i)$$

and the weighted control deviation

$$\triangle u_i \overset{\triangle}{=} \mathcal{V}_i^{\frac{1}{2}} u_i + \mathcal{V}_i^{\frac{1}{2}} R^{-1} B^T M_{i+1} [I + (BR^{-1}B^T - \gamma^{-2} B_w B_w^T) M_{i+1}]^{-1} Ax_i$$

is minimized. By using $\triangle w_i$, we obtain the following state-space model:

$$x_{i+1} = A_{a,i} x_i + B_{a,i} u_i + B_w \mathcal{W}_i^{-\frac{1}{2}} \triangle w_i$$
$$y_i = C x_i + D_w \mathcal{W}_i^{-\frac{1}{2}} \triangle w_i \tag{2.245}$$

where $A_{a,i} = A + B_w(\gamma^2 I - B_w^T M_{i+1} B_w)^{-1} B_w^T M_{i+1} A$, and $B_{a,i} = B + B_w(\gamma^2 I - B_w^T M_{i+1} B_w)^{-1} B_w^T M_{i+1} B$. Note that $D_w B_w^T = 0$ is assumed as mentioned before.

The performance criterion (2.244) for the state-space model (2.245) is just one for the H_∞ filter that estimates u_i^* with respect to $\triangle w_i$ by using measurements y_i. This is a similar structure to (2.183). Note that the variable z_i in (2.183) to be estimated corresponds to u_i^*. Using the result of H_∞ filters, we can think of this as finding out the output feedback H_∞ control u_i by obtaining the estimator of u_i^*.

All derivations require long and tedious algebraic calculations. In this book, we just summarize the final result. The output feedback H_∞ control is given by

$$u_i = -K_{of,i} \hat{x}_i$$
$$K_{of,i} = R^{-1} B^T M_{i+1} [I + (BR^{-1}B^T - \gamma^{-2} B_w B_w^T) M_{i+1}]^{-1} A$$
$$\hat{x}_{i+1} = A_{a,i} \hat{x}_i + L_{of,i} \begin{bmatrix} 0 \\ y_i - C\hat{x}_i \end{bmatrix} + Bu_i$$

where M_i, i.e. M_{i,i_f}, is given in (2.140) and $L_{of,i}$ is defined as

$$L_{of,i} = \left(A_{a,i} S_{of,i} \begin{bmatrix} -K_{of,i}^T & C^T \end{bmatrix} - \gamma^2 B_w \begin{bmatrix} \bar{S}_i & 0 \end{bmatrix} \right) R_{of,i}^{-1}$$
$$S_{of,i+1} = A_{a,i} S_{of,i} A_{a,i}^T - \gamma^2 B_w W_i^{-1} B_w^T - L_{of,i} R_{of,i} L_{of,i}^T$$
$$R_{of,i} = \begin{bmatrix} -\gamma^2 Z_i^{-1} & 0 \\ 0 & I \end{bmatrix} + \begin{bmatrix} -K_{of,i} \\ C \end{bmatrix} S_{of,i} \begin{bmatrix} -K_{of,i}^T & C^T \end{bmatrix}$$
$$\begin{bmatrix} W_i^{-1} & \bar{S}_i \\ \bar{S}_i^T & Z_i^{-1} \end{bmatrix} = \left(\begin{bmatrix} -\gamma^2 I & 0 \\ 0 & R \end{bmatrix} + \begin{bmatrix} B_w^T \\ B^T \end{bmatrix} M_{i+1} \begin{bmatrix} B_w & B \end{bmatrix} \right)^{-1}$$

with the initial conditions $M_{i_f,i_f} = Q_f$ and $S_{i_0} = 0$.

Infinite Horizon Output Feedback H_∞ Control

Now we introduce the infinite horizon H_∞ output feedback control in the following theorem:

Theorem 2.12. *Infinite horizon H_∞ output feedback control is composed of*

$$u_i = -K_{of,\infty}\hat{x}_i$$

$$K_{of,\infty} = R^{-1}B^T M_\infty [I + (BR^{-1}B^T - \gamma^{-2}B_w B_w^T)M_\infty]^{-1}A$$

$$\hat{x}_{i+1} = A_{a,\infty}\hat{x}_i + L_{of,\infty}\begin{bmatrix} y_i - C\hat{x}_i \\ 0 \end{bmatrix} + Bu_i$$

where

$$A_{a,\infty} = A + B_w(\gamma^2 I - B_w^T M_\infty B_w)^{-1}B_w^T M_\infty A$$

$$L_{of,\infty} = \left(A_{a,\infty}S_{of,\infty}\left[-K_{of,\infty}^T \ C^T\right] - \gamma^2 B_w \left[\bar{S}_\infty \ 0\right] \right) R_{of,\infty}^{-1}$$

$$S_{of,\infty} = A_{a,\infty}S_{of,\infty}A_{a,\infty}^T - \gamma^2 B_w W_\infty^{-1}B_w^T - L_{of,\infty}R_{of,\infty}L_{of,\infty}^T$$

$$R_{of,\infty} = \begin{bmatrix} -\gamma^2(Z_\infty)^{-1} & 0 \\ 0 & I_p \end{bmatrix} + \begin{bmatrix} -K_{of,\infty} \\ C \end{bmatrix} P_i \left[-K_{of,\infty}^T \ C^T\right]$$

$$\begin{bmatrix} W_\infty^{-1} & \bar{S}_\infty \\ \bar{S}_\infty^T & Z_\infty^{-1} \end{bmatrix} = \left(\begin{bmatrix} -\gamma^2 I_l & 0 \\ 0 & R \end{bmatrix} + \begin{bmatrix} B_w^T \\ B^T \end{bmatrix} M_\infty \left[B_w \ B\right] \right)^{-1}$$

and achieves the following specification

$$\frac{\sum_{i=i_0}^\infty x_i^T R x_i + u_i^T Q u_i}{\sum_{i=i_0}^\infty w_i^T w_i} < \gamma^2$$

if and only if there exists solutions $M_\infty \geq 0$ satisfying (2.159) and $S_{of,\infty} \geq 0$ such that

(1) $A - \left[B \ B_w\right] \left(\begin{bmatrix} R & 0 \\ 0 & -\gamma^2 I \end{bmatrix} + \begin{bmatrix} B^T \\ B_w^T \end{bmatrix} M_\infty \left[B \ B_w\right] \right)^{-1} \begin{bmatrix} B^T \\ B_w^T \end{bmatrix} M_\infty A$ *is stable.*

(2) The numbers of the positive and negative eigenvalues of the two following matrices are the same:

$$\begin{bmatrix} R & 0 \\ 0 & -\gamma^2 I_l \end{bmatrix}, \quad \begin{bmatrix} R & 0 \\ 0 & -\gamma^2 I_l \end{bmatrix} + \begin{bmatrix} B^T \\ B_w^T \end{bmatrix} M_\infty \left[B \ B_w\right] \quad (2.246)$$

(3) $A_{a,\infty} - L_{of,\infty}\begin{bmatrix} C \\ (I + B^T M_\infty B)^{\frac{1}{2}}K_{of,\infty} \end{bmatrix}$ *is stable.*

(4) The numbers of the positive and negative eigenvalues of $\begin{bmatrix} I_p & 0 \\ 0 & -\gamma^2 I_m \end{bmatrix}$ are the same as those of the following matrix:

$$\begin{bmatrix} I_p & 0 \\ 0 & -\gamma^2 I_m + T \end{bmatrix} + \begin{bmatrix} C \\ X^{\frac{1}{2}}K_{of,\infty} \end{bmatrix} S_{of,\infty} \left[C^T \ K_{of,\infty}^T X^{\frac{1}{2}}\right] \quad (2.247)$$

where $T = X^{-\frac{1}{2}} B^T M_\infty Z^{-1} M_\infty B X^{-\frac{1}{2}}$, $Z = I - B^T M_\infty (I + BB^T M_\infty)^{-1} B$, *and* $X = I + B^T M_\infty B$.

We can see a proof of Theorem 2.12 in [HSK99]. It is shown in [Bur98] that output feedback H_∞ control can be obtained from a solution to an estimation problem.

2.6 Linear Optimal Controls via Linear Matrix Inequality

In this section, optimal control problems for discrete linear time-invariant systems are reformulated in terms of linear matrix inequalities (LMIs). Since LMI problems are convex, it can be solved very efficiently and the global minimum is always found. We first consider the LQ control and then move to H_∞ control.

2.6.1 Infinite Horizon Linear Quadratic Control via Linear Matrix Inequality

Let us consider the infinite horizon LQ cost function as follows:

$$J_\infty = \sum_{i=0}^{\infty} \left\{ x_i^T Q x_i + u_i^T R u_i \right\},$$

where $Q > 0, R > 0$. It is noted that, unlike the standard LQ control, Q is positive-definite. The nonsingularity of Q is required to solve an LMI problem. We aim to find the control u_i which minimizes the above cost function. The main attention is focused on designing a linear optimal state-feedback control, $u_i = H x_i$. Assume that $V(x_i)$ has the form

$$V(x_i) = x_i^T K x_i, \quad K > 0$$

and satisfies the following inequality:

$$V(x_{i+1}) - V(x_i) \leq -[x_i^T Q x_i + u_i^T R u_i] \tag{2.248}$$

Then, the system controlled by u_i is asymptotically stable and $J_\infty \leq V(x_0)$. With $u_i = H x_i$, the inequality (2.248) is equivalently rewritten as

$$x_i^T (A + BH)^T K(A + BH) x_i - x_i^T K x_i \leq -x_i^T [Q + H^T RH] x_i \tag{2.249}$$

From (2.249), it is clear that (2.248) is satisfied if there exists H and K such that

$$(A + BH)^T K(A + BH) - K + Q + H^T RH \leq 0 \tag{2.250}$$

Instead of directly minimizing $x_0^T K x_0$, we take an approach where its upper bound is minimized. For this purpose, assume that there exists $\gamma_2 > 0$ such that

$$x_0^T K x_0 \leq \gamma_2 \tag{2.251}$$

Now the optimal control problem for given x_0 can be formulated as follows:

$$\min_{\gamma_2, K, H} \gamma_2 \quad \text{subject to (2.250) and (2.251)}$$

However, the above optimization problem does not seem easily solvable because the matrix inequalities (2.250) and (2.251) are not of LMI forms. In the following, matrix inequalities (2.250) and (2.251) are converted to LMI conditions. First, let us turn to the condition in (2.250), which can be rewritten as follows:

$$-K + \begin{bmatrix} (A+BH)^T & H^T & I \end{bmatrix} \begin{bmatrix} K^{-1} & 0 & 0 \\ 0 & R^{-1} & 0 \\ 0 & 0 & Q^{-1} \end{bmatrix}^{-1} \begin{bmatrix} (A+BH) \\ H \\ I \end{bmatrix} \leq 0$$

From the Schur complement, the above inequality is equivalent to

$$\begin{bmatrix} -K & (A+BH)^T & H^T & I \\ (A+BH) & -K^{-1} & 0 & 0 \\ H & 0 & -R^{-1} & 0 \\ I & 0 & 0 & -Q^{-1} \end{bmatrix} \leq 0 \tag{2.252}$$

Also from the Schur complement, (2.251) is converted to

$$\begin{bmatrix} \gamma_2 & x_0^T \\ x_0 & K^{-1} \end{bmatrix} \geq 0 \tag{2.253}$$

Pre- and post-multiply (2.252) by $\text{diag}\{K^{-1}, I, I, I\}$. It should be noted that this operation does not change the inequality sign. Introducing new variables $Y \triangleq HK^{-1}$ and $S \triangleq K^{-1}$, (2.252) is equivalently changed into

$$\begin{bmatrix} -S & (AS+BY)^T & Y^T & S \\ (AS+BY) & -S & 0 & 0 \\ Y & 0 & -R^{-1} & 0 \\ S & 0 & 0 & -Q^{-1} \end{bmatrix} \leq 0 \tag{2.254}$$

Furthermore, (2.253) is converted to

$$\begin{bmatrix} \gamma_2 & x_0^T \\ x_0 & S \end{bmatrix} \geq 0 \tag{2.255}$$

Now that (2.254) and (2.255) are LMI conditions, the resulting optimization problem is an infinite horizon control, which is represented as follows:

$$\min_{\gamma_2, Y, S} \gamma_2$$
$$\text{subject to (2.254) and (2.255)}$$

Provided that the above optimization problem is feasible, then $H = YS^{-1}$ and $K = S^{-1}$.

2.6.2 Infinite Horizon H_∞ Control via Linear Matrix Inequality

Consider the system

$$x_{i+1} = Ax_i + Bu_i \tag{2.256}$$

$$z_i = C_z x_i + D_{zu} u_i \tag{2.257}$$

where A is a stable matrix. For the above system, the well-known bounded real lemma (BRL) is stated as follows:

Lemma 2.13 (Bounded Real Lemma). *Let $\gamma > 0$. If there exists $X > 0$ such that*

$$\begin{bmatrix} -X^{-1} & A & B & 0 \\ A^T & -X & 0 & C_z^T \\ B^T & 0 & -\gamma W_u & D^T \\ 0 & C & D_{zu} & -\gamma W_z^{-1} \end{bmatrix} < 0 \tag{2.258}$$

then

$$\frac{\sum_{i=i_0}^{\infty} z_i^T W_z z_i}{\sum_{i=i_0}^{\infty} u_i^T W_u u_i} < \gamma^2 \tag{2.259}$$

where u_i and z_i are governed by the system (2.256) and (2.257).

Proof. The inequality (2.259) is equivalent to

$$J_{zu} = \sum_{i=0}^{\infty} \left\{ z_i^T W_z z_i - \gamma^2 u_i^T W_u u_i \right\} < 0 \tag{2.260}$$

Let us take $V(x)$ as follows:

$$V(x) = x^T K x, \quad K > 0$$

Respectively adding and subtracting $\sum_{i=0}^{\infty} \left\{ V(x_{i+i}) - V(x_i) \right\}$ to and from J_{zu} in (2.260), does not make any difference to J_{zu}. Hence, it follows that

$$J_{zu} = \sum_{i=0}^{\infty} \left\{ z_i^T W_z z_i - \gamma^2 u_i^T W_u u_i + V(x_{i+1}) - V(x_i) \right\} + V(x_0) - V(x_\infty)$$

Since x_0 is assumed to be zero and $V(x_\infty) \geq 0$, we have

$$J_{zu} \leq \sum_{i=0}^{\infty} \left\{ z_i^T W_z z_i - \gamma^2 u_i^T W_u u_i + V(x_{i+1}) - V(x_i) \right\}$$

Furthermore,

$$\sum_{i=0}^{\infty} \{z_i^T W_z z_i - \gamma^2 u_i^T W_u u_i + V(x_{i+1}) - V(x_i)\}$$

$$= \sum_{i=0}^{\infty} \left\{ [C_z x_i + D_{zu} u_i]^T W_z [C_z x_i + D_{zu} u_i] - \gamma^2 u_i^T W_u u_i \right.$$

$$\left. + [Ax_i + Bu_i]^T K[Ax_i + Bu_i] - x_i^T K x_i \right\}$$

$$= \sum_{i=0}^{\infty} \left\{ \begin{bmatrix} x_i \\ u_i \end{bmatrix}^T \Lambda \begin{bmatrix} x_i \\ u_i \end{bmatrix} \right\}$$

where

$$\Lambda \triangleq \begin{bmatrix} -K + A^T K A + C_z^T W_z C_z & A^T K B + C_z^T W_z D_{zu} \\ B^T K A + D_{zu}^T W_z C_z & B^T K B + D_{zu}^T W_z D_{zu} - \gamma^2 W_u \end{bmatrix} \quad (2.261)$$

Hence, if the 2-by-2 block matrix Λ in (2.261) is negative definite, then $J_{zu} < 0$ and equivalently the inequality (2.259) holds.

The 2-by-2 block matrix Λ can be rewritten as follows:

$$\Lambda = \begin{bmatrix} -K & 0 \\ 0 & -\gamma^2 W_u \end{bmatrix} + \begin{bmatrix} A^T & C_z^T \\ B^T & D_{zu}^T \end{bmatrix} \begin{bmatrix} K & 0 \\ 0 & W_z \end{bmatrix} \begin{bmatrix} A & B \\ C_z & D_{zu} \end{bmatrix}$$

From the Schur complement, the negative definiteness of Λ is guaranteed if the following matrix equality holds:

$$\begin{bmatrix} -K & 0 & A^T & C_z^T \\ 0 & -\gamma^2 W_u & B^T & D_{zu}^T \\ A & B & -K^{-1} & 0 \\ C_z & D_{zu} & 0 & -W_z^{-1} \end{bmatrix} < 0 \quad (2.262)$$

Define Π as follows:

$$\Pi \triangleq \begin{bmatrix} 0 & I & 0 & 0 \\ 0 & 0 & I & 0 \\ I & 0 & 0 & 0 \\ 0 & 0 & 0 & I \end{bmatrix}$$

Pre- and post-multiplying (2.262) by Π^T and Π respectively does not change the inequality sign. Hence, the condition in (2.262) is equivalently represented by

$$\begin{bmatrix} -K^{-1} & A & B & 0 \\ A^T & -K & 0 & C_z^T \\ B^T & 0 & -\gamma^2 W_u & D_{zu}^T \\ 0 & C_z & D_{zu} & -W_z^{-1} \end{bmatrix} < 0 \quad (2.263)$$

Pre- and post-multiplying (2.263) by $\text{diag}\{\sqrt{\gamma}I, \sqrt{\gamma}^{-1}I, \sqrt{\gamma}^{-1}I, \sqrt{\gamma}I\}$ and introducing a change of variables such that $X \triangleq \frac{1}{\sqrt{\gamma}} K$, the condition in (2.263) is equivalently changed to (2.258). This completes the proof. ∎

Using the BRL, the LMI-based H_∞ control problem can be formulated. Let us consider the system

$$x_{i+1} = Ax_i + B_w w_i + B u_i, \quad x_0 = 0$$
$$z_i = C_z x_i + D_{zu} u_i$$

As in the LMI-based LQ problem, the control is constrained to have a state-feedback, $u_i = Hx_i$. With $u_i = Hx_i$, the above system is rewritten as follows:

$$x_{i+1} = [A + BH]x_i + B_w w_i, \quad x_0 = 0$$
$$z_i = [C_z + D_{zu}H]x_i$$

According to the BRL, H which guarantees $\|G_{cl}(z)\|_\infty < \gamma_\infty$ should satisfy, for some $X > 0$,

$$\begin{bmatrix} -X^{-1} & (A+BH) & B_w & 0 \\ (A+BH)^T & -X & 0 & (C_z + D_{zu}H)^T \\ B_w^T & 0 & -\gamma_\infty I & 0 \\ 0 & (C_z + D_{zu}H) & 0 & -\gamma_\infty I \end{bmatrix} < 0 \qquad (2.264)$$

where $G_{cl}(z) = [C_z + D_{zu}H](zI - A - BH)^{-1}B_w$. Pre- and post-multiplying (2.264) by $\mathrm{diag}\{I, X^{-1}, I, I\}$ and introducing a change of variables such that $S_\infty \triangleq X^{-1}$ and $Y \triangleq HX^{-1}$ lead to

$$\begin{bmatrix} -S_\infty & (AS_\infty + BY) & B_w & 0 \\ (AS_\infty + BY)^T & -S_\infty & 0 & (C_z S_\infty + D_{zm}Y)^T \\ B_w^T & 0 & -\gamma_\infty I & 0 \\ 0 & (C_z S_\infty + D_{zu}Y) & 0 & -\gamma_\infty I \end{bmatrix} < 0 \quad (2.265)$$

Provided that the above LMI is feasible for some given γ_∞, H_∞ state-feedback control guaranteeing $\|G_{cl}(z)\|_\infty < \gamma_\infty$ is given by

$$H = YS_\infty^{-1}$$

In this case, we can obtain the infinite horizon H_∞ control via LMI, which minimizes γ_∞ by solving the following optimization problem:

$$\min_{\gamma_\infty, Y, S_\infty} \gamma_\infty \quad \text{subject to (2.265)}$$

2.7 * H_2 Controls

Since LQ regulator and LQG control problems are studied extensively in this book, H_2 controls and H_2 filters are introduced in limited problems and are only briefly summarized without proofs in this section.

To manipulate more general problems, it is very useful to have a general system with the input, the disturbance, the controlled output, and the measure output given by

$$x_{i+1} = Ax_i + B_w w_i + Bu_i$$
$$y_i = Cx_i + D_w w_i$$
$$z_i = C_z x_i + D_{zw} w_i + D_{zu} u_i \qquad (2.266)$$

The standard H_2 problem is to find a proper, real rational controller $u(z) = K(z)y(z)$ which stabilizes the closed-loop system internally and minimizes the H_2 norm of the transfer matrix T_{zw} from w_i to z_i.

The H_2 norm can be represented as

$$\|T_{zw}(e^{jw})\|_2^2 = \frac{1}{2\pi} \int_{-\pi}^{\pi} \text{tr}\{T_{zw}^*(e^{jw})T_{zw}(e^{jw})\}dw = \sum_{k=i_0}^{\infty} \text{tr}\{H_{k-i_0}^* H_{k-i_0}\}$$

$$= \sum_{l=1}^{m} \sum_{i=-\infty}^{\infty} z_i^{lT} z_i^l \qquad (2.267)$$

where

$$T_{zw}(e^{jw}) = \sum_{k=0}^{\infty} H_k e^{-jwk}$$

and A^* is a complex conjugate transpose of A and z^l is an output resulting from applying unit impulses to lth input. From (2.267), we can see that the H_2 norm can be obtained from applying unit impulses to each input. We should require the output to settle to zero before applying an impulse to the next input. In the case of single input systems, the H_2 norm is obtained by the driving unit impulse once, i.e. $\|T_{zw}(e^{jw})\|_2 = \|z\|_2$.

The H_2 norm can be given another interpretation for stochastic systems. The expected power in the error signal z_i is then given by

$$E\{z_i^T z_i\} = \text{tr}[E\{z_i z_i^T\}] = \frac{1}{2\pi} \int_{-\pi}^{\pi} \text{tr}\{T_{zw}(e^{jw})T_{zw}^*(e^{jw})\} \, dw$$

$$= \frac{1}{2\pi} \int_{-\pi}^{\pi} \text{tr}\{T_{zw}^*(e^{jw})T_{zw}(e^{jw})\} \, dw$$

where the second equality comes from Theorem C.4 in Appendix C.

Thus, by minimizing the H_2 norm, the output (or error) power of the generalized system, due to a unit intensity white noise input, is minimized.

It is noted that for the given system transfer function

$$G(z) \triangleq \left[\begin{array}{c|c} A & B \\ \hline C & D \end{array}\right] = C(zI - A)^{-1}B + D$$

$\|G(z)\|_2$ is obtained by

$$\|G(z)\|_2^2 = \text{tr}(D^T D + B^T L_o B) = \text{tr}(DD^T + CL_c C^T) \qquad (2.268)$$

where L_c and L_o are the controllability and observability Gramians

$$AL_cA^T - L_c + BB^T = 0 \qquad (2.269)$$
$$A^TL_oA - L_o + C^TC = 0 \qquad (2.270)$$

The H_2 norm has a number of good mathematical and numerical properties, and its minimization has important engineering implications. However, the H_2 norm is not an induced norm and does not satisfy the multiplicative property.

It is assumed that the following things are satisfied for the system (2.266):

(i) (A, B) is stabilizable and (C, A) is detectable,

(ii) D_{zu} is of full column rank with $\begin{bmatrix} D_{zu} & D_\perp \end{bmatrix}$ unitary and D_w is full row with $\begin{bmatrix} D_w \\ \tilde{D}_\perp \end{bmatrix}$ unitary,

(iii) $\begin{bmatrix} A - e^{j\theta}I & B \\ C_z & D_{zu} \end{bmatrix}$ has full column rank for all $\theta \in [0 \ 2\pi]$,

(iv) $\begin{bmatrix} A - e^{j\theta}I & B_w \\ C & D_w \end{bmatrix}$ has full rank for all $\theta \in [0 \ 2\pi]$.

Let $X_2 \geq 0$ and $Y_2 \geq 0$ be the solutions to the following Riccati equations:

$$A_x^*(I + X_2BB^T)^{-1}X_2A_x - X_2 + C_z^TD_\perp D_\perp^TC_z = 0 \qquad (2.271)$$
$$A_y(I + Y_2C^TC)^{-1}Y_2A_y^T - Y_2 + B_wD_\perp^TD_\perp B_w^T = 0 \qquad (2.272)$$

where

$$A_x = A - BD_{zu}^TC_z, \ A_y = A - B_wC_w^TC \qquad (2.273)$$

Note that the stabilizing solutions exist by the assumptions (iii) and (iv). The solution to the standard H_2 problem is given by

$$\hat{x}_{i+1} = (\hat{A}_2 - BL_0C)\hat{x}_i - (L_2 - BL_0)y_i \qquad (2.274)$$
$$u_i = (F_2 - L_0C)\hat{x}_i + L_0y_i \qquad (2.275)$$

where

$$F_2 = -(I + B^TX_2B)^{-1}(B^TX_2A + D_{zu}^TC_z)$$
$$L_2 = -(AY_2C^T + B_wC_w^T)(I + CY_2C^T)^{-1}$$
$$L_0 = (F_2Y_2C^T + F_2C_w^T)(I + CY_2C^T)^{-1}$$
$$\hat{A}_2 = A + BF_2 + L_2C$$

The well-known LQR control problems can be seen as a special H_2 problem. The standard LQR control problem is to find an optimal control law $u \in l_2[0 \ \infty]$ such that the performance criterion $\sum_{i=0}^{\infty} z_i^T z_i$ is minimized in the following system with the impulse input $w_i = \delta_i$:

$$x_{i+1} = Ax_i + x_0 w_i + Bu_i$$
$$z_i = C_z x_i + D_{zu} u_i$$
$$y_i = x_i \tag{2.276}$$

where w_i is a scalar value, $C_z^T D_{zu} = O$, $C_z^T C_z = Q$, and $D_{zu}^T D_{zu} = R$. Note that B_w in (2.266) corresponds to x_0 in (2.276). Here, the H_2 performance criterion becomes an LQ criterion.

The LQG control problem is an important special case of the H_2 optimal control for the following system:

$$x_{i+1} = Ax_i + \left[GQ_w^{\frac{1}{2}} \; O \right] \begin{bmatrix} w_i \\ v_i \end{bmatrix} + Bu_i$$
$$y_i = Cx_i + \left[O \; R_v^{\frac{1}{2}} \right] \begin{bmatrix} w_i \\ v_i \end{bmatrix} \tag{2.277}$$
$$z_i = C_z x_i + D_{zu} u_i$$

where $C_z^T D_{zu} = O$, $C_z^T C_z = Q$, and $D_{zu}^T D_{zu} = R$. The LQG control is obtained so that the H_2 norm of the transfer function from w and v_i to z is minimized. It is noted that, according to (C.14) in Appendix C, the performance criterion (2.267) for the system (2.277) can be considered by observing the steady-state mean square value of the controlled output

$$E\left[\lim_{N\to\infty} \frac{1}{N} \sum_{i=0}^{N-1} z_i^T z_i \right] = E\left[\lim_{N\to\infty} \frac{1}{N} \sum_{i=0}^{N-1} \left(x_i^T Q x_i + u_i^T R u_i \right) \right]$$

when the white Gaussian noises with unit power are applied. It is noted that w_i and v_i can be combined into one disturbance source w_i as in (2.266).

The H_2 filter problem can be solved as a special case of the H_2 control problem. Suppose a state-space model is described by the following:

$$x_{i+1} = Ax_i + B_w w_i$$
$$y_i = Cx_i + D_w w_i \tag{2.278}$$

The H_2 filter problem is to find an estimate $\hat{x}_i$ of x_i using the measurement of y_i so that the H_2 norm from w_i to $x_i - \hat{x}_i$ is minimized. The filter has to be causal so that it can be realized.

The H_2 filter problem can be regarded as the following control problem:

$$x_{i+1} = Ax_i + B_w w_i + 0 \times \hat{x}_i$$
$$z_i = x_i - \hat{x}_i \tag{2.279}$$
$$y_i = Cx_i + D_w w_i$$

where the following correspondences to (2.266) hold

$$u_i \longleftarrow \hat{x}_i$$
$$B \longleftarrow 0$$
$$C_z \longleftarrow I$$
$$D_{zu} \longleftarrow -I$$

H_2/H_∞ Controls Based on Mixed Criteria

Each performance criterion has its own advantages and disadvantages, so that there are trade-offs between them. In some cases we want to adopt two or more performance criteria simultaneously in order to satisfy specifications. In this section, we introduce two kinds of controls based on mixed criteria. It is noted that an LQ control is a special case of H_2 controls. Here, the LQ control is used for simplicity.

1. Minimize the H_2 norm for a fixed guaranteed H_∞ norm such that

$$\min_{\gamma_2, Y, S} \gamma_2 \tag{2.280}$$

subject to

$$\begin{bmatrix} \gamma_2 & x_0^T \\ x_0 & S \end{bmatrix} \geq 0 \tag{2.281}$$

$$\begin{bmatrix} -S & (AS+BY)^T & Y^T & S \\ (AS+BY) & -S & 0 & 0 \\ Y & 0 & -R^{-1} & 0 \\ S & 0 & 0 & -Q^{-1} \end{bmatrix} \leq 0 \tag{2.282}$$

$$\begin{bmatrix} -S & (AS+BY) & B_w & 0 \\ (AS+BY)^T & -S & 0 & (C_zS+D_{zu}Y)^T \\ B_w^T & 0 & -\gamma_\infty I & 0 \\ 0 & (C_zS+D_{zu}Y) & 0 & -\gamma_\infty I \end{bmatrix} < 0 \tag{2.283}$$

From Y and S, the state feedback gain is obtained, i.e. $H = YS^{-1}$.

2. Minimize the H_∞ norm for a fixed guaranteed H_2 norm such that

$$\min_{\gamma_\infty, Y, S} \gamma_\infty \tag{2.284}$$

subject to
(2.281), (2.282), (2.283).

The state feedback gain is obtained from Y and S, i.e. $H = YS^{-1}$.

2.8 References

The material presented in this chapter has been established for a long time and is covered in several excellent books. The subject is so large that it would be a considerable task to provide comprehensive references.

Therefore, in this chapter, some references will be provided so that it is enough to understand the contents.

Dynamic programming and the minimum principle of Section 2.2 are discussed in many places. For general systems, dynamic programming and the minimum principle of Pontryagin appeare in [BD62, BK65, Bel57] and in [PBGM62] respectively. For a short review, [Kir70] is a useful reference for the minimum criterion in both dynamic programming and the minimum principle. For a treatment of the minimax criterion, see [Bur98] for the minimax principle and [BB91] [BLW91] [KIF93] for dynamic programming. Readers interested in rigorous mathematics are referred to [Str68].

The literature on LQ controls is vast and old. LQR for tracking problems as in Theorems 2.1, 2.2, and 2.4 is via dynamic programming [AM89] and via the minimum principle [BH75] [KS72]. In the case of a fixed terminal with a reference signal, the closed-loop solution is first introduced in this book as in Theorem 2.3.

The H_∞ control in Section 2.3.2 is closely related to an LQ difference game. The books by [BH75] and [BO82] are good sources for results on game theories. [BB91] is a book on game theories that deals explicitly with the connections between game theories and H_∞ control. The treatment of the finite horizon state feedback H_∞ control in this book is based on [LAKG92].

The Kalman filters as in Theorem 2.6 can be derived in many ways for stochastic systems. The seminal papers on the optimal estimation are [KB60] and [KB61].

The perfect square expression of the quadratic cost function in Lemmas 2.8 and 2.9 appeared in [Lew86b, Lew86a]. A bibliography of LQG controls is compiled by [MG71]. A special issue of the *IEEE Transactions on Automatic Control* was devoted to LQG controls in 1971 [Ath71]. Most of the contents about LQG in this book originate from the text of [Lew86b]. The LQG separation theorem appeared in [Won68]. The perfect square expression of the H_∞ cost function in Theorem 2.11 appeared in [GL95].

Even though finite horizon and infinite horizon LQ controls are obtained analytically from a Riccati approach, we also obtain them numerically from an LMI in this book, which can be useful for constrained systems. Detailed treatments of LMI can be found in [BGFB94, GNLC95]. The LMIs for *LQ* and H_∞ controls in Sections 2.6.1 and 2.6.2 appear in [GNLC95]. The bounded real lemma in Section 2.6.2 is investigated in [Yak62], [Kal63], and [Pop64], and also in a book by [Bur98].

The general H_2 problem and its solution are considered well in the frequency domain in [ZDG96]. This work is based on the infinite horizon. In [BGFB94], H_2 and H_∞ controls are given in LMI form, which can be used for the H_2/H_∞ mixed control. The work by [BH89] deals with the problem requiring the minimization of an upper bound on the H_2 norm under an H_∞ norm constraint.

2.9 Problems

2.1. Consider the system

$$x_{i+1} = \alpha x_i + u_i - \frac{u_i^2}{M - x_i} \tag{2.285}$$

where $0 < \alpha < 1$ and $x_{i_0} < M$. In particular, we want to maximize

$$J = \beta^{i_f - i_0} c x_{i_f} + \sum_{i=i_0}^{i_f - 1} [px_i - u_i] \beta^{i - i_0} \tag{2.286}$$

where $p > 0$, $0 < \beta < 1$, and $c > 0$. Find an optimal control u_i so that (2.286) is maximized.

2.2. Consider a nonlinear system $x_{k+1} = f(x_k, u_k)$ with constraints given by $\phi(x_k) \geq 0$ and the performance criterion (2.2).

(1) Show that above $\phi(x_k) \geq 0$ can be represented by an extra state variable $x_{n+1,k+1}$ such as

$$x_{n+1,k+1} = x_{n+1,k} + \{[\phi_1(x_k)]^2 \tilde{u}(-\phi_1(x_k)) + \cdots + [\phi_l(x_k)]^2 \tilde{u}(-\phi_l(x_k))\}$$

where $\tilde{u}(\cdot)$ is a unit function given by $\tilde{u}(x) = 1$ only for $x > 0$ and 0 otherwise with

$$x_{n+1,i_0} = 0, \quad x_{n+1,i_f} = 0$$

(2) Using the minimum principle, find the optimal control so that the system

$$x_{1,k+1} = 0.4 x_{2,k} \tag{2.287}$$
$$x_{2,k+1} = -0.2 x_{2,k} + u_k \tag{2.288}$$

is to be controlled to minimize the performance criterion

$$J = \sum_{k=0}^{3} 0.5[x_{1,k}^2 + x_{2,k}^2 + u_k^2] \tag{2.289}$$

The control and states are constrained by

$$-1 \leq u_k \leq 1 \tag{2.290}$$
$$-2 \leq x_{2,k} \leq 2 \tag{2.291}$$

2.3. Suppose that a man has his initial savings S and lives only on interest that comes from his savings at a fixed rate. His current savings x_k are therefore governed by the equation

$$x_{k+1} = \alpha x_k - u_k \tag{2.292}$$

where $\alpha > 1$ and u_k denotes his expenditure. His immediate enjoyment due to expenditure is $u_k^{\frac{1}{2}}$. As time goes on, the enjoyment is diminished as fast as β^k, where $|\beta| < 1$. Thus, he wants to maximize

$$J = \sum_{k=0}^{N} \beta^k u_k^{\frac{1}{2}} \tag{2.293}$$

where S, α, and β are set to 10, 1.8, and 0.6, respectively. Make simulations for three kinds of planning based on Table 1.1. For the long-term planning, use $N = 100$. For the periodic and short-term plannings, use $N = 5$ and the simulation time is 100. Using the minimum principle, find optimal solutions u_i analytically, not numerically.

2.4. Consider the following general nonlinear system:

$$x_{i+1} = f(x_i) + g(x_i)w_i, \quad z_i = h(x_i) + J(x_i)w_i \tag{2.294}$$

(1) If there exists a nonnegative function $V : \Re^n \to \Re$ with $V(0) = 0$ such that for all $w \in \Re^p$ and $k = 0, 1, 2 \cdots$

$$V(x_{k+1}) - V(x_k) \leq \gamma^2 \|w_k\|^2 - \|z_k\|^2 \tag{2.295}$$

show that the following inequality is satisfied:

$$\sum_{k=0}^{N} \|z_k\|^2 \leq \gamma^2 \sum_{k=0}^{N} \|w_k\|^2$$

Conversely, show that a nonnegative function $V : \Re^n \to \Re$ with $V(0) = 0$ exists if the H_∞ norm of the system is less than γ^2.

(2) Suppose that there exists a positive definite function $V(x)$ satisfying

$$g^T(0)\frac{\partial^2 V}{\partial^2 x}(0)g(0) + J^T(0)J(0) - \gamma^2 < 0 \tag{2.296}$$

$$0 = V(f(x) + g(x)\alpha(x)) - V(x)$$
$$+ \frac{1}{2}(\|h(x) + J(x)\alpha(x)\| - \gamma^2\|\alpha(x)\|) \tag{2.297}$$

where $\alpha(x)$ is a unique solution of

$$\frac{\partial V}{\partial \alpha(x)}(x)g(x) + \alpha(x)^T(J^T(x)J(x) - \gamma^2) = -h^T(x)J(x) \tag{2.298}$$

and the systems is observable, i.e.

$$z_k|_{w_k=0} = h(x_k) = 0 \to \lim_{k \to \infty} x_k = 0 \tag{2.299}$$

Show that the system $x_{i+1} = f(x_i)$ is stable.

2.5. We consider a minimum time performance criterion in which the objective is to steer a current state into a specific target set in minimum time.

For the system

$$x_{i+1} = \begin{bmatrix} 1 & 1 \\ 0 & 1 \end{bmatrix} x_i + \begin{bmatrix} 0.5 \\ 1 \end{bmatrix} u_i \qquad (2.300)$$

the performance criterion is given as

$$J = \sum_{i=i_0}^{i_f-1} 1 = i_f - i_0 \qquad (2.301)$$

where $|u_i| \leq 1$. Find the control u_i to bring the state from the initial point $x_{i_0} = [1 \ 4]^T$ to the origin in the minimum time.

2.6. An optimal investment plan is considered here. Without any external investment, the manufacturing facilities at the next time $k + 1$ decrease in proportion to the manufacturing facilities at the current time k. In order to increase the manufacturing facilities, we should invest money. Letting x_k and u_k be the manufacturing facilities at time k and the investment at the time k respectively, we can construct the following model:

$$x_{k+1} = \alpha x_k + \gamma u_k \qquad (2.302)$$

where $|\alpha| < 1$, $\gamma > 0$, and x_0 are given. Assume that manufacturing facilities are worth the value proportional to the investment and the product at the time k is proportional to the manufacturing facilities at time k. Then, the profit can be represented as

$$J = \beta x_N + \sum_{i=0}^{N-1} (\beta x_i - u_i) \qquad (2.303)$$

The investment is assumed to be nonnegative and bounded above, i.e. $0 \leq u_i \leq \bar{u}$. Obtain the optimal investment with respect to α, β, γ, and N.

2.7. Let a free body obey the following dynamics:

$$y_{i+1} = y_i + v_i \qquad (2.304)$$
$$v_{i+1} = v_i + u_i \qquad (2.305)$$

with y_i the position and v_i the velocity. The state is $x_i = [y_i \ v_i]^T$. Let the acceleration input u_i be constrained in magnitude by

$$|u_i| \leq 1 \qquad (2.306)$$

Suppose the objective is to determine a control input to bring any given initial state (y_{i_0}, v_{i_0}) to the origin so that

$$\begin{bmatrix} y_{i_f} \\ v_{i_f} \end{bmatrix} = 0 \tag{2.307}$$

The control should use minimum fuel, so let

$$J(i_0) = \sum_{i=i_0}^{i_f} |u_i| \tag{2.308}$$

(1) Find the minimum-fuel control law to drive any x_{i_0} to the origin in a given time $N = i_f - i_0$ if $|u_i| \le 1$.

(2) Draw the phase-plane trajectory. N, y_{i_0}, and v_{i_0} are set to 35, 10, and 10 respectively.

2.8. Consider a performance criterion with $Q = R = 0$ and a positive definite Q_f. The control can be given (2.43)–(2.45) with the inverse replaced by a pseudo inverse.

(1) Show that the solution to Riccati Equation (3.47) can be represented as

$$K_{i_f - k} = (Q_f^{\frac{1}{2}} A^k)^T \left[I - Q_f^{\frac{1}{2}} W_k (Q_f^{\frac{1}{2}} W_k)^T [Q_f^{\frac{1}{2}} W_k (Q_f^{\frac{1}{2}} W_k)^T]^\dagger \right] (Q_f^{\frac{1}{2}} A^k). \tag{2.309}$$

where $W_k = [B \ \ AB \ \ A^2 B \ \ \cdots \ \ A^{k-1} B]$ for $k = 1, 2, ..., N-1$ with $W_0 = 0$ and $A^\dagger$ is a pseudo inverse of A.

(2) In the deadbeat control problem, we desire that $x_{i_f} = 0$; this can happen only if a performance criterion is equal to zero, i.e. if $K_{i_0} = 0$. Show that K_{i_0} can be zero if the following condition holds for some k:

$$\text{Im}(A^k) \subset \text{Im}(W_k) \tag{2.310}$$

where $\text{Im}(M)$ is the image of the matrix M.

(3) Show that (2.310) is satisfied if the system is controllable.

2.9. Consider the following performance criterion for the system (2.27):

$$J(x^r, u) = \sum_{i=i_0}^{i_f - 1} \begin{bmatrix} x_i \\ u_i \end{bmatrix}^T \begin{bmatrix} Q & M \\ M^T & R \end{bmatrix} \begin{bmatrix} x_i \\ u_i \end{bmatrix} + x_{i_f}^T Q_f x_{i_f}$$

where

$$Q = Q^T \ge 0, \quad \begin{bmatrix} Q & M \\ M^T & R \end{bmatrix} \ge 0, \quad R = R^T > 0$$

Show that the optimal control is given as

$$u_i = -K_i x_i$$
$$K_i = (B^T S_{i+1} B + R)^{-1} (B^T S_{i+1} A + M^T)$$
$$S_i = A^T S_{i+1} A - (B^T S_{i+1} A + M^T)^T (B^T S_{i+1} B + R)^{-T} (B^T S_{i+1} A + M^T)$$
$$\quad + Q$$

where $S_{i_f} = Q_f$.

2.10. Show that the general tracking control (2.103) is reduced to the simpler tracking control (2.79) if Q in (2.103) becomes zero.

2.11. Consider the minimum energy performance criterion given by

$$J = \sum_{i=i_0}^{i_f-1} u_i^T R u_i \tag{2.311}$$

for the system (2.27). Find the control u_i that minimizes (2.311) and satisfies the constraints $|u_i| \leq \bar{u}$ and $x_{i_f} = 0$.

2.12. Consider an optimal control problem on $[i_0 \ i_f]$ for the system (2.27) with the LQ performance criterion

$$J(x, u) = \sum_{i=i_0}^{i_f-1} (x_i^T Q x_i + u_i^T R u_i) \tag{2.312}$$

(1) Find the optimal control u_k subject to $C x_{i_f} + b = 0$.
(2) Find the optimal control u_k subject to $x_{i_f}^T P x_{i_f} \leq 1$, where P is a symmetric positive definite matrix.

2.13. Consider the following performance criterion for the system (2.120):

$$J(x^r, u) = \sum_{i=i_0}^{i_f-1} \left\{ \begin{bmatrix} x_i \\ u_i \end{bmatrix}^T \begin{bmatrix} Q & M \\ M^T & R \end{bmatrix} \begin{bmatrix} x_i \\ u_i \end{bmatrix} - \gamma^2 w_i^T w_i \right\} + x_{i_f}^T Q_f x_{i_f},$$

where

$$Q = Q^T \geq 0, \quad \begin{bmatrix} Q & M \\ M^T & R \end{bmatrix} \geq 0, \quad R = R^T > 0. \tag{2.313}$$

Derive the H_∞ control.

2.14. Derive the last term h_{i,i_f} in the optimal cost (2.148) associated with the H_∞ control.

2.15. Consider the stochastic model (2.161) and (2.162) where w_i and v_i are zero-mean, white noise sequences with variance given by

$$E\left\{ \begin{bmatrix} w_i \\ v_i \end{bmatrix} \begin{bmatrix} w_j^T & v_j^T \end{bmatrix} \right\} = \begin{bmatrix} \Xi_{11} & \Xi_{12} \\ \Xi_{12} & \Xi_{22} \end{bmatrix} \delta_{i-j}$$

(1) Show that (2.161) and (2.162) are equivalent to the following model:

$$x_{i+1} = \bar{A} x_i + B u_i + G \Xi_{12} \Xi_{22}^{-1} y_i + G \xi_i$$
$$y_i = C x_i + v_i$$

where

$$\bar{A} = A - G \Xi_{12} \Xi_{22}^{-1} C, \quad \xi_i = w_i - \Xi_{12} \Xi_{22}^{-1} v_i$$

(2) Find $E\{\xi_i v_i^T\}$.

(3) Show that the controllability and observability of the pairs $\{\bar{A}, B\}$ and $\{\bar{A}, C\}$ are guaranteed by the controllability and observability of the pairs $\{A, B\}$ and $\{A, C\}$, respectively.

(4) Find the Kalman filter $\hat{x}_{k+1|k}$.

2.16. Consider the following system:

$$x_{k+1} = \begin{bmatrix} 0 & 1 & 0 \\ 0 & 0 & 1 \\ 0 & 0 & 0 \end{bmatrix} x_k + \begin{bmatrix} 0 \\ 0 \\ 1 \end{bmatrix} \xi_k$$

$$y_k = \begin{bmatrix} 1 & 0 & 0 \end{bmatrix} x_k + \theta_k$$

where ξ_k and θ_k are zero-mean Gaussian white noises with covariance 1 and μ.

(1) Express the covariance of the state estimation error P_k as a function of μ.

(2) Calculate the gain matrix of the Kalman filter.

(3) Calculate and plot the poles and zeros of the closed-loop system.

2.17. Consider the LQG problem

$$\begin{bmatrix} x_{1,i+1} \\ x_{2,i+1} \end{bmatrix} = \begin{bmatrix} 1 & 1 \\ 0 & 1 \end{bmatrix} \begin{bmatrix} x_{1,i} \\ x_{2,i} \end{bmatrix} + \sqrt{\rho} \begin{bmatrix} 1 & 0 \\ 1 & 0 \end{bmatrix} w + \begin{bmatrix} 0 \\ 1 \end{bmatrix} u \qquad (2.314)$$

$$y = \begin{bmatrix} 1 & 0 \end{bmatrix} \begin{bmatrix} x_1 \\ x_2 \end{bmatrix} \qquad (2.315)$$

for the following performance criterion:

$$J = \sum_{i=i_0}^{i_f-1} \rho(x_1 + x_2)^2 + u_i^2 \qquad (2.316)$$

Discuss the stability margin, such as gain and phase margins, for steady-state control.

2.18. Consider a controllable pair $\{A, B\}$ and assume A does not have unit-circle eigenvalues. Consider also arbitrary matrices $\{Q, S, R\}$ of appropriate dimensions and define a Popov function

$$S_y(z) = \begin{bmatrix} B^T(zI - A^T)^{-1} & I \end{bmatrix} \begin{bmatrix} Q & S \\ S^T & R \end{bmatrix} \begin{bmatrix} (z^{-1}I - A)^{-1}B \\ I \end{bmatrix} \qquad (2.317)$$

where the central matrix is Hermitian but may be indefinite. The KYP lemma[KSH00] can be stated as follows.

The following three statements are all equivalent:

(a) $S_y(e^{jw}) \geq 0$ for all $w \in [-\pi, \pi]$.

(b) There exists a Hermitian matrix P such that

$$\begin{bmatrix} Q - P + A^T P A & S + A^T P B \\ S^T + B^T P A & R + B^T P B \end{bmatrix} \geq 0 \tag{2.318}$$

(c) There exist an $n \times n$ Hermitian matrix P, a $p \times p$ matrix $R_e \geq 0$, and an $n \times p$ matrix K_p, such that

$$\begin{bmatrix} Q - P + A^T P A & S + A^T P B \\ S^T + B^T P A & R + B^T P B \end{bmatrix} = \begin{bmatrix} K_p \\ I \end{bmatrix} R_e \begin{bmatrix} K_p^T & I \end{bmatrix} \tag{2.319}$$

Derive the bounded real lemma from the above KYP lemma and compare it with the LMI result in this section.

3

State Feedback Receding Horizon Controls

3.1 Introduction

In this chapter, state feedback receding horizon controls for linear systems will be given for both quadratic and H_∞ performance criteria.

The state feedback receding horizon LQ controls will be extensively investigated because they are bases for the further developments of other receding controls. The receding horizon control with the quadratic performance criterion will be derived with detailed procedures. Time-invariant systems are dealt with with simple notations. The important monotonicity of the optimal cost will be introduced with different conditions, such as a free terminal state and a fixed terminal state. A nonzero terminal cost for the free terminal state is often termed a *free terminal state* thereafter, and a fixed terminal state as a *terminal equality constraint* thereafter. Stability of the receding horizon controls is proved under cost monotonicity conditions. Horizon sizes for guaranteeing the stability are determined regardless of terminal weighting matrices. Some additional properties of the receding horizon controls are presented.

Similar results are given for the H_∞ controls that are obtained from the minimax criterion. In particular, monotonicity of the saddle-point value and stability of the state feedback receding horizon H_∞ controls are discussed.

Since cost monotonicity conditions look difficult to obtain, we introduce easy computation of receding horizon LQ and H_∞ controls by the LMI.

In order to explain the concept of a receding horizon, we introduce the predictive form, say x_{k+j}, and the referenced predictive form, say $x_{k+j|k}$, in this chapter. Once the concept is clearly understood by using the reference predictive form, we will use the predictive form instead of the reference predictive form.

The organization of this chapter is as follows. In Section 3.2, predictive forms for systems and performance criteria are introduced. In Section 3.3, receding horizon LQ controls are extensively introduced with cost monotonicity, stability, and internal properties. A special case of input–output systems is investigated for GPC. In Section 3.4, receding horizon H_∞ controls are dealt with with cost monotonicity, stability, and internal properties. In Section 3.5, receding horizon LQ control and H_∞ control are represented via batch and LMI forms.

3.2 Receding Horizon Controls in Predictive Forms

3.2.1 Predictive Forms

Consider the following state-space model:

$$x_{i+1} = Ax_i + Bu_i \qquad (3.1)$$
$$z_i = C_z x_i \qquad (3.2)$$

where $x_i \in \Re^n$ and $u_i \in \Re^m$ are the state and the input respectively. z_i in (3.2) is called a controlled output. Note that the time index i is an arbitrary time point. This time variable will also be used for recursive equations.

With the standard form (3.1) and (3.2) it is not easy to represent the future time from the current time. In order to represent the future time from the current time, we can introduce a *predictive form*

$$x_{k+j+1} = Ax_{k+j} + Bu_{k+j} \qquad (3.3)$$
$$z_{k+j} = C_z x_{k+j} \qquad (3.4)$$

where k and j indicate the current time and the time distance from it respectively. Note that x_{k+j}, z_{k+j}, and u_{k+j} mean future state, future output, and future input at time $k + j$ respectively. In the previous chapter it was not necessary to identify the current time. However, in the case of the RHC the current time and the specific time points on the horizon should be distinguished. Thus, k is used instead of i for RHC, which offers a clarification during the derivation procedure. The time on the horizon denoted by $k + j$ means the time after j from the current time. This notation is depicted in Figure 3.1. However, the above predictive form also does not distinguish the current time if they are given as numbers. For example, $k = 10$ and $j = 3$ give $x_{k+j} = x_{13}$. When x_{13} is given, there is no way to know what the current time is. Therefore, in order to identify the current time we can introduce a *referenced predictive form*

$$x_{k+j+1|k} = Ax_{k+j|k} + Bu_{k+j|k} \qquad (3.5)$$
$$z_{k+j|k} = C_z x_{k+j|k} \qquad (3.6)$$

with the initial condition $x_{k|k} = x_k$. In this case, when $k = 10$ and $j = 3$, $x_{k+j|k}$ can be represented as $x_{13|10}$. We can see that the current time k is 10 and the distance j from the current time is 3. A referenced predictive form

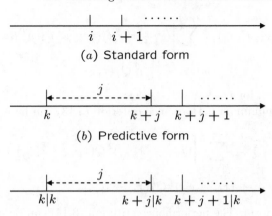

(a) Standard form

(b) Predictive form

(c) Referenced predictive form

Fig. 3.1. Times in predictive forms

improves understanding. However, a predictive form will often be used in this book because the symbol k indicates the current time.

For a minimax problem, the following system is considered:

$$x_{i+1} = Ax_i + Bu_i + B_w w_i \qquad (3.7)$$
$$z_i = C_z x_i \qquad (3.8)$$

where w_i is a disturbance. In order to represent the future time we can introduce a *predictive form*

$$x_{k+j+1} = Ax_{k+j} + Bu_{k+j} + B_w w_{k+j} \qquad (3.9)$$
$$z_{k+j} = C_z x_{k+j} \qquad (3.10)$$

and a *referenced predictive form*

$$x_{k+j+1|k} = Ax_{k+j|k} + Bu_{k+j|k} + B_w w_{k+j} \qquad (3.11)$$
$$z_{k+j|k} = C_z x_{k+j|k} \qquad (3.12)$$

with the initial condition $x_{k|k} = x_k$.

In order to explain the concept of a receding horizon, we introduce the predictive form and the referenced predictive form. Once the concept is clearly understood by using the referenced predictive form, we will use the predictive form instead of the referenced predictive form for notational simplicity.

3.2.2 Performance Criteria in Predictive Forms

In the minimum performance criterion (2.31) for the free terminal cost, i_0 can be arbitrary and is set to k so that we have

$$J(x_k, x^r_., u.) = \sum_{i=k}^{i_f-1} \left[(x_i - x^r_i)^T Q(x_i - x^r_i) + u_i^T R u_i \right]$$
$$+ (x_{i_f} - x^r_{i_f})^T Q_f(x_{i_f} - x^r_{i_f}) \qquad (3.13)$$

where x^r_i is given for $i = k, k+1, \cdots, i_f$.

The above minimum performance criterion (3.13) can be represented by

$$J(x_k, x^r_{k+.}, u_{k+.}) = \sum_{j=0}^{i_f-k-1} \left[(x_{k+j} - x^r_{k+j})^T Q(x_{k+j} - x^r_{k+j}) + u_{k+j}^T R u_{k+j} \right]$$
$$+ (x_{i_f} - x^r_{i_f})^T Q_f(x_{i_f} - x^r_{i_f}) \qquad (3.14)$$

in a predictive form. The performance criterion (3.14) can be rewritten as

$$J(x_{k|k}, x^r, u_{k+\cdot|k}) = \sum_{j=0}^{i_f-k-1} \left[(x_{k+j|k} - x^r_{k+j|k})^T Q(x_{k+j|k} - x^r_{k+j|k}) \right.$$
$$\left. + u_{k+j|k}^T R u_{k+j|k} \right] + (x_{i_f|k} - x^r_{i_f|k})^T Q_f(x_{i_f|k} - x^r_{i_f|k}) \qquad (3.15)$$

in a referenced predictive form, where x^r is used instead of $x^r_{k+\cdot|k}$ for simplicity.

As can be seen in (3.13), (3.14), and (3.15), the performance criterion depends on the initial state, the reference trajectory, and the input on the horizon. If minimizations are taken for the performance criteria, then we denote them by $J^*(x_k, x^r)$ in a predictive form and $J^*(x_{k|k}, x^r)$ in a referenced predictive form. We can see that the dependency of the input disappears for the optimal performance criterion.

The performance criterion for the terminal equality constraint can be given as in (3.13), (3.14), and (3.15) without terminal costs, i.e. $Q_f = 0$. The terminal equality constraints are represented as $x_{i_f} = x^r_{i_f}$ in (3.13) and (3.14) and $x_{i_f|k} = x^r_{i_f|k}$ in (3.15).

In the minimax performance criterion (2.128) for the free terminal cost, i_0 can be arbitrary and is set to k so that we have

$$J(x_k, x^r, u., w.) = \sum_{i=k}^{i_f-1} \left[(x_i - x^r_i)^T Q(x_i - x^r_i) + u_i^T R u_i - \gamma^2 w_i^T R_w w_i \right]$$
$$+ (x_{i_f} - x^r_{i_f})^T Q_f(x_{i_f} - x^r_{i_f}) \qquad (3.16)$$

where x^r_i is given for $i = k, k+1, \cdots, i_f$.

The above minimax performance criterion (3.16) can be represented by

$$J(x_k, x^r, u_{k+.}, w_{k+.}) = \sum_{j=0}^{i_f-k-1} \left[(x_{k+j} - x^r_{k+j})^T Q(x_{k+j} - x^r_{k+j}) + u_{k+j}^T R u_{k+j} \right.$$
$$\left. - \gamma^2 w_{k+j}^T R_w w_{k+j} \right] + (x_{i_f} - x^r_{i_f})^T Q_f(x_{i_f} - x^r_{i_f}) \qquad (3.17)$$

in a predictive form. The performance criterion (3.17) can be rewritten as

$$J(x_{k|k}, x^r, u_{k+\cdot|k}, w_{k+\cdot|k}) = \sum_{j=0}^{i_f-k-1} \left[(x_{k+j|k} - x^r_{k+j|k})^T Q (x_{k+j|k} - x^r_{k+j|k}) \right.$$

$$\left. + u^T_{k+j|k} R u_{k+j|k} - \gamma^2 w^T_{k+j|k} R_w w_{k+j|k} \right]$$

$$+ (x_{i_f|k} - x^r_{i_f|k})^T Q_f (x_{i_f|k} - x^r_{i_f|k}) \tag{3.18}$$

in a referenced predictive form.

Unlike the minimization problem, the performance criterion for the minimaxization problem depends on the disturbance. Taking the minimization and the maximization with respect to the input and the disturbance respectively yields the optimal performance criterion that depends only on the initial state and the reference trajectory. As in the minimization problem, we denote the optimal performance criterion by $J^*(x_k, x^r)$ in a predictive form and $J^*(x_{k|k}, x^r)$ in a referenced predictive form.

3.3 Receding Horizon Control Based on Minimum Criteria

3.3.1 Receding Horizon Linear Quadratic Control

Consider the following discrete time-invariant system of a referenced predictive form:

$$x_{k+j+1|k} = Ax_{k+j|k} + Bu_{k+j|k} \tag{3.19}$$
$$z_{k+j|k} = C_z x_{k+j|k} \tag{3.20}$$

A state feedback RHC for the system (3.19) and (3.20) is introduced in a tracking form. As mentioned before, the current time and the time distance from the current time are denoted by k and j for clarification. The time variable j is used for the derivation of the RHC.

Free Terminal Cost

The optimal control for the system (3.19) and (3.20) and the free terminal cost (3.18) can be rewritten in a referenced predictive form as

$$u^*_{k+j|k} = -R^{-1}B^T[I + K_{k+j+1,i_f|k}BR^{-1}B^T]^{-1}$$
$$\times [K_{k+j+1,i_f|k}Ax_{k+j|k} + g_{k+j+1,i_f|k}] \tag{3.21}$$

where

$$K_{k+j,i_f|k} = A^T[I + K_{k+j+1,i_f|k}BR^{-1}B^T]^{-1}K_{k+j+1,i_f|k}A + Q$$

$$g_{k+j,i_f|k} = A^T[I + K_{k+j+1,i_f|k}BR^{-1}B^T]^{-1}g_{k+j+1,i_f|k} - Qx^r_{k+j|k}$$

with $K_{i_f,i_f|k} = Q_f$ and $g_{i_f,i_f|k} = -Q_f x^r_{i_f|k}$.

The receding horizon concept was introduced in the introduction chapter and is depicted in Figure 3.2. The optimal control is obtained first on the horizon $[k,\ k+N]$. Here, k indicates the current time and $k+N$, is the final time on the horizon. Therefore, $i_f = k + N$, where N is the horizon size. The

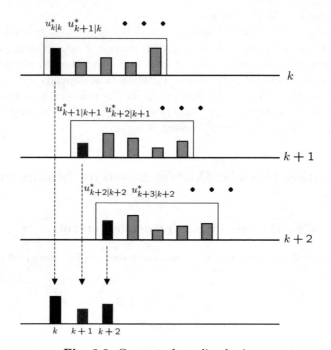

Fig. 3.2. Concept of receding horizon

performance criterion can be given in a referenced predictive form as

$$J(x_{k|k}, x^r, u_{k+\cdot|k}) = \sum_{j=0}^{N-1}\left[(x_{k+j|k} - x^r_{k+j|k})^T Q(x_{k+j|k} - x^r_{k+j|k})\right.$$

$$\left. +u^T_{k+j|k}Ru_{k+j|k}\right] + (x_{k+N|k} - x^r_{k+N|k})^T Q_f(x_{k+N|k} - x^r_{k+N|k}) \quad (3.22)$$

The optimal control on the interval $[k,\ k+N]$ is given in a referenced predictive form by

$$u^*_{k+j|k} = -R^{-1}B^T[I + K_{k+j+1,k+N|k}BR^{-1}B^T]^{-1}$$

$$\times [K_{k+j+1,k+N|k}Ax_{k+j|k} + g_{k+j+1,k+N|k}] \quad (3.23)$$

where $K_{k+j+1,k+N|k}$ and $g_{k+j+1,k+N|k}$ are given by

$$K_{k+j,k+N|k} = A^T[I + K_{k+j+1,k+N|k}BR^{-1}B^T]^{-1}K_{k+j+1,k+N|k}A$$
$$+ Q \qquad (3.24)$$
$$g_{k+j,k+N|k} = A^T[I + K_{k+j+1,k+N|k}BR^{-1}B^T]^{-1}g_{k+j+1,k+N|k}$$
$$- Qx^r_{k+j|k} \qquad (3.25)$$

with

$$K_{k+N,k+N|k} = Q_f \qquad (3.26)$$
$$g_{k+N,k+N|k} = -Q_f x^r_{k+N|k} \qquad (3.27)$$

The receding horizon LQ control at time k is given by the first control $u_{k|k}$ among $u_{k+i|k}$ for $i = 0, 1, \cdots, k + N - 1$ as in Figure 3.2. It can be obtained from (3.23) with $j = 0$ as

$$u^*_{k|k} = -R^{-1}B^T[I + K_{k+1,k+N|k}BR^{-1}B^T]^{-1}$$
$$\times [K_{k+1,k+N|k}Ax_k + g_{k+1,k+N|k}] \qquad (3.28)$$

where $K_{k+1,k+N|k}$ and $g_{k+1,k+N|k}$ are computed from (3.24) and (3.25).

The above notation in a referenced predictive form can be simplified to a predictive form by dropping the reference value.

It simply can be represented by a predictive form

$$u^*_{k+j} = -R^{-1}B^T[I + K_{k+j+1,i_f}BR^{-1}B^T]^{-1}$$
$$\times [K_{k+j+1,i_f}Ax_{k+j} + g_{k+j+1,i_f}] \qquad (3.29)$$

where

$$K_{k+j,i_f} = A^T[I + K_{k+j+1,i_f}BR^{-1}B^T]^{-1}K_{k+j+1,i_f}A + Q \qquad (3.30)$$
$$g_{k+j,i_f} = A^T[I + K_{k+j+1,i_f}BR^{-1}B^T]^{-1}g_{k+j+1,i_f} - Qx^r_{k+j} \qquad (3.31)$$

with $K_{i_f,i_f} = Q_f$ and $g_{i_f,i_f} = -Q_f x^r_{i_f}$. Thus, $u_{k|k}$ and $K_{k+1,k+N|k}$ are replaced by u_k and $K_{k+1,k+N}$ so that we have

$$u^*_k = -R^{-1}B^T[I + K_{k+1,k+N}BR^{-1}B^T]^{-1}[K_{k+1,k+N}Ax_k + g_{k+1,k+N}] \quad (3.32)$$

where $K_{k+1,k+N}$ and $g_{k+1,k+N}$ are computed from

$$K_{k+j,k+N} = A^T[I + K_{k+j+1,k+N}BR^{-1}B^T]^{-1}K_{k+j+1,k+N}A + Q \quad (3.33)$$
$$g_{k+j,k+N} = A^T[I + K_{k+j+1,k+N}BR^{-1}B^T]^{-1}g_{k+j+1,k+N} - Qx^r_{k+j} \quad (3.34)$$

with

$$K_{k+N,k+N} = Q_f \qquad (3.35)$$
$$g_{k+N,k+N} = -Q_f x^r_{k+N} \qquad (3.36)$$

Note that $I + K_{k+j,k+N} BR^{-1} B^T$ is nonsingular since $K_{k+j,k+N}$ is guaranteed to be positive semidefinite and the nonsingularity of $I + MN$ implies that of $I + NM$ for any matrices M and N.

For the zero reference signal x_i^r becomes zero, so that for the free terminal state, we have

$$u_k^* = -R^{-1} B^T [I + K_{k+1,k+N} BR^{-1} B^T]^{-1} K_{k+1,k+N} Ax_k \qquad (3.37)$$

from (3.32).

Terminal Equality Constraint

So far, the free terminal costs are utilized for the receding horizon tracking control (RHTC). The terminal equality constraint can also be considered for the RHTC. In this case, the performance criterion is written as

$$J(x_k, x^r, u_{k+\cdot|k}) = \sum_{j=0}^{N-1} \left[(x_{k+j|k} - x_{k+j|k}^r)^T Q(x_{k+j|k} - x_{k+j|k}^r) \right.$$
$$\left. + u_{k+j|k}^T R u_{k+j|k} \right] \qquad (3.38)$$

where

$$x_{k+N|k} = x_{k+N|k}^r \qquad (3.39)$$

The condition (3.39) is often called the terminal equality condition. The RHC for the terminal equality constraint with a nonzero reference signal is obtained by replacing i and i_f by k and $k + N$ in (2.103) as follows:

$$u_k = -R^{-1} B^T (I + K_{k+1,k+N} BR^{-1} B^T)^{-1} \left[K_{k+1,k+N} Ax_k + M_{k+1,k+N} \right.$$
$$\left. \times S_{k+1,k+N}^{-1} (x_{k+N}^r - M_{k,k+N}^T x_k - h_{k,k+N}) + g_{k+1,k+N} \right] \qquad (3.40)$$

where $K_{k+\cdot,k+N}$, $M_{k+\cdot,k+N}$, $S_{k+\cdot,k+N}$, $g_{k+\cdot,k+N}$, and $h_{k+\cdot,k+N}$ are as follows:

$$K_{k+j,k+N} = A^T K_{k+j+1,k+N} (I + BR^{-1} B^T K_{k+j+1,k+N})^{-1} A + Q$$

$$M_{k+j,k+N} = (I + BR^{-1} B^T K_{k+j+1,k+N})^{-T} M_{k+j+1,k+N}$$

$$S_{k+j,k+N} = S_{k+j+1,k+N}$$
$$\quad - M_{k+j+1,k+N}^T B(B^T K_{k+j+1,k+N} B + R)^{-1} B^T M_{k+j+1,k+N}$$

$$g_{k+j,k+N} = A^T g_{k+j+1,k+N}$$
$$\quad - A^T K_{k+j+1,k+N} (I + BR^{-1} B^T K_{k+j+1,k+N})^{-1} BR^{-1} B^T$$
$$\quad \times g_{k+j+1,k+N} - Qx_{k+j}^r$$

$$h_{k+j,k+N} = h_{k+j+1,k+N}$$
$$\quad - M_{k+j+1,k+N}^T (I + BR^{-1} B^T K_{k+j+1,k+N})^{-1} BR^{-1} B^T g_{k+j+1,k+N}$$

The boundary conditions are given by

$$K_{k+N,k+N} = 0, M_{k+N,k+N} = I, S_{k+N,k+N} = 0, g_{k+N,k+N} = 0, h_{k+N,k+N} = 0$$

For the regulation problem, (3.40) is reduced to

$$u_k^* = -R^{-1}B^T(I + K_{k+1,k+N}BR^{-1}B^T)^{-1}[K_{k+1,k+N}A$$
$$- M_{k+1,k+N}S_{k+1,k+N}^{-1}M_{k,k+N}^T]x_k \tag{3.41}$$

From (2.68), u_k^* in (3.41) is represented in another form

$$u_k^* = -R^{-1}B^T P_{k+1,k+N+1}^{-1} A x_k \tag{3.42}$$

where $P_{k+1,k+N+1}$ is computed from (2.65)

$$P_{k+j,k+N+1} = A^{-1}[I + P_{k+j+1,k+N+1}A^{-T}QA^{-1}]^{-1}P_{k+j+1,k+N+1}A$$
$$+ BR^{-1}B^T \tag{3.43}$$

with

$$P_{k+N+1,k+N+1} = 0 \tag{3.44}$$

Note that the system matrix A should be nonsingular in Riccati Equation (3.43). However, this requirement can be relaxed in the form of (3.41) or with the batch form, which is left as a problem at the end of this chapter.

3.3.2 Simple Notation for Time-invariant Systems

In previous sections the Riccati equations have had two arguments, one of which represents the terminal time. However, only one argument is used for time-invariant systems in this section for simplicity. If no confusion arises, then one argument will be used for Riccati equations throughout this book, particularly for Riccati equations for time-invariant systems.

Time-invariant homogeneous systems such as $x_{i+1} = f(x_i)$ have a special property known as shift invariance. If the initial condition is the same, then the solution depends on the distance from the initial time. Let x_{i,i_0} denote the solution at i with the initial i_0, as can be seen in Figure 3.3. That is

$$x_{i,i_0} = x_{i+N,i_0+N} \tag{3.45}$$

for any N with $x_{i_0,i_0} = x_{i_0+N,i_0+N}$.

Free Terminal Cost

Since (3.33) is also a time-invariant system, the following equation is satisfied:

$$K_{k+j,k+N} = K_{j,N} \tag{3.46}$$

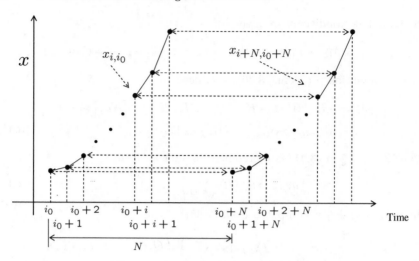

x

x_{i,i_0}

x_{i+N,i_0+N}

i_0 i_0+2 i_0+i i_0+N i_0+2+N

i_0+1 i_0+i+1 i_0+1+N

Time

N

Fig. 3.3. Property of shift invariance

with $K_{N,N} = K_{k+N,k+N} = Q_f$.

Since N is fixed, we denote $K_{j,N}$ by simply K_j and K_j satisfies the following equation:

$$K_j = A^T K_{j+1} A - A^T K_{i+1} B[R + B^T K_{j+1} B]^{-1} B^T K_{j+1} A + Q$$
$$= A^T K_{j+1}[I + BR^{-1}B^T K_{j+1}]^{-1} A + Q \tag{3.47}$$

with the boundary condition

$$K_N = Q_f \tag{3.48}$$

Thus, the receding horizon control (3.32) can be represented as

$$u_k^* = -R^{-1}B^T[I + K_1 BR^{-1}B^T]^{-1}[K_1 A x_k + g_{k+1,k+N}] \tag{3.49}$$

where K_1 is obtained from (3.47) and $g_{k+1,k+N}$ is computed from

$$g_{k+j,k+N} = A^T[I + K_{j+1}BR^{-1}B^T]^{-1} g_{k+j+1,k+N} - Q x_{k+j}^r \tag{3.50}$$

with the boundary condition

$$g_{k+N,k+N} = -Q_f x_{k+N}^r \tag{3.51}$$

It is noted that (3.34) is not a time-invariant system due to a time-varying signal, x_{k+j}^r. If x_{k+j}^r is a constant signal denoted by $\bar{x}^r$, then

$$g_j = A^T[I + K_{j+1}BR^{-1}B^T]^{-1} g_{j+1} - Q\bar{x}^r \tag{3.52}$$

with the boundary condition

$$g_N = -Q_f \bar{x}^r \tag{3.53}$$

The control can be written as

$$u_k = -R^{-1}B^T[I + K_1BR^{-1}B^T]^{-1}(K_1Ax_k + g_1) \tag{3.54}$$

It is noted that from shift invariance with a new boundary condition

$$K_{N-1} = Q_f \tag{3.55}$$

K_1 in (3.49) and (3.54) becomes K_0.

For the zero reference signal x_i^r becomes zero, so that for the free terminal cost we have

$$u_k^* = -R^{-1}B^T[I + K_1BR^{-1}B^T]^{-1}K_1Ax_k \tag{3.56}$$

from (3.32).

Terminal Equality Constraint

The RHC (3.40) for the terminal equality constraint with a nonzero reference can be represented as

$$u_k = -R^{-1}B^T[I + K_1BR^{-1}B^T]^{-1}\bigg[K_1Ax_k$$

$$+ M_1S_1^{-1}(x_{k+N}^r - M_0^Tx_k - h_{k,k+N}) + g_{k+1,k+N}\bigg] \tag{3.57}$$

where K_j, M_j, S_j, $g_{k+j,k+N}$, and $h_{k+j,k+N}$ are as follows:

$$K_j = A^TK_{j+1}(I + BR^{-1}B^TK_{j+1})^{-1}A + Q$$
$$M_j = (I + BR^{-1}B^TK_{j+1})^{-T}M_{j+1}$$
$$S_j = S_{j+1} - M_{j+1}^TB(B^TK_{j+1}B + R)^{-1}B^TM_{j+1}$$
$$g_{k+j,k+N} = A^Tg_{k+j+1,k+N}$$
$$\quad - A^TK_{j+1}(I + BR^{-1}B^TK_{j+1})^{-1}BR^{-1}B^Tg_{k+j+1,k+N}$$
$$\quad - Qx_{k+j}^r$$
$$h_{k+j,k+N} = h_{k+j+1,k+N}$$
$$\quad - M_{k+j+1,k+N}^T(I + BR^{-1}B^TK_{j+1})^{-1}BR^{-1}B^Tg_{k+j+1,k+N}$$

The boundary conditions are given by

$$K_N = 0,\ M_N = I,\ S_N = 0,\ g_{k+N,k+N} = 0,\ h_{k+N,k+N} = 0$$

For the regulation problem, (3.57) is reduced to

$$u_k^* = -R^{-1}B^T[I + K_1BR^{-1}B^T]^{-1}[K_1A - M_1S_1^{-1}M_0^T]x_k \tag{3.58}$$

From (3.42), u_k^* in (3.58) is represented in another form

$$u_k^* = -R^{-1}B^T P_1^{-1} A x_k \qquad (3.59)$$

where P_1 is computed from

$$P_j = A^{-1}\left[I + P_{j+1}A^{-T}QA^{-1}\right]^{-1}P_{j+1}A + BR^{-1}B^T \qquad (3.60)$$

with

$$P_{N+1} = 0 \qquad (3.61)$$

Note that it is assumed that the system matrix A is nonsingular.

Forward Computation

The computation of (3.47) is made in a backward way and the following forward computation can be introduced by the transformation

$$\overrightarrow{K}_j = K_{N-j+1} \qquad (3.62)$$

Thus, K_1 starting from $K_N = Q_f$ is obtained as

$$Q_f = \overrightarrow{K}_1 = K_N, \ \overrightarrow{K}_2 = K_{N-1}, \ \cdots, \ \overrightarrow{K}_N = K_1$$

The Riccati equation can be written as

$$\overrightarrow{K}_{j+1} = A^T \overrightarrow{K}_j A - A^T \overrightarrow{K}_j B[R + B^T \overrightarrow{K}_j B]^{-1} B^T \overrightarrow{K}_j A + Q$$
$$\qquad (3.63)$$

$$= A^T \overrightarrow{K}_j [I + BR^{-1}B^T \overrightarrow{K}_j]^{-1} A + Q, \qquad (3.64)$$

with the initial condition

$$\overrightarrow{K}_1 = Q_f \qquad (3.65)$$

In the same way as the Riccati equation, g_1 starting from $g_N = -Q_f \bar{x}^r$ is obtained as

$$\overrightarrow{g}_{j+1} = A^T[I + \overrightarrow{K}_j BR^{-1}B^T]^{-1}\overrightarrow{g}_j - Q\bar{x}^r \qquad (3.66)$$

where $\overrightarrow{g}_1 = -Q_f \bar{x}^r$. The relation and dependency among K_j, $\overrightarrow{K}_j$, g_j, and $\overrightarrow{g}_j$ are shown in Figure 3.4 and Figure 3.5.

The control is represented by

$$u_k = -R^{-1}B^T[I + \overrightarrow{K}_N BR^{-1}B^T]^{-1}(\overrightarrow{K}_N A x_k + \overrightarrow{g}_N) \qquad (3.67)$$

For forward computation, the RHC (3.42) and Riccati Equation (3.43) can be written as

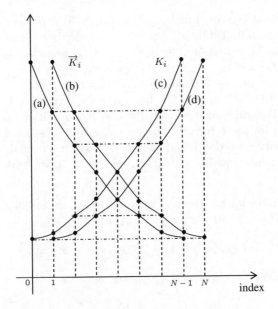

Fig. 3.4. Computation of K_i and $\hat{K}_i$. Initial conditions $i = 0$ in (a), $i = 1$ in (b), $i = N - 1$ in (c), and $i = N$ in (d)

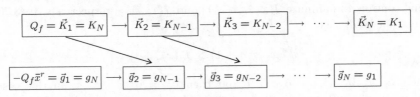

Fig. 3.5. Relation between K_i and g_i

$$u_k^* = -R^{-1}B^T \overrightarrow{P}_N^{-1} A x_k \tag{3.68}$$

where $\overrightarrow{P}_N$ is computed by

$$\overrightarrow{P}_{j+1} = A^{-1}\overrightarrow{P}_j[I + A^{-T}QA^{-1}\overrightarrow{P}_j]^{-1}A^{-T} + BR^{-1}B^T \tag{3.69}$$

with $\overrightarrow{P}_1 = 0$.

3.3.3 Monotonicity of the Optimal Cost

In this section, some conditions are proposed for time-invariant systems which guarantee the monotonicity of the optimal cost. In the next section, under the proposed cost monotonicity conditions, the closed-loop stability of the RHC is shown. Since the closed-loop stability can be treated with the regulation problem, the g_i can be zero in this section.

It is noted that the cost function J (3.13)–(3.15) depends on several variables, such as the initial state x_i, input $u_{i+.}$, initial time i, and terminal time i_f. Thus, it can be represented as $J(x_i, u_{i+.}, i, i_f)$ and the optimal cost can be given as $J^*(x_i, i, i_f)$.

Define $\delta J^*(x_\tau, \sigma)$ as $\delta J^*(x_\tau, \sigma) = J^*(x_\tau, \tau, \sigma + 1) - J^*(x_\tau, \tau, \sigma)$. If $\delta J^*(x_\tau, \sigma) \leq 0$ or $\delta J^*(x_\tau, \sigma) \geq 0$ for any $\sigma > \tau$, then it is called a cost monotonicity. We will show first that the cost monotonicity condition can be easily achieved by the terminal equality condition. Then, the more general cost monotonicity condition is introduced by using a terminal cost.

For the terminal equality condition, i.e. $x_{i_f} = 0$, we have the following result.

Theorem 3.1. *For the terminal equality constraint, the optimal cost $J^*(x_i, i, i_f)$ satisfies the following cost monotonicity relation:*

$$J^*(x_\tau, \tau, \sigma + 1) \leq J^*(x_\tau, \tau, \sigma), \quad \tau \leq \sigma \tag{3.70}$$

If the Riccati solution exists for (3.70), then we have

$$K_{\tau, \sigma+1} \leq K_{\tau, \sigma} \tag{3.71}$$

Proof. This can be proved by contradiction. Assume that u_i^1 and u_i^2 are optimal controls to minimize $J(x_\tau, \tau, \sigma + 1)$ and $J(x_\tau, \tau, \sigma)$ respectively. If (3.70) does not hold, then

$$J^*(x_\tau, \tau, \sigma + 1) > J^*(x_\tau, \tau, \sigma)$$

Replace u_i^1 by u_i^2 up to $\sigma - 1$ and then $u_i^1 = 0$ at $i = \sigma$. In this case, $x_\sigma = 0$, $u_\sigma^1 = 0$, and thus $x_{\sigma+1}^1 = 0$. Therefore, the cost for this control is $\bar{J}(x_\tau, \tau, \sigma + 1) = J^*(x_\tau, \tau, \sigma)$. Since this control may not be optimal for $J(x_\tau, \tau, \sigma + 1)$, we have $\bar{J}(x_\tau, \tau, \sigma + 1) \geq J^*(x_\tau, \tau, \sigma + 1)$, which implies that

$$J^*(x_\tau, \tau, \sigma) \geq J^*(x_\tau, \tau, \sigma + 1) \tag{3.72}$$

This is a contradiction to (3.70). This completes the proof. ∎

For the time-invariant systems we have

$$K_\tau \leq K_{\tau+1}$$

The above result is for the terminal equality condition. Next, the cost monotonicity condition using a free terminal cost is introduced.

Theorem 3.2. *Assume that Q_f in (3.15) satisfies the following inequality:*

$$Q_f \geq Q + H^T R H + (A - BH)^T Q_f (A - BH) \tag{3.73}$$

for some $H \in \Re^{m \times n}$.

For the free terminal cost, the optimal cost $J^*(x_i, i, i_f)$ *then satisfies the following monotonicity relation:*

$$J^*(x_\tau, \tau, \sigma + 1) \leq J^*(x_\tau, \tau, \sigma), \quad \tau \leq \sigma \tag{3.74}$$

and thus

$$K_{\tau, \sigma+1} \leq K_{\tau, \sigma} \tag{3.75}$$

Proof. u_i^1 and u_i^2 are the optimal controls to minimize $J(x_\tau, \tau, \sigma + 1)$ and $J(x_\tau, \tau, \sigma)$ respectively. If we replace u_i^1 by u_i^2 up to $\sigma - 1$ and $u_\sigma^1 = -Hx_\sigma$, then the cost for this control is given by

$$\bar{J}(x_\tau, \sigma + 1) \triangleq \sum_{i=\tau}^{\sigma-1} [x_i^{2T} Q x_i^2 + u_i^{2T} R u_i^2] + x_\sigma^{2T} Q x_\sigma^2 + x_\sigma^{2T} H^T R H x_\sigma^2$$
$$+ x_\sigma^{2T} (A - BH)^T Q_f (A - BH) x_\sigma^2$$
$$\geq J^*(x_\tau, \sigma + 1) \tag{3.76}$$

where the last inequality comes from the fact that this control may not be optimal. The difference between the adjacent optimal cost is less than or equal to zero as

$$J^*(x_\tau, \sigma + 1) - J^*(x_\tau, \sigma) \leq \bar{J}(x_\tau, \sigma + 1) - J^*(x_\tau, \sigma)$$
$$= x_\sigma^{2T} Q x_\sigma^2 + x_\sigma^{2T} H^T R H x_\sigma^2$$
$$+ x_\sigma^{2T} (A - BH)^T Q_f (A - BH) x_\sigma^2 - x_\sigma^{2T} Q_f x_\sigma^2$$
$$= x_\sigma^{2T} \{Q + H^T R H + (A - BH)^T Q_f (A - BH) - Q_f\} x_\sigma^2$$
$$\leq 0 \tag{3.77}$$

where

$$J^*(x_\tau, \sigma) = \sum_{i=\tau}^{\sigma-1} [x_i^{2T} Q x_i^2 + u_i^{2T} R u_i^2] + x_\sigma^{2T} Q_f x_\sigma^2 \tag{3.78}$$

From (3.77) we have

$$\triangle J^*(x_\tau, \sigma) = x_\tau^T [K_{\tau, \sigma+1} - K_{\tau, \sigma}] x_\tau \leq 0 \tag{3.79}$$

for all x_τ, and thus $K_{\tau, \sigma+1} - K_{\tau, \sigma} \leq 0$. This completes the proof. ∎

It looks difficult to find out Q_f and H satisfying (3.73). One approach is as follows: if H that makes $A - BH$ Hurwitz is given, then Q_f can be systematically obtained. First choose one matrix $M > 0$ such that $M \geq Q + H^T R H$. Then, calculate the solution Q_f to the following Lyapunov equation:

$$(A - BH)^T Q_f (A - BH) - Q_f = -M \tag{3.80}$$

It can be easily seen that Q_f obtained from (3.80) satisfies (3.73). Q_f can be explicitly expressed as

$$Q_f = \sum_{i=0}^{\infty} (A - BH)^{Ti} M (A - BH)^i \tag{3.81}$$

Another approach to find out Q_f and H satisfying (3.73) is introduced in Section 3.5.1, where LMIs are used.

It is noted that for time-invariant systems the inequality (3.75) implies

$$K_{\tau,\sigma+1} \le K_{\tau+1,\sigma+1} \tag{3.82}$$

which leads to

$$K_\tau \le K_{\tau+1} \tag{3.83}$$

The monotonicity of the optimal cost is presented in Figure 3.6. There are

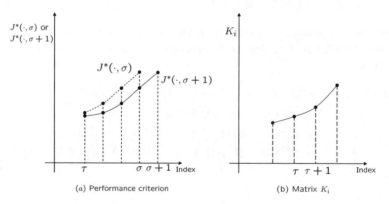

(a) Performance criterion   (b) Matrix K_i

Fig. 3.6. Cost monotonicity of Theorem 3.1

several cases that satisfy the condition of Theorem 3.2.

Case 1:

$$Q_f \ge A^T Q_f [I + BR^{-1} B^T Q_f]^{-1} A + Q \tag{3.84}$$

If H is replaced by a matrix $H = [R + B^T Q_f B]^{-1} B^T Q_f A$, then we have

$$
\begin{aligned}
Q_f &\ge Q + H^T R H + (A - BH)^T Q_f (A - BH) \\
&= Q + A^T Q_f B [R + B^T Q_f B]^{-1} R [R + B^T Q_f B]^{-1} B^T Q A \\
&\quad + (A - B[R + B^T Q_f B]^{-1} B^T Q_f A)^T Q_f (A - B[R + B^T Q_f B]^{-1} B^T Q_f A) \\
&= Q + A^T Q_f A - A^T Q_f B [R + B^T Q_f B]^{-1} B^T Q_f A \\
&= A^T Q_f [I + BR^{-1} B^T Q_f]^{-1} A + Q
\end{aligned} \tag{3.85}
$$

which is a special case of (3.73). All Q_f values satisfying the inequality (3.85) are a subset of all Q_f values satisfying (3.73).

It can be seen that (3.73) implies (3.85) regardless of H as follows:

$$
\begin{aligned}
& Q_f - Q - A^T Q_f [I + BR^{-1}B^T Q_f]^{-1} A \\
& \geq -A^T Q_f [I + BR^{-1}B^T Q_f]^{-1} A + H^T R H + (A - BH)^T Q_f (A - BH) \\
& = [A^T Q_f B (R + B^T Q_f B)^{-1} - H]^T (R + B^T Q_f B) \\
& \quad \times [(R + B^T Q_f B)^{-1} B^T Q_f A - H] \\
& \geq 0
\end{aligned}
\tag{3.86}
$$

Therefore, all Q_f values satisfying (3.73) also satisfy (3.85), and thus are a subset of all Q_f values satisfying (3.85). Thus, (3.73) is surprisingly equivalent to (3.85).

Q_f that satisfies (3.85) can also be obtained explicitly from the solution to the following Riccati equation:

$$
Q_f^* = \beta^2 A^T Q_f^* [I + \gamma BR^{-1}B^T Q_f^*]^{-1} A + \alpha Q
\tag{3.87}
$$

with $\alpha \geq 1$, $\beta \geq 1$, and $0 \leq \gamma \leq 1$. It can be easily seen that Q_f^* satisfies (3.85), since

$$
\begin{aligned}
Q_f^* &= \beta^2 A^T Q_f^* [I + \gamma BR^{-1}B^T Q_f^*]^{-1} A + \alpha Q \\
&\geq A^T Q_f^* [I + BR^{-1}B^T Q_f^*]^{-1} A + Q
\end{aligned}
$$

Case 2:

$$
Q_f = Q + H^T R H + (A - BH)^T Q_f (A - BH)
\tag{3.88}
$$

This Q_f is a special case of (3.73) and has the following meaning. Note that u_i is unknown for the interval $[\tau\ \sigma - 1]$ and defined as $u_i = -Hx_i$ on the interval $[\sigma,\ \infty]$. If a pair (A, B) is stabilizable and $u_i = -Hx_i$ is a stabilizing control, then

$$
\begin{aligned}
J &= \sum_{i=\tau}^{\infty} [x_i^T Q x_i + u_i^T R u_i] \\
&= \sum_{i=\tau}^{\sigma-1} [x_i^T Q x_i + u_i^T R u_i] \\
&\quad + x_\sigma^T \sum_{i=\sigma}^{\infty} (A^T - H^T B^T)^{i-\sigma} [Q + H^T R H] (A - BH)^{i-\sigma} x_\sigma \\
&= \sum_{i=\tau}^{\sigma-1} [x_i^T Q x_i + u_i^T R u_i] + x_\sigma^T Q_f x_\sigma
\end{aligned}
\tag{3.89}
$$

where Q_f satisfies $Q_f = Q + H^T R H + (A - BH)^T Q_f (A - BH)$. Therefore, Q_f is related to the steady-state performance with the control $u_i = -Hx_i$.

It is noted that, under (3.73), $u_i = -Hx_i$ will be proved to be a stabilizing control in Section 3.3.4.

Case 3:

$$Q_f = A^T Q_f [I + BR^{-1}B^T Q_f]^{-1} A + Q \tag{3.90}$$

This is actually the steady-state Riccati equation and is a special case of (3.85), and thus of (3.73). This Q_f is related to the steady-state optimal performance with the optimal control.

Case 4:

$$Q_f = Q + A^T Q_f A \tag{3.91}$$

If the system matrix A is stable and u_i is identically equal to zero for $i \geq \sigma \geq \tau$, then Q_f satisfies $Q_f = Q + A^T Q_f A$, which is also a special case of (3.73).

Proposition 3.3. Q_f *satisfying (3.73) has the following lower bound:*

$$Q_f \geq \bar{K} \tag{3.92}$$

where $\bar{K}$ is the steady-state solution to the Riccati equation in (3.90) and assumed to exist uniquely.

Proof. By the cost monotonicity condition, the solution to the recursive Riccati equation starting from Q_f satisfying Case 3 can be ordered

$$Q_f = K_{i_0} \geq K_{i_0+1} \geq K_{i_0+2} \geq \cdots \tag{3.93}$$

where

$$K_{i+1} = A^T K_i [I + BR^{-1}B^T K_i]^{-1} A + Q \tag{3.94}$$

with $K_{i_0} = Q_f$.

Since K_i is composed of two positive semidefinite matrices, K_i is also positive semidefinite, or bounded below, i.e. $K_i \geq 0$.

K_i is decreasing and bounded below, so that K_i has a limit value, which is denoted by $\bar{K}$. Clearly, it can be easily seen that

$$Q_f \geq K_i \geq \bar{K} \tag{3.95}$$

for any $i \geq i_0$.

The only thing we have to do is to guarantee that $\bar{K}$ satisfies the condition corresponding to Case 3. Taking the limitation on both sides of (3.94), we have

$$\lim_{i \to \infty} K_{i+1} = \lim_{i \to \infty} A^T K_i [I + BR^{-1}B^T K_i]^{-1} A + Q \tag{3.96}$$

$$\bar{K} = A^T \bar{K} [I + BR^{-1}B^T \bar{K}]^{-1} A + Q \tag{3.97}$$

The uniqueness of the solution to the Riccati equation implies that $\bar{K}$ is the solution that satisfies Case 3. This completes the proof. ∎

Theorem 3.2 discusses the nonincreasing monotonicity for the optimal cost. In the following, the nondecreasing monotonicity of the optimal cost can be obtained.

Theorem 3.4. *Assume that Q_f in (3.15) satisfies the inequality*

$$Q_f \leq A^T Q_f [I + BR^{-1}B^T Q_f]^{-1} A + Q \qquad (3.98)$$

The optimal cost $J^*(x_i, i, i_f)$ then satisfies the relation

$$J^*(x_\tau, \tau, \sigma + 1) \geq J^*(x_\tau, \tau, \sigma), \quad \tau \leq \sigma \qquad (3.99)$$

and thus

$$K_{\tau, \sigma+1} \geq K_{\tau, \sigma} \qquad (3.100)$$

Proof. u_i^1 and u_i^2 are the optimal controls to minimize $J(x_\tau, \tau, \sigma + 1)$ and $J(x_\tau, \tau, \sigma)$ respectively. If we replace u_i^2 by u_i^1 up to $\sigma - 1$, then by the optimal principle we have

$$J^*(x_\tau, \sigma) = \sum_{i=\tau}^{\sigma-1} [x_i^{2T} Q x_i^2 + u_i^{2T} R u_i^2] + x_\sigma^{2T} Q_f x_\sigma^2 \qquad (3.101)$$

$$\leq \sum_{i=\tau}^{\sigma-1} [x_i^{1T} Q x_i^1 + u_i^{1T} R u_i^1] + x_\sigma^{1T} Q_f x_\sigma^1 \qquad (3.102)$$

The difference between the adjacent optimal cost can be expressed as

$$\delta J^*(x_\tau, \sigma) = \sum_{i=\tau}^{\sigma} [x_i^{1T} Q x_i^1 + u_i^{1T} R u_i^1] + x_{\sigma+1}^{1T} Q_f x_{\sigma+1}^1$$

$$- \sum_{i=\tau}^{\sigma-1} [x_i^{2T} Q x_i^2 + u_i^{2T} R u_i^2] - x_\sigma^{2T} Q_f x_\sigma^2 \qquad (3.103)$$

Combining (3.102) and (3.103) yields

$$\delta J^*(x_\tau, \sigma) \geq x_\sigma^{1T} Q x_\sigma^1 + u_\sigma^{1T} R u_\sigma^1 + x_{\sigma+1}^{1T} Q_f x_{\sigma+1}^1 - x_\sigma^{1T} Q_f x_\sigma^1$$

$$= x_\sigma^{1T} \{Q + A^T Q_f [I + BR^{-1}B^T Q_f]^{-1} A - Q_f\} x_\sigma^1$$

$$\geq 0 \qquad (3.104)$$

where $u_\sigma^1 = -H x_\sigma^1$, $x_{\sigma+1}^1 = (A - BH) x_\sigma^1$ and $H = -(R + B^T Q_f B)^{-1} B^T Q_f A$. The second equality of (3.104) comes from the following fact:

$$H^T R H + (A - BH)^T Q_f (A - BH) =$$
$$A^T Q_f [I + BR^{-1}B^T Q_f]^{-1} A \qquad (3.105)$$

as can be seen in (3.85). The last inequality of (3.104) comes from (3.98). From (2.49) and (3.99) we have

$$\delta J^*(x_\tau, \sigma) = x_\tau^T [K_{\tau, \sigma+1} - K_{\tau, \sigma}] x_\tau \geq 0 \qquad (3.106)$$

for all x_τ. Thus, $K_{\tau, \sigma+1} - K_{\tau, \sigma} \geq 0$. This completes the proof. ∎

It is noted that the relation (3.100) on the Riccati equation can be represented simply by one argument as

$$K_\tau \geq K_{\tau+1} \tag{3.107}$$

for time-invariant systems.

The monotonicity of the optimal cost in Theorem 3.4 is presented in Figure 3.7.

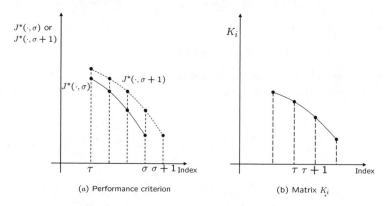

(a) Performance criterion   (b) Matrix K_i

Fig. 3.7. Cost monotonicity of Theorem 3.2

We have at least one important case that satisfies the condition of Theorem 3.4.

Case 1: $Q_f = 0$

It is noted that the free terminal cost with the zero terminal weighting, $Q_f = 0$, satisfies (3.98). Thus, Theorem 3.4 includes the monotonicity of the optimal cost of the free terminal cost with the zero terminal weighting.

The terminal equality condition is more conservative than the free terminal cost. Actually, it is a strong requirement that the nonzero state must go to zero within a finite time. Thus, the terminal equality constraint has no solution for the small horizon size N, whereas the free terminal cost always gives a solution for any horizon size N. The free terminal cost requires less computation than the terminal equality constraint. However, the terminal equality constraint provides a simple approach for guaranteeing stability.

It is noted that the cost monotonicity in Theorems 3.1, 3.2 and 3.4 are obtained from the optimality. Thus, the cost monotonicity may hold even for nonlinear systems, which will be explained later.

In the following theorem, it will be shown that when the monotonicity of the optimal cost or the Riccati equations holds once, it holds for all subsequent times.

Theorem 3.5.

(a) If

$$J^*(x_{\tau'}, \tau', \sigma+1) \leq J^*(x_{\tau'}, \tau', \sigma) \quad (or \geq J^*(x_{\tau'}, \tau', \sigma)) \quad (3.108)$$

for some τ', then

$$J^*(x_{\tau''}, \tau'', \sigma+1) \leq J^*(x_{\tau''}, \tau'', \sigma) \quad (or \geq J^*(x_{\tau''}, \tau'', \sigma)) \quad (3.109)$$

where $\tau_0 \leq \tau'' \leq \tau'$.

(b) If

$$K_{\tau', \sigma+1} \leq K_{\tau', \sigma} \quad (or \geq K_{\tau', \sigma}) \quad (3.110)$$

for some τ', then

$$K_{\tau'', \sigma+1} \leq K_{\tau'', \sigma} \quad (or \geq K_{\tau'', \sigma}) \quad (3.111)$$

where $\tau_0 \leq \tau'' \leq \tau'$.

Proof. We first prove the part (a). u_i^1 and u_i^2 are the optimal controls to minimize $J(x_{\tau''}, \tau'', \sigma+1)$ and $J(x_{\tau''}, \tau'', \sigma)$ respectively. If we replace u_i^1 by u_i^2 up to $\tau' - 1$, then by the optimal principle we have

$$J^*(x_{\tau''}, \sigma+1) = \sum_{i=\tau''}^{\tau'-1} [x_i^{1T} Q x_i^1 + u_i^{1T} R u_i^1] + J^*(x_{\tau'}^1, \tau', \sigma+1)$$

$$\leq \sum_{i=\tau''}^{\tau'-1} [x_i^{2T} Q x_i^2 + u_i^{2T} R u_i^2] + J^*(x_{\tau'}^2, \tau', \sigma+1) \quad (3.112)$$

The difference between the adjacent optimal cost can be expressed as

$$\delta J^*(x_{\tau''}, \sigma) = \sum_{i=\tau''}^{\tau'-1} [x_i^{1T} Q x_i^1 + u_i^{1T} R u_i^1] + J^*(x_{\tau'}^1, \tau', \sigma+1)$$

$$- \sum_{i=\tau''}^{\tau'-1} [x_i^{2T} Q x_i^2 + u_i^{2T} R u_i^2] - J^*(x_{\tau'}^2, \tau', \sigma) \quad (3.113)$$

Combining (3.112) and (3.113) yields

$$\delta J^*(x_{\tau''}, \sigma) \leq \sum_{i=\tau''}^{\tau'-1} [x_i^{2T} Q x_i^2 + u_i^{2T} R u_i^2] + J^*(x_{\tau'}^2, \tau', \sigma+1)$$

$$- \sum_{i=\tau''}^{\tau'-1} [x_i^{2T} Q x_i^2 + u_i^{2T} R u_i^2] - J^*(x_{\tau'}^2, \tau', \sigma)$$

$$= J^*(x_{\tau'}^2, \tau', \sigma+1) - J^*(x_{\tau'}^2, \tau', \sigma)$$

$$= \delta J^*(x_{\tau'}^2, \sigma) \leq 0 \quad (3.114)$$

Replacing u_i^2 by u_i^1 up to $\tau' - 1$ and taking similar steps we have

$$\delta J^*(x_{\tau''}, \sigma) \geq \delta J^*(x_{\tau'}^1, \sigma) \tag{3.115}$$

from which $\delta J^*(x_{\tau'}^1, \sigma) \geq 0$ implies $\delta J^*(x_{\tau''}, \sigma) \geq 0$.

Now we prove the second part of the theorem. From (2.49), (3.108), and (3.109), the monotonicities of the Riccati equations hold. From the inequality

$$J^*(x_{\tau''}, \tau'', \sigma + 1) - J^*(x_{\tau''}, \tau'', \sigma) = x_{\tau''}^T [K_{\tau'', \sigma+1} - K_{\tau'', \sigma}] x_{\tau''}$$
$$\leq (\geq) \; 0$$

$K_{\tau'', \sigma+1} \leq (\geq) K_{\tau'', \sigma}$ is satisfied. This completes the proof.   ∎

For time-invariant systems the above relations can be simplified. If

$$K_{\tau'} \leq K_{\tau'+1} \; (\text{or} \geq K_{\tau'+1}) \tag{3.116}$$

for some τ', then

$$K_{\tau''} \leq K_{\tau''+1} \; (\text{or} \geq K_{\tau''+1}) \tag{3.117}$$

for $\tau_0 < \tau'' < \tau'$.

Part (a) of Theorem 3.5 may be extended to constrained and nonlinear systems, whereas part (b) is only for linear systems.

Computation of the solutions of cost monotonicity conditions (3.73), (3.84), and (3.98) looks difficult to solve, but it can be easily solved by using LMI, as seen in Section 3.5.1.

3.3.4 Stability of Receding Horizon Linear Quadratic Control

For the existence of a stabilizing feedback control, we assume that the pair (A, B) is stabilizable. In this section it will be shown that the RHC is a stable control under cost monotonicity conditions.

Theorem 3.6. *Assume that the pairs (A, B) and $(A, Q^{\frac{1}{2}})$ are stabilizable and observable respectively, and that the receding horizon control associated with the quadratic cost $J(x_i, i, i+N)$ exists. If $J^*(x_i, i, i+N+1) \leq J^*(x_i, i, i+N)$, then asymptotical stability is guaranteed.*

Proof. For time-invariant systems, the system is asymptotically stable if the zero state is attractive. We show that the zero state is attractive. Since $J^*(x_i, i, \sigma + 1) \leq J^*(x_i, i, \sigma)$,

$$J^*(x_i, i, i+N) = x_i^T Q x_i + u_i^{*T} R u_i^* + J^*(x(i+1; x_i, i, u_i^*), i+1, i+N)$$
$$\geq x_i^T Q x_i + u_i^{*T} R u_i^*$$
$$+ J^*(x(i+1; x_i, i, u_i^*), i+1, i+N+1) \tag{3.118}$$

Note that u_i is the receding horizon control since it is the first control on the interval $[i, i + N]$. From (3.118) we have

$$J^*(x_i, i, i + N) \geq J^*(x(i + 1; x_i, i, u_i^*), i + 1, i + N + 1) \qquad (3.119)$$

Recall that a nonincreasing sequence bounded below converges to a constant. Since $J^*(x_i, i, i + N)$ is nonincreasing and $J^*(x_i, i, i + N) \geq 0$, we have

$$J^*(x_i, i, i + N) \to c \qquad (3.120)$$

for some nonnegative constant c as $i \to \infty$. Thus, as $i \to \infty$,

$$u_i^{*T} R u_i^* + x_i^T Q x_i \to 0 \qquad (3.121)$$

and

$$\sum_{j=i}^{i+l-1} x_j^T Q x_j + u_j^{*T} R u_j^* = x_i^T \sum_{j=i}^{i+l-1} (A^T - L_f^T B^T)^{j-i} (Q + L_f^T R L_f)$$

$$\times (A - B L_f)^{j-i} x_i = x_i^T G_{i,i+l}^o x_i \longrightarrow 0,$$

where L_f is the feedback gain of the RHC and $G_{i,i+l}^o$ is an observability Gramian of $(A - BL_f, (Q + L_f^T R L_f)^{\frac{1}{2}})$. However, since the pair $(A, Q^{\frac{1}{2}})$ is observable, it is guaranteed that $G_{i,i+l}^o$ is nonsingular for $l \geq n_c$ by Theorem B.5 in Appendix B. This means that $x_i \to 0$ as $i \to \infty$, independently of i_0. Therefore, the closed-loop system is asymptotically stable. This completes the proof.   ∎

Note that if the condition $Q > 0$ is added in the condition of Theorem 3.6, then the horizon size N could be greater than or equal to 1.

The observability in Theorem 3.6 can be weakened with the detectability similarly as in [KK00].

It was proved in the previous section that the optimal cost with the terminal equality constraint has a nondecreasing property. Therefore, we have the following result.

Theorem 3.7. *Assume that the pairs (A, B) and $(A, Q^{\frac{1}{2}})$ are controllable and observable respectively. The receding horizon control (3.42) obtained from the terminal equality constraint is asymptotically stable for $n_c \leq N < \infty$.*

Proof. The controllability and $n_c \leq N < \infty$ are required for the existence of the optimal control, as seen in Figure 2.3. Then it follows from Theorem 3.1 and Theorem 3.6.   ∎

Note that if the condition $Q > 0$ is added in the condition of Theorem 3.7, then the horizon size N could be $\max(n_c) \leq N < \infty$.

So far, we have discussed a terminal equality constraint. For the free terminal cost we have a cost monotonicity condition in Theorem 3.2 for the stability.

Theorem 3.8. *Assume that the pairs (A, B) and $(A, Q^{\frac{1}{2}})$ are stabilizable and observable respectively. For $Q_f \geq 0$ satisfying (3.73) for some H, the system (3.4) with the receding horizon control (3.56) is asymptotically stable for $1 \leq N < \infty$.* ∎

Theorem 3.8 follows from Theorems 3.2 and 3.6. Q_f in the four cases in Section 3.3.3 satisfies (3.73) and thus the condition of Theorem 3.8.

What we have talked about so far can be seen from a different perspective. The difference Riccati equation (3.47) for $j = 0$ can be represented as

$$K_1 = A^T K_1 A - A^T K_1 B [R + B^T K_1 B]^{-1} B^T K_1 A + \bar{Q} \qquad (3.122)$$

$$\bar{Q} = Q + K_1 - K_0 \qquad (3.123)$$

Equation (3.122) no longer looks like a recursion, but rather an algebraic equation for K_1. Therefore, Equation (3.122) is called the fake ARE (FARE).

The closed-loop stability of the RHC obtained from (3.122) and (3.123) requires the condition $\bar{Q} \geq 0$ and the detectability of the pair $(A, \bar{Q}^{\frac{1}{2}})$. The pair $(A, \bar{Q}^{\frac{1}{2}})$ is detectable if the pair $(A, Q^{\frac{1}{2}})$ is detectable and $K_1 - K_0 \geq 0$. The condition $K_1 - K_0 \geq 0$ is satisfied under the terminal inequality (3.73).

The free parameter H obtained in Theorem 3.8 is combined with the performance criterion to guarantee the stability of the closed-loop system. However, the free parameter H can be used itself as a stabilizing control gain.

Theorem 3.9. *Assume that the pairs (A, B) and $(A, Q^{\frac{1}{2}})$ are stabilizable and observable respectively. The system (3.4) with the control $u_i = -Hx_i$ is asymptotically stable where H is obtained from the inequality (3.73).*

Proof. Let

$$V(x_i) = x_i^T Q_f x_i \qquad (3.124)$$

where we can show that Q_f is positive definite as follows. As just said before, Q_f of (3.73) satisfies the inequality (3.84). If $\triangle$ is defined as

$$\triangle = Q_f - A^T Q_f [I + BR^{-1}B^T Q_f]^{-1} A - Q \geq 0 \qquad (3.125)$$

we can consider the following Riccati equation:

$$Q_f = A^T Q_f [I + BR^{-1}B^T Q_f]^{-1} A + Q + \triangle \qquad (3.126)$$

The observability of $(A, Q^{\frac{1}{2}})$ implies the observability of $(A, (Q + \triangle)^{\frac{1}{2}})$, so that the unique positive solution Q_f comes from (3.126). Therefore, $V(x_i)$ can be considered to be a candidate of Lyapunov functions.

Subtracting $V(x_i)$ from $V(x_{i+1})$ yields

$$V(x_{i+1}) - V(x_i) = x_i^T[(A - BH)^T Q_f(A - BH) - Q_f]x_i$$
$$\leq x_i^T[-Q - H^T RH]x_i \leq 0$$

In order to use LaSalle's theorem, we try to find out the set $S = \{x_i|V(x_{i+l+1}) - V(x_{i+l}) = 0, l = 0, 1, 2, \cdots\}$. Consider the following equation:

$$x_i^T(A - BH)^{lT}[Q + H^T RH](A - BH)^l x_i = 0 \qquad (3.127)$$

for all $l \geq 0$. According to the observability of $(A, Q^{\frac{1}{2}})$, the only solution that can stay identically in S is the trivial solution $x_i = 0$. Thus, the system driven by $u_i = -Hx_i$ is asymptotically stable. This completes the proof.   ∎

Note that the control in Theorem 3.8 considers both the performance and the stability, whereas the one in Theorem 3.9 considers only the stability.

These results of Theorems 3.7 and 3.8 can be extended further. The matrix Q in Theorems 3.7 and 3.8 must be nonzero. However, it can even be zero in the extended result.

Let us consider the receding horizon control introduced in (3.59)

$$u_i = -R^{-1}B^T P_1^{-1} A x_i \qquad (3.128)$$

where P_1 is computed from

$$P_i = A^{-1}P_{i+1}[I + A^{-T}QA^{-1}P_{i+1}]^{-1}A^{-T} + BR^{-1}B^T \qquad (3.129)$$

with the boundary condition for the free terminal cost

$$P_N = Q_f^{-1} + BR^{-1}B^T \qquad (3.130)$$

and $P_N = BR^{-1}B^T$ for the terminal equality constraint. However, we will assume that P_N can be arbitrarily chosen from now on and is denoted by P_f, $P_N = P_f$.

In the theorem to follow, we will show the stability of the closed-loop systems with the receding horizon control (3.128) under a certain condition that includes the well-known condition $P_f = 0$.

In fact, Riccati Equation (3.129) with the condition $P_f \geq 0$ can be obtained from the following system:

$$\hat{x}_{i+1} = A^{-T}\hat{x}_i + A^{-T}Q^{\frac{1}{2}}\hat{u}_i \qquad (3.131)$$

with a performance criterion

$$\hat{J}(\hat{x}_{i_0}, i_0, i_f) = \sum_{i=i_0}^{i_f-1}[\hat{x}_i^T BR^{-1}B^T\hat{x}_i + \hat{u}_i^T\hat{u}_i] + \hat{x}_{i_f}^T P_f\hat{x}_{i_f} \qquad (3.132)$$

The optimal performance criterion (3.132) for the system (3.131) is given by $\hat{J}^*(\hat{x}_i, i, i_f) = \hat{x}_i^T P_{i,i_f}\hat{x}_i$.

The nondecreasing monotonicity of (3.132) is given in the following corollary by using Theorem 3.4.

Assume that P_f in (3.132) satisfies the following inequality:

$$P_f \leq A^{-1}P_f[I + A^{-T}QA^{-1}P_f]^{-1}A^{-T} + BR^{-1}B^T \qquad (3.133)$$

From Theorem 3.4 we have

$$P_{\tau,\sigma+1} \geq P_{\tau,\sigma} \qquad (3.134)$$

For time-invariant systems we have

$$P_\tau \geq P_{\tau+1} \qquad (3.135)$$

It is noted that Inequality (3.134) is the same as (3.71). The well-known condition for the terminal equality constraint $P_f = 0$ satisfies (3.133), and thus (3.134) holds.

Before investigating the stability under the condition (3.133), we need knowledge of an adjoint system. The two systems $x_{1,i+1} = Ax_{1,i}$ and $x_{2,i+1} = A^{-T}x_{2,i}$ are said to be adjoint to each other. They generate state trajectories while making the value of $x_i^T x_i$ fixed. If one system is shown to be unstable for any initial state the other system is guaranteed to be stable. Note that all eigenvalues of the matrix A are located outside the unit circle if and only if the system is unstable for any initial state. Additionally, it is noted that the eigenvalues of A are inverse to those of A^{-T}.

Now we are in a position to investigate the stability of the closed-loop systems with the control (3.128) under the condition (3.133) that includes the well-known condition $P_f = 0$.

Theorem 3.10. *Assume that the pair (A, B) is controllable and A is nonsingular.*

(1) If $P_{i+1} \leq P_i$ for some i, then the system (3.4) with the receding horizon control (3.128) is asymptotically stable for $n_c + 1 \leq N < \infty$.

(2) If $P_f \geq 0$ satisfies (3.133), then the system (3.4) with the receding horizon control (3.128) is asymptotically stable for $n_c + 1 \leq N < \infty$.

Proof. Consider the adjoint system of the system (3.4) with the control (3.128)

$$\hat{x}_{i+1} = [A - BR^{-1}B^T P_1^{-1}A]^{-T}\hat{x}_i \qquad (3.136)$$

and the associated scalar-valued function

$$V(\hat{x}_i) = \hat{x}_i^T A^{-1}P_1 A^{-T}\hat{x}_i \qquad (3.137)$$

Note that the inverse of (3.136) is guaranteed to exist since, from (3.129), we have

$$P_1 = A^{-1}P_2[I + A^{-T}QA^{-1}P_2]^{-1}A^{-T} + BR^{-1}B^T$$
$$> BR^{-1}B^T$$

for nonsingular, A and P_2. Note that $P_1 > 0$ and $(P_1 - BR^{-1}B^T)P_1^{-1}$ is nonsingular so that $A - BR^{-1}B^T P_1^{-1}A$ is nonsingular.

Taking the subtraction of functions at time i and $i + 1$ yields

$$
\begin{aligned}
V(\hat{x}_i) &- V(\hat{x}_{i+1}) \\
&= \hat{x}_i^T A^{-1}P_1 A^{-T}\hat{x}_i - \hat{x}_{i+1}^T A^{-1}P_1 A^{-T}\hat{x}_{i+1} \\
&= \hat{x}_{i+1}^T \Big[(A - BR^{-1}B^T P_1^{-1}A)A^{-1}P_1 A^{-T}(A - BR^{-1}B^T P_1^{-1}A)^T \\
&\quad - A^{-1}P_1 A^{-T}\Big]\hat{x}_{i+1}^T \\
&= -\hat{x}_{i+1}^T\Big[P_1 - 2BR^{-1}B^T + BR^{-1}B^T P_1^{-1}BR^{-1}B^T \\
&\quad - A^{-1}P_1 A^{-T}\Big]\hat{x}_{i+1}^T
\end{aligned}
$$
(3.138)

We have

$$
\begin{aligned}
P_1 &= (A^T P_2^{-1}A + Q)^{-1} + BR^{-1}B^T \\
&= A^{-1}(P_2^{-1} + A^{-T}QA^{-1})^{-1}A^{-T} + BR^{-1}B^T \\
&= A^{-1}\Big[P_2 - P_2 A^{-T}Q^{\frac{1}{2}}(Q^{\frac{1}{2}}A^{-1}P_2 A^{-T}Q^{\frac{1}{2}} + I)^{-1}Q^{\frac{1}{2}}A^{-1}P_2\Big] \\
&\quad \times A^{-T} + BR^{-1}B^T \\
&= A^{-1}P_2 A^{-T} + BR^{-1}B^T - Z
\end{aligned}
$$
(3.139)

where

$$Z = A^{-1}P_2 A^{-T}Q^{\frac{1}{2}}(Q^{\frac{1}{2}}A^{-1}P_2 A^{-T}Q^{\frac{1}{2}} + I)^{-1}Q^{\frac{1}{2}}A^{-1}P_2 A^{-T}$$

Substituting (3.139) into (3.138) we have

$$V(\hat{x}_i) - V(\hat{x}_{i+1}) = \hat{x}_{i+1}^T[-BR^{-1}B^T + BR^{-1}B^T P_1^{-1}BR^{-1}B^T]\hat{x}_{i+1}$$

$$+ \hat{x}_{i+1}^T[A^{-1}(P_2 - P_1)A^{-T} - Z]\hat{x}_{i+1}$$

Since $P_2 < P_1$ and $Z \geq 0$ we have

$$
\begin{aligned}
V(\hat{x}_i) - V(\hat{x}_{i+1}) &\leq \hat{x}_{i+1}^T[-BR^{-1}B^T + BR^{-1}B^T P_1^{-1}BR^{-1}B^T]\hat{x}_{i+1} \\
&= \hat{x}_{i+1}^T BR^{-\frac{1}{2}}[-I + R^{-\frac{1}{2}}B^T P_1^{-1}BR^{-\frac{1}{2}}]R^{-\frac{1}{2}}B^T\hat{x}_{i+1} \\
&= -\hat{x}_{i+1}^T BR^{-\frac{1}{2}}SR^{-\frac{1}{2}}B^T\hat{x}_{i+1}
\end{aligned}
$$
(3.140)

where $S = I - R^{-\frac{1}{2}} B^T P_1^{-1} B R^{-\frac{1}{2}}$. Note that S is positive definite, since the following equality holds:

$$S = I - R^{-\frac{1}{2}} B^T P_1^{-1} B R^{-\frac{1}{2}} = I - R^{-\frac{1}{2}} B^T [\hat{P}_1 + B R^{-1} B^T]^{-1} B R^{-\frac{1}{2}}$$
$$= [I + R^{-\frac{1}{2}} B^T \hat{P}_1^{-1} B R^{-\frac{1}{2}}]^{-1}$$

where the second equality comes from the relation $P_1 = \hat{P}_1 + B R^{-1} B^T$. Note that $\hat{P}_1 > 0$ if $N \geq n_c + 1$. Summing both sides of (3.140) from i to $i + n_c - 1$, we have

$$\sum_{k=i}^{i+n_c-1} \left[V(\hat{x}_{k+1}) - V(\hat{x}_k) \right] \geq \sum_{k=i}^{i+n_c-1} \hat{x}_{k+1}^T B R^{-\frac{1}{2}} S R^{-\frac{1}{2}} B^T \hat{x}_{k+1} \quad (3.141)$$

$$V(\hat{x}_{i+n_c}) - V(\hat{x}_i) \geq \hat{x}_i^T \Theta \hat{x}_i \quad (3.142)$$

where

$$\Theta = \sum_{k=i}^{i+n_c-1} \left[\Psi^{(i-k-1)} W \Psi^{T(i-k-1)} \right]$$
$$\Psi = A - B R^{-1} B^T P_1^{-1} A$$
$$W = B R^{-\frac{1}{2}} S R^{-\frac{1}{2}} B^T$$

Recalling that $\lambda_{\max}(A^{-1} P_1 A^{-1}) |\hat{x}_i| \geq V(\hat{x}_i)$ and using (3.142), we obtain

$$V(\hat{x}_{i+n_c}) \geq \hat{x}_i^T \Theta \hat{x}_i + V(\hat{x}_i)$$
$$\geq \lambda_{\min}(\Theta) |\hat{x}_i|^2 + V(\hat{x}_i)$$
$$\geq \varpi V(\hat{x}_i) + V(\hat{x}_i) \quad (3.143)$$

where

$$\varpi = \lambda_{\min}(\Theta) \frac{1}{\lambda_{\max}(A^{-1} P_1 A^{-1})} \quad (3.144)$$

Note that if (A, B) is controllable, then $(A - BH, B)$ and $((A - BH)^{-1}, B)$ are controllable. Thus, Θ is positive definite and its minimum eigenvalue is positive. ϖ is also positive. Therefore, from (3.143), the lower bound of the state is given as

$$\| \hat{x}_{i_0 + m \times n_c} \|^2 \geq \frac{1}{\lambda_{\max}(A^{-1} P_1 A^{-1})} (\varpi + 1)^m V(\hat{x}_{i_0}) \quad (3.145)$$

The inequality (3.145) implies that the closed-loop system (3.136) is exponentially increasing, i.e. the closed-loop system (3.4) with (3.128) is exponentially decreasing. The second part of this theorem can be easily proved from the first part. This completes the proof. ∎

It is noted that the receding horizon control (3.59) is a special case of controls in Theorem 3.10.

Theorem 3.7 requires the observability condition, whereas Theorem 3.10 does not. Theorem 3.10 holds for arbitrary $Q \geq 0$, including the zero matrix. When $Q = 0$, P_1 can be expressed as the following closed form:

$$P_1 = \sum_{j=i+1}^{i+N} A^{j-i-1} B R^{-1} B^T A^{(j-i-1)T} + A^N P_f (A^N)^T \qquad (3.146)$$

where A is nonsingular. It is noted that, in the above equation, P_f can even be zero. This is the *simplest RHC*

$$u_i = -R^{-1} B^T \left[\sum_{j=i+1}^{i+N} A^{j-i-1} B R^{-1} B^T A^{(j-i-1)T} \right]^{-1} A x_i \qquad (3.147)$$

that guarantees the closed-loop stability.

It is noted that P_f satifying (3.133) is equivalent to Q_f satisfying (3.84) in the relation of $P_f = Q_f^{-1} + B R^{-1} B^T$. Replacing P_f with $Q_f^{-1} + B R^{-1} B^T$ in (3.133) yields the following inequality:

$$Q_f^{-1} + B R^{-1} B^T \leq A^{-1} [Q_f^{-1} + B R^{-1} B^T + A^{-T} Q A^{-1}]^{-1} A^{-T} + B R^{-1} B^T$$
$$= [A^T (Q_f^{-1} + B R^{-1} B^T)^{-1} A + Q]^{-1} + B R^{-1} B^T$$

Finally, we have

$$Q_f \geq A^T (Q_f^{-1} + B R^{-1} B^T)^{-1} A + Q \qquad (3.148)$$

Therefore, if Q_f satisfies (3.148), P_f also satisfies (3.133).

Theorem 3.11. *Assume that the pair (A, B) is controllable and A is nonsingular.*

(1) *If $K_{i+1} \geq K_i > 0$ for some i, then the system (3.4) with receding horizon control (3.56) is asymptotically stable for $1 \leq N < \infty$.*

(2) *For $Q_f > 0$ satisfying (3.73) for some H, the system (3.4) with receding horizon control (3.56) is asymptotically stable for $1 \leq N < \infty$.*

Proof. The first part is proved as follows. $K_{i+1} \geq K_i > 0$ implies $0 < K_{i+1}^{-1} \leq K_i^{-1}$, from which we have $0 < P_{i+1} \leq P_i$ satisfying the inequality (3.135). Thus, the control (3.128) is equivalent to (3.56). The second part is proved as follows. Inequalities $K_{i+1} \geq K_i > 0$ are satisfied for K_i generated from $Q_f > 0$ satisfying (3.73) for some H. Thus, the second result can be seen from the first one. This completes the proof. ∎

It is noted that (3.148) is equivalent to (3.73), as mentioned before.

So far, the cost monotonicity condition has been introduced for stability. Without this cost monotonicity condition, there still exists a finite horizon such that the resulting receding horizon control stabilizes the closed-loop systems.

Before proceeding to the main theorem, we introduce a matrix norm $\|A\|_{\rho,\epsilon}$, which satisfies the properties of the norm and $\rho(A) \leq \|A\|_{\rho,\epsilon} \leq \rho(A) + \epsilon$. Here, ϵ is a design parameter and $\rho(A)$ is the spectral radius of A, i.e. $\rho(A) = \max_{1 \leq i \leq n} |\lambda_i|$.

Theorem 3.12. *Suppose that $Q \geq 0$ and $R > 0$. If the pairs (A, B) and $(A, Q^{\frac{1}{2}})$ are controllable and observable respectively, then the receding horizon control (3.56) for the free terminal cost stabilizes the systems for the following horizon:*

$$N > 1 + \frac{1}{\ln \|A_c^T\|_{\rho,\epsilon} + \ln \|A_c\|_{\rho,\epsilon}} \ln \left[\frac{1}{\beta \|BR^{-1}B^T\|_{\rho,\epsilon}} \left\{ \frac{1}{\|A_c\|_{\rho,\epsilon}} - 1 - \epsilon \right\} \right]$$

$$(3.149)$$

where $\beta = \|Q_f - K_\infty\|_{\rho,\epsilon}$, $A_c = A - BR^{-1}B^T[I + K_\infty BR^{-1}B^T]^{-1}K_\infty A$, and K_∞ is the solution to the steady-state Riccati equation.

Proof. Denote $\triangle K_{i,N}$ by $K_{i,N} - K_\infty$, where $K_{i,N}$ is the solution at time i to the Riccati equation starting from time N, and K_∞ is the steady-state solution to the Riccati equation which is given by (2.108). In order to enhance the clarification, $K_{i,N}$ is used instead of K_i. $K_{N,N} = Q_f$ and $K_{1,N}$ of $i = 1$ are involved with the control gain of the RHC with a horizon size N. From properties of the Riccati equation, we have the following inequality:

$$\triangle K_{i,N} \leq A_c^T \triangle K_{i+1,N} A_c \tag{3.150}$$

Taking the norm $\| \cdot \|_{\rho,\epsilon}$ on both sides of (3.150), we obtain

$$\|\triangle K_{i,N}\|_{\rho,\epsilon} \leq \|A_c^T\|_{\rho,\epsilon} \|\triangle K_{i+1,N}\|_{\rho,\epsilon} \|A_c\|_{\rho,\epsilon} \tag{3.151}$$

where a norm $\| \cdot \|_{\rho,\epsilon}$ is defined just before this theorem. From (3.151), it can be easily seen that $\|\triangle K_{1,N}\|_{\rho,\epsilon}$ is bounded below as follows:

$$\|\triangle K_{1,N}\|_{\rho,\epsilon} \leq \|A_c^T\|_{\rho,\epsilon}^{N-1} \|\triangle K_{N,N}\|_{\rho,\epsilon} \|A_c\|_{\rho,\epsilon}^{N-1} = \|A_c^T\|_{\rho,\epsilon}^{N-1} \beta \|A_c\|_{\rho,\epsilon}^{N-1} \tag{3.152}$$

The closed-loop system matrix $A_{c,N}$ from the control gain $K_{1,N}$ is given by

$$A_{c,N} = A - BR^{-1}B^T[I + K_{1,N}BR^{-1}B^T]^{-1}K_{1,N}A \tag{3.153}$$

It is known that the steady-state closed-loop system matrices A_c and $A_{c,N}$ in (3.153) are related to each other as follows:

$$A_{c,N} = \left[I + BR_{o,N}^{-1}B^T \triangle K_{1,N} \right] A_c \tag{3.154}$$

where $R_{o,N} = R + B^T K_{1,N} B$. Taking the norm $\|\cdot\|_{\rho,\epsilon}$ on both sides of (3.154) and using the inequality (3.152), we have

$$\|A_{c,N}\|_{\rho,\epsilon} \le \left[1 + \epsilon + \|BR^{-1}B^T\|_{\rho,\epsilon}\|\Delta K_{1,N}\|_{\rho,\epsilon}\right]\|A_c\|_{\rho,\epsilon}$$

$$\le \left[1 + \epsilon + \|BR^{-1}B^T\|_{\rho,\epsilon}\|A_c^T\|_{\rho,\epsilon}^{N-1}\beta\|A_c\|_{\rho,\epsilon}^{N-1}\right]\|A_c\|_{\rho,\epsilon} \quad (3.155)$$

where ϵ should be chosen so that $\epsilon < \frac{1}{\|A_c\|_{\rho,\epsilon}} - 1$. In order to guarantee $\|A_{c,N}\|_{\rho,\epsilon} < 1$, we have only to make the right-hand side in (3.155) less than 1. Therefore, we have

$$\|A_c^T\|_{\rho,\epsilon}^{N-1}\|A_c\|_{\rho,\epsilon}^{N-1} \le \frac{1}{\beta\|BR^{-1}B^T\|_{\rho,\epsilon}}\left[\frac{1}{\|A_c\|_{\rho,\epsilon}} - 1 - \epsilon\right] \quad (3.156)$$

It is noted that if the right-hand side of (3.156) is greater than or equal to 1, then the inequality (3.156) always holds due to the Hurwitz matrix A_c. Taking the logarithm on both sides of (3.156), we have (3.149). This completes the proof. ∎

Theorem 3.12 holds irrespective of Q_f. The determination of a suitable N is an issue.

The case of zero terminal weighting leads to generally large horizons and large terminal weighting to small horizons, as can be seen in the next example.

Example 3.1

We consider a scalar, time-invariant system and the quadratic cost

$$x_{i+1} = ax_i + bu_i \quad (3.157)$$

$$J = \sum_{j=0}^{N-1}[qx_{k+j}^2 + ru_{k+j}^2] + fx_{k+N}^2 \quad (3.158)$$

where $b \ne 0$, $r > 0$ and $q > 0$. It can be easily seen that (a, b) in (3.157) is stabilizable and $(a, \sqrt{q})$ is observable. In this case, the Riccati equation is simply represented as

$$p_k = a^2 p_{k+1} - \frac{a^2 b^2 p_{k+1}^2}{b^2 p_{k+1} + r} + q = \frac{a^2 r p_{k+1}}{b^2 p_{k+1} + r} + q \quad (3.159)$$

with $p_N = f$. The RHC with a horizon size N is obtained as

$$u_k = -Lx_k \quad (3.160)$$

where

$$L = \frac{abp_1}{b^2 p_1 + r} \tag{3.161}$$

Now, we investigate the horizon of the RHC for stabilizing the closed-loop system. The steady-state solution to the ARE and the system matrix of the closed-loop system can be written as

$$p_\infty = \frac{q}{2\Pi}\left[\pm\sqrt{(1-\Pi)^2 + \frac{4\Pi}{1-a^2}} - (1-\Pi)\right] \tag{3.162}$$

$$a^{cl} = a - bL = \frac{a}{1 + \frac{b^2}{r}p_\infty} \tag{3.163}$$

where

$$\Pi = \frac{b^2 q}{(1-a^2)r} \tag{3.164}$$

We will consider two cases. One is for a stable system and the other for an unstable system.

(1) Stable system ($|a| < 1$)

Since $|a| < 1$, $1 - a^2 > 0$ and $\Pi > 0$. In this case, we have the positive solution as

$$p_\infty = \frac{q}{2\Pi}\left[\sqrt{(1-\Pi)^2 + \frac{4\Pi}{1-a^2}} - (1-\Pi)\right] \tag{3.165}$$

From (3.163), we have $|a^{cl}| < |a| < 1$. So, the asymptotical stability is guaranteed for the closed-loop system.

(2) Unstable system ($|a| > 1$)

Since $|a| > 1$, $1 - a^2 < 0$ and $\Pi < 0$. In this case, we have the positive solution given by

$$p_\infty = -\frac{q}{2\Pi}\left[\sqrt{(1-\Pi)^2 + \frac{4\Pi}{1-a^2}} + (1-\Pi)\right] \tag{3.166}$$

The system matrix a^{cl} of the closed-loop system can be represented as

$$a^{cl} = \frac{a}{1 - \frac{1-a^2}{2}\left[\sqrt{(1-\Pi)^2 + \frac{4\Pi}{1-a^2}} + 1 - \Pi\right]} \tag{3.167}$$

We have $|a^{cl}| < 1$ from the following relation:

$$|2 + \sqrt{((a^2-1)(1-\Pi) + 2)^2 - 4a^2} + (a^2-1)(1-\Pi)|$$
$$> |2 + \sqrt{(a^2+1)^2 - 4a^2} + (a^2-1)|$$
$$= |2a^2| > 2|a|.$$

From a^{cl}, the lower bound of the horizon size guaranteeing the stability is obtained as

$$N > 1 + \frac{1}{2\ln|a^{cl}|}\ln\left[\frac{r}{b^2|f - p_\infty|}\left\{\frac{1}{|a^{cl}|} - 1\right\}\right] \qquad (3.168)$$

where $\epsilon = 0$ and absolute values of scalar values are used for $\|\cdot\|_{\rho,\epsilon}$ norm. ∎

As can be seen in this example, the gain and phase margins of the RHC are greater than those of the conventional steady-state LQ control. For the general result on multi-input–multi-output systems, this is left as an exercise.

3.3.5 Additional Properties of Receding Horizon Linear Quadratic Control

A Prescribed Degree of Stability

We introduce another performance criterion to make closed-loop eigenvalues smaller than a specific value. Of course, as closed-loop eigenvalues get smaller, the closed-loop system becomes more stable, probably with an excessive control energy cost.

Consider the following performance criterion:

$$J = \sum_{j=0}^{N-1} \alpha^{2j}(u_{k+j}^T R u_{k+j} + x_{k+j}^T Q x_{k+j}) + \alpha^{2N} x_{k+N}^T Q_f x_{k+N} \quad (3.169)$$

where $\alpha > 1$ and the pair (A, B) is stabilizable.

The first thing we have to do for dealing with (3.169) is to make transformations that convert the given problem to a standard LQ problem. Therefore, we introduce new variables such as

$$\hat{x}_{k+j} \stackrel{\triangle}{=} \alpha^j x_{k+j} \qquad (3.170)$$

$$\hat{u}_{k+j} \stackrel{\triangle}{=} \alpha^j u_{k+j} \qquad (3.171)$$

Observing that

$$\hat{x}_{k+j+1} = \alpha^{j+1} x_{k+j+1} = \alpha\alpha^j[Ax_{k+j} + Bu_{k+j}] = \alpha A\hat{x}_{k+j} + \alpha B\hat{u}_{k+j} \quad (3.172)$$

we may associate the system (3.172) with the following performance criterion:

$$J = \sum_{j=0}^{N-1}(\hat{u}_{k+j}^T R\hat{u}_{k+j} + \hat{x}_{k+j}^T Q\hat{x}_{k+j}) + \hat{x}_{k+N}^T Q_f\hat{x}_{k+N} \qquad (3.173)$$

The receding horizon control for (3.172) and (3.173) can be written as

$$\hat{u}_k = -R^{-1}\alpha B^T[I + K_1\alpha BR^{-1}\alpha B^T]^{-1}K_1\alpha A\hat{x}_k \qquad (3.174)$$

where K_1 is obtained from

$$K_j = \alpha A^T [I + K_{j+1}\alpha BR^{-1}\alpha B^T]^{-1} K_{j+1}\alpha A + Q \qquad (3.175)$$

with $K_N = Q_f$. The RHC u_k can be written as

$$u_k = -R^{-1}B^T[I + K_1\alpha BR^{-1}\alpha B^T]^{-1}K_1\alpha Ax_k \qquad (3.176)$$

Using the RHC u_k in (3.176), we introduce a method to stabilize systems with a high degree of closed-loop stability. If α is chosen to satisfy the following cost monotonicity condition:

$$Q_f \geq Q + H^T RH + \alpha(A - BH)^T Q_f (A - BH)\alpha \qquad (3.177)$$

then the RHC (3.176) stabilizes the closed-loop system. Note that since α is assumed to be greater than 1, the cost monotonicity condition holds even by replacing αA with A.

In order to check the stability of the RHC (3.176), the time index k is replaced with the arbitrary time point i and the closed-loop systems are constructed. The RHC (3.176) satisfying (3.177) makes $\hat{x}_i$ approach zero according to the following state-space model:

$$\hat{x}_{i+1} = \alpha(A\hat{x}_i + B\hat{u}_i) \qquad (3.178)$$
$$= \alpha(A + BR^{-1}B^T[I + K_1\alpha BR^{-1}\alpha B^T]^{-1}K_1\alpha A)\hat{x}_i \qquad (3.179)$$

where

$$\alpha\rho(A + BR^{-1}B^T[I + K_1\alpha BR^{-1}\alpha B^T]^{-1}K_1\alpha A) \leq 1 \qquad (3.180)$$

From (3.178) and (3.179), the real state x_k and control u_k can be written as

$$x_{i+1} = Ax_i + Bu_i \qquad (3.181)$$
$$= (A + BR^{-1}B^T[I + K_1\alpha BR^{-1}\alpha B^T]^{-1}K_1\alpha A)x_i \qquad (3.182)$$

The spectral radius of the closed-loop eigenvalues for (3.181) and (3.182) is obtained from (3.180) as follows:

$$\rho(A + BR^{-1}B^T[I + K_1\alpha BR^{-1}\alpha B^T]^{-1}K_1\alpha A) \leq \frac{1}{\alpha} \qquad (3.183)$$

Then, we can see that it is possible to define a modified receding horizon control problem which achieves a closed-loop system with a prescribed degree of stability α. That is, for a prescribed α, the state x_i approaches zero at least as fast as $|\frac{1}{\alpha}|^i$. The smaller that α is, the more stable is the closed-loop system. The same goes for the terminal equality constraint.

From now on we investigate the optimality of the RHC. The receding horizon control is optimal in the sense of the receding horizon concept. But this may not be optimal in other senses, such as the finite or infinite horizon

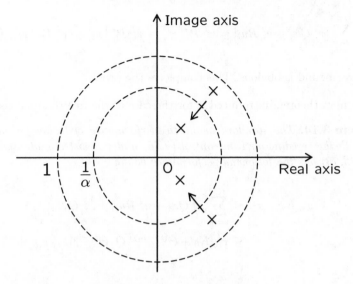

Fig. 3.8. Effect of parameter α

optimal control concept. Likewise, standard finite or infinite optimal control may not be optimal in the sense of the receding horizon control, whereas it is optimal in the sense of the standard optimal control. Therefore, it will be interesting to compare between them.

For simplicity we assume that there is no reference signal to track.

Theorem 3.13. *The standard quadratic performance criterion for the systems with the receding horizon control (3.59) under a terminal equality constraint has the following performance bounds:*

$$x_{i_0} K_{i_0, i_f} x_{i_0} \leq \sum_{i=i_0}^{i_f - 1} [x_i^T Q x_i + u_i^T R u_i] \leq x_{i_0}^T P_0^{-1} x_{i_0} \tag{3.184}$$

Proof. We have the following inequality:

$$x_i^T P_0^{-1} x_i - x_{i+1}^T P_0^{-1} x_{i+1} = x_i^T P_{-1}^{-1} x_i - x_{i+1}^T P_0^{-1} x_{i+1} + x_i^T [P_0^{-1} - P_{-1}^{-1}] x_i$$
$$\geq x_i^T P_{-1}^{-1} x_i - x_{i+1}^T P_0^{-1} x_{i+1} \tag{3.185}$$

which follows from the fact that $P_0^{-1} \geq P_{-1}^{-1}$. By using the optimality, we have

$$x_i^T P_{-1}^{-1} x_i - x_{i+1}^T P_0^{-1} x_{i+1} = J^*(x_i, i, i+N+1) - J^*(x_{i+1}, i+1, i+N+1)$$
$$\geq J^*(x_i, i, i+N+1) - J(x_{i+1}, i+1, i+N+1)$$
$$\geq x_i^T Q x_i + u_i R u_i \tag{3.186}$$

where $J(x_{i+1}, i+1, i+N+1)$ is a cost function generated from the state driven by the optimal control that is based on the interval $[i, i+N+1]$. From (3.186) we have

$$\sum_{i=i_0}^{i_f-1} [x_i^T Q x_i + u_i^T R u_i] \leq x_{i_0}^T P_0^{-1} x_{i_0} - x_{i_f}^T P_0^{-1} x_{i_f} \leq x_{i_0}^T P_0^{-1} x_{i_0} \quad (3.187)$$

The lower bound is obvious. This completes the proof. ∎

The next theorem introduced is for the case of the free terminal cost.

Theorem 3.14. *The standard quadratic performance criterion for the systems with the receding horizon control (3.56) under a cost monotonicity condition (3.73) has the following performance bounds:*

$$x_{i_0} K_{i_0,i_f} x_{i_0} \leq \sum_{i=i_0}^{i_f-1} [x_i^T Q x_i + u_i^T R u_i] + x_{i_f}^T Q_f x_{i_f}$$

$$\leq x_{i_0}^T [K_0 + \Theta^{(i_f-i_0)T} Q_f \Theta^{i_f-i_0}] x_{i_0}$$

where

$$\Theta \stackrel{\triangle}{=} A - BR^{-1}B^T K_1[I + BR^{-1}B^T K_1]^{-1} A$$

K_0 *is obtained from (3.47) starting from* $K_N = Q_f$, *and* K_{i_0,i_f} *is obtained by starting from* $K_{i_f,i_f} = Q_f$.

Proof. The lower bound is obvious, since K_{i_0,i_f} is the cost incurred by the standard optimal control law. The gain of the receding horizon control is given by

$$L \stackrel{\triangle}{=} R^{-1}B^T K_1[I + BR^{-1}B^T K_1]^{-1} A$$
$$= [R + B^T K_1 B]^{-1} B^T K_1 A.$$

As is well known, the quadratic cost for the feedback control $u_i = -L x_i$ is given by

$$\sum_{i=i_0}^{i_f-1} [x_i^T Q x_i + u_i^T R u_i] + x_{i_f}^T Q_f x_{i_f} = x_{i_0}^T N_{i_0} x_{i_0}$$

where N_i is the solution of the following difference equation:

$$N_i = [A - BL]^T N_{i+1}[A - BL] + L^T R L + Q$$
$$N_{i_f} = Q_f$$

From (3.47) and (3.48) we have

$$K_i = A^T K_{i+1} A - A^T K_{i+1} B[R + B^T K_{i+1} B]^{-1} B^T K_{i+1} A + Q$$

where $K_N = Q_f$. If we note that, for $i = 0$ in (3.188),

$$A^T K_1 B[R + B^T K_1 B]^{-1} B^T K_1 A = A^T K_1 BL = L^T B^T K_1 A$$
$$= L^T [R + B^T K_1 B]L$$

we can easily have

$$K_0 = [A - BL]^T K_1 [A - BL] + L^T RL + Q$$

Let

$$T_i \triangleq N_i - K_0$$

then T_i satisfies

$$T_i = [A - BL]^T [T_{i+1} - \tilde{T}_i][A - BL] \le [A - BL]^T T_{i+1}[A - BL]$$

with the boundary condition $T_{i_f} = Q_{i_f} - K_0$, where $\tilde{T}_i = K_1 - K_0 \ge 0$ under a cost monotonicity condition. We can obtain T_{i_0} by evaluating recursively, and finally we have

$$T_{i_0} \le \Theta^{(i_f - i_0)T} T_{i_f} \Theta^{i_f - i_0}$$

where $\Theta = A - BL$. Thus, we have

$$N_{i_0} \le K_0 + \Theta^{(i_f - i_0)T} [Q_{i_f} - K_0] \Theta^{i_f - i_0}$$

from which follows the result. This completes the proof.   ∎

Since $\Theta^{(i_f - i_0)T} \to 0$, the infinite time cost has the following bounds:

$$x_{i_0}^T K_{i_0, \infty} x_{i_0} \le \sum_{i=0}^{\infty} x_i^T Q x_i + u_i^T R u_i \qquad (3.188)$$

$$\le x_{i_0}^T K_0 x_{i_0} \qquad (3.189)$$

The receding horizon control is optimal in its own right. However, the receding horizon control can be used for a suboptimal control for the standard regulation problem. In this case, Theorem 3.14 provides a degree of suboptimality.

Example 3.2

In this example, it is shown via simulation that the RHC has good tracking ability for the nonzero reference signal. For simulation, we consider a two-dimensional free body system. This free body is moved by two kinds of forces, i.e. a horizontal force and a vertical force. According to Newton's laws, the following dynamics are obtained:

$$M\ddot{x} + B\dot{x} = u_x$$
$$M\ddot{x} + B\dot{y} = u_y$$

where M, B, x, y, u_x, and u_y are the mass of the free body, the friction coefficient, the horizontal position, the vertical position, the horizontal force, and the vertical force respectively. Through plugging the real values into the parameters and taking the discretization procedure, we have

$$x_{k+1} = \begin{bmatrix} 1 & 0.0550 & 0 & 0 \\ 0 & 0.9950 & 0 & 0 \\ 0 & 0 & 1 & 0.0550 \\ 0 & 0 & 0 & 0.9995 \end{bmatrix} x_k + \begin{bmatrix} 0.0015 & 0 \\ 0.0550 & 0 \\ 0 & 0.0015 \\ 0 & 0.0550 \end{bmatrix} u_k$$

$$y_k = \begin{bmatrix} 1 & 0 & 0 & 0 \\ 0 & 0 & 1 & 0 \end{bmatrix} x_k$$

where the first and second components of x_i denote the positions of x and y respectively, and the two components of u_i are for the horizontal and vertical forces.

The sampling time and the horizon size are taken as 0.055 and 3. The reference signal is given by

$$x_i^r = \begin{cases} 1 - \frac{i}{100} & 0 \le i < 100 \\ 0 & 100 \le i < 200 \\ \frac{i}{100} - 2 & 200 \le i < 300 \\ 1 & 300 \le i < 400 \\ 1 & i \ge 400 \end{cases} \qquad y_i^r = \begin{cases} 1 & 0 \le i < 100 \\ 2 - \frac{i}{100} & 100 \le i < 200 \\ 0 & 200 \le i < 300 \\ \frac{i}{100} - 3 & 300 \le i < 400 \\ 1 & i \ge 400 \end{cases}$$

Q and R for the LQ and receding horizon controls are chosen to be unit matrices. The final weighting matrix for the RHC is set to $10^5 I$. In Figure 3.9, we can see that the RHC has the better performance for the given reference trajectory. Actually, the trajectory for the LQTC is way off the reference signal. However, one for the RHC keeps up with the reference well. ∎

Prediction Horizon

In general, the horizon N in the performance criterion (3.22) is divided into two parts, $[k,\ k+N_c-1]$ and $[k+N_c,\ k+N]$. The control on $[k,\ k+N_c-1]$ is obtained optimally to minimize the performance criterion on $[k,\ k+N_c-1]$, while the control on $[k+N_c,\ k+N]$ is usually a given control, say a linear control $u_i = Hx_i$ on this horizon. In this case, the horizon or horizon size N_c is called the control horizon and N is called the prediction horizon, the performance horizon, or the cost horizon. Here, N can be denoted as N_p to indicate the prediction horizon. In the previous problem we discussed so far, the control horizon N_c was the same as the prediction horizon N_p. In this case, we will use the term control horizon instead of prediction horizon. We consider the following performance criterion:

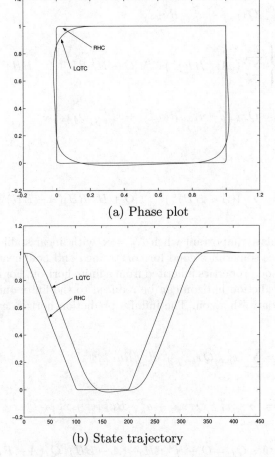

(a) Phase plot

(b) State trajectory

Fig. 3.9. Comparison RHC and LQTC

$$J = \sum_{j=0}^{N_c-1} (u_{k+j}^T R u_{k+j} + x_{k+j}^T Q x_{k+j}) + \sum_{j=N_c}^{N_p} (u_{k+j}^T R u_{k+j}$$
$$+ x_{k+j}^T Q x_{k+j}) \tag{3.190}$$

where the control horizon and the prediction horizon are $[k,\ k+N_c-1]$ and $[k,\ k+N_p]$ respectively. If we apply a linear control $u_i = H x_i$ on $[k+N_c,\ k+N_p]$, we have

$$J = \sum_{j=0}^{N_c-1} [x_{k+j}^T Q x_{k+j} + u_{k+j}^T R u_{k+j}]$$

$$+ x_{k+N_c}^T \left\{ \sum_{j=N_c}^{N_p} ((A - HB)^T)^{j-N_c} [Q + H^T R H](A - BH)^{j-N_c} \right\} x_{k+N_c}$$

$$= \sum_{j=0}^{N_c-1} [x_{k+j}^T Q x_{k+j} + u_{k+j}^T R u_{k+j}] + x_{k+N_c}^T Q_f x_{k+N_c} \qquad (3.191)$$

where

$$Q_f = \sum_{j=N_c}^{N_p} ((A - HB)^T)^{j-N_c} [Q + H^T R H](A - BH)^{j-N_c} \qquad (3.192)$$

This is particularly important when $N_p = \infty$ with linear stable control, since this approach is sometimes good for constrained and nonlinear systems. But we may lose good properties inherited from a finite horizon. For linear systems, the infinite prediction horizon can be reduced to the finite one, which is the same as the control horizon. The infinite prediction horizon can be changed as

$$J = \sum_{j=0}^{\infty} [x_{k+j}^T Q x_{k+j} + u_{k+j}^T R u_{k+j}]$$

$$= \sum_{j=0}^{N_c-1} [x_{k+j}^T Q x_{k+j} + u_{k+j}^T R u_{k+j}] + x_{k+N_c}^T Q_f x_{k+N_c} \qquad (3.193)$$

where Q_f satisfies $Q_f = Q + H^T R H + (A - BH)^T Q_f (A - BH)$. Therefore, Q_f is related to the terminal weighting matrix.

3.3.6 A Special Case of Input–Output Systems

In addition to the state-space model (3.1) and (3.2), GPC has used the following CARIMA model:

$$A(q^{-1}) y_i = B(q^{-1}) \triangle u_{i-1} \qquad (3.194)$$

where

$$A(q^{-1}) = 1 + a_1 q^{-1} + \cdots + a_n q^{-n}, \quad a_n \neq 0 \qquad (3.195)$$
$$B(q^{-1}) = b_1 + b_2 q^{-1} + \cdots + b_m q^{-m+1} \qquad (3.196)$$

where q^{-1} is the unit delay operator, such as $q^{-1} y_i = y_{i-1}$, and $\triangle u_i = (1 - q^{-1}) u_i = u_i - u_{i-1}$. It is noted that $B(q^{-1})$ can be

$$b_1 + b_2 q^{-1} + \cdots + b_n q^{-n+1}, \quad m \leq n \tag{3.197}$$

where $b_i = 0$ for $i > m$. It is noted that (3.194) can be represented as

$$A(q^{-1})y_i = \widetilde{B}(q^{-1})\triangle u_i \tag{3.198}$$

where

$$\widetilde{B}(q^{-1}) = b_1 q^{-1} + b_2 q^{-2} \cdots + b_n q^{-n} \tag{3.199}$$

The above model (3.198) in an input–output form can be transformed to the state-space model

$$\begin{aligned} x_{i+1} &= \bar{A}x_i + \bar{B}\triangle u_i \\ y_i &= \bar{C}x_i \end{aligned} \tag{3.200}$$

where $x_i \in R^n$ and

$$\bar{A} = \begin{bmatrix} -a_1 & 1 & 0 & \cdots & 0 \\ -a_2 & 0 & 1 & \cdots & 0 \\ \vdots & \vdots & \vdots & \ddots & \vdots \\ -a_{n-1} & 0 & 0 & \cdots & 1 \\ -a_n & 0 & 0 & \cdots & 0 \end{bmatrix} \quad \bar{B} = \begin{bmatrix} b_1 \\ b_2 \\ \vdots \\ b_{n-1} \\ b_n \end{bmatrix} \tag{3.201}$$

$$\bar{C} = \begin{bmatrix} 1 & 0 & 0 & \cdots & 0 \end{bmatrix}$$

It is clear that $y_i = x_i{}^{(1)}$, where $x_i{}^{(1)}$ indicates the first element of x_i.

The common performance criterion for the CARIMA model (3.194) is given as

$$\sum_{j=1}^{N_c} \left[q(y_{k+j} - y_{k+j}^r)^2 + r(\triangle u_{k+j-1})^2 \right] \tag{3.202}$$

Here, N_c is the control horizon.

Since y_k is given, the optimal control for (3.202) is also optimal for the following performance index:

$$\sum_{j=0}^{N_c-1} \left[q(y_{k+j} - y_{k+j}^r)^2 + r(\triangle u_{k+j})^2 \right] + q(y_{k+N_c} - y_{k+N_c}^r)^2 \tag{3.203}$$

The performance index (3.202) can be extended to include a free terminal cost such as

$$\sum_{j=1}^{N_c} \left[q(y_{k+j} - y_{k+j}^r)^2 + r(\triangle u_{k+j-1})^2 \right] + \sum_{j=N_c+1}^{N_p} q_f(y_{k+j} - y_{k+j}^r)^2 \tag{3.204}$$

We can consider a similar performance that generates the same optimal control for (3.204), such as

$$\sum_{j=0}^{N_c-1} \left[q(y_{k+j} - y_{k+j}^r)^2 + r(\triangle u_{k+j})^2 \right] + \sum_{j=N_c}^{N_p} q_f^{(j)}(y_{k+j} - y_{k+j}^r)^2 \quad (3.205)$$

where

$$q_f^{(j)} = \begin{cases} q, & j = N_c \\ q_f, & j > N_c \end{cases}$$

For a given $\bar{C}$, there exists always an L such that $\bar{C}L = I$. Let $x_i^r = Ly_i^r$. The performance criterion (3.202) becomes

$$\sum_{j=0}^{N_c-1} \left[(x_{k+j} - x_{k+j}^r)^T Q(x_{k+j} - x_{k+j}^r) + \triangle u_{k+j}^T R \triangle u_{k+j} \right]$$

$$+ \sum_{j=N_c}^{N_p} (x_{k+j} - x_{k+j}^r)^T Q_f^{(j)}(x_{k+j} - x_{k+j}^r) \quad (3.206)$$

where $Q = q\bar{C}^T\bar{C}$, $Q_f^{(j)} = q_f^{(j)}\bar{C}^T\bar{C}$, and $R = r$. GPC can be obtained using the state model (3.200) with the performance criterion (3.206), whose solutions are described in detail in this book. It is noted that the performance criterion (3.206) has two values in the time-varying state and input weightings. The optimal control is given in a state feedback form. From the special structure of the CARIMA model

$$x_i = \tilde{A}Y_{i-n} + \tilde{B}\triangle U_{i-n} \quad (3.207)$$

where

$$Y_{i-n} = \begin{bmatrix} y_{i-n} \\ \vdots \\ y_{i-1} \end{bmatrix} \quad \triangle U_{i-n} = \begin{bmatrix} \triangle u_{i-n} \\ \vdots \\ \triangle u_{i-1} \end{bmatrix} \quad (3.208)$$

$$\tilde{A} = \begin{bmatrix} -a_n & -a_{n-1} & \cdots & -a_2 & -a_1 \\ 0 & -a_n & \cdots & -a_3 & -a_2 \\ \vdots & \vdots & \ddots & \vdots & \vdots \\ 0 & 0 & \cdots & -a_n & -a_{n-1} \\ 0 & 0 & \cdots & 0 & -a_n \end{bmatrix} \quad (3.209)$$

$$\tilde{B} = \begin{bmatrix} b_n & b_{n-1} & \cdots & b_2 & b_1 \\ 0 & b_n & \cdots & b_3 & b_2 \\ \vdots & \vdots & \ddots & \vdots & \vdots \\ 0 & 0 & \cdots & b_n & b_{n-1} \\ 0 & 0 & \cdots & 0 & b_n \end{bmatrix} \quad (3.210)$$

the state can be computed with input control and measured output. The optimal control can be given as an output feedback control.

If $\triangle u_{k+N_c} = \ldots = \triangle u_{k+N_p-1} = 0$ is assumed, for $N_p = N_c + n - 1$, then the terminal cost can be represented as

$$\sum_{j=N_c}^{N_p} q_f^{(j)} (y_{k+j} - y_{k+j}^r)^2$$
$$= (Y_{k+N_c} - Y_{k+N_c}^r)^T \bar{Q}_f (Y_{k+N_c} - Y_{k+N_c}^r) \qquad (3.211)$$

where

$$\bar{Q}_f = [\text{diag}(\overbrace{q\ q_f\ \cdots\ q_f}^{N_p-N_c+1})], \quad Y_{k+N_c}^r = \begin{bmatrix} y_{k+N_c}^r \\ \vdots \\ y_{k+N_c+n-1}^r \end{bmatrix} \qquad (3.212)$$

In this case the terminal cost becomes

$$(x_{k+N_c+n} - \tilde{A}Y_{k+N_c}^r)^T (\tilde{A}^T)^{-1} \bar{Q}_f \tilde{A}^{-1} (x_{k+N_c+n} - \tilde{A}Y_{k+N_c}^r)$$
$$= (x_{k+N_c} - x_{k+N_c}^r)^T Q_f (x_{k+N_c} - x_{k+N_c}^r)$$

where

$$Q_f = q_f (\bar{A}^T)^n (\tilde{A}^T)^{-1} \bar{Q}_f \tilde{A}^{-1} \bar{A}^n \qquad (3.213)$$
$$x_{k+N_c}^r = (\bar{A}^n)^{-1} \tilde{A}Y_{k+N_c}^r \qquad (3.214)$$

It is noted that $\bar{A}$ and $\tilde{A}$ are all nonsingular.

The performance criterion (3.204) is for the free terminal cost. We can now introduce a terminal equality constraint, such as

$$y_{k+j} = y_{k+j}^r, \quad j = N_c, \cdots, N_p \qquad (3.215)$$

which is equivalent to $x_{k+N_c} = x_{k+N_c}^r$. GPC can be obtained from the results in state-space forms that have already been discussed.

3.4 Receding Horizon Control Based on Minimax Criteria

3.4.1 Receding Horizon H_∞ Control

In this section, a receding horizon H_∞ control in a tracking form for discrete time-invariant systems is obtained.

Based on the following system in a predictive form:

$$x_{k+j+1} = Ax_{k+j} + B_w w_{k+j} + Bu_{k+j} \qquad (3.216)$$
$$z_{k+j} = C_z x_{k+j} \qquad (3.217)$$

with the initial state x_k, the optimal control and the worst-case disturbance can be written in a predictive form as

$$u^*_{k+j} = -R^{-1}B^T \Lambda^{-1}_{k+j+1,i_f}[M_{k+j+1,i_f}Ax_{k+j} + g_{k+j+1,i_f}]$$

$$w^*_{k+j} = \gamma^{-2}R_w^{-1}B_w^T \Lambda^{-1}_{k+j+1,i_f}[M_{k+j+1,i_f}Ax_{k+j} + g_{k+j+1,i_f}]$$

M_{k+j,i_f} and g_{k+j,i_f} can be obtained from

$$M_{k+j,i_f} = A^T \Lambda^{-1}_{k+j+1,i_f}M_{k+j+1,i_f}A + Q, \quad i = i_0,\cdots,i_f - 1 \quad (3.218)$$

$$M_{i_f,i_f} = Q_f \quad (3.219)$$

and

$$g_{k+j,i_f} = -A^T \Lambda^{-1}_{k+j+1,i_f}g_{k+j+1,i_f} - Qx^r_{k+j} \quad (3.220)$$

$$g_{i_f,i_f} = -Q_f x^r_{i_f} \quad (3.221)$$

where

$$\Lambda_{k+j+1,i_f} = I + M_{k+j+1,i_f}(BR^{-1}B^T - \gamma^{-2}B_wR_w^{-1}B_w^T)$$

If we replace i_f with $k + N$, the optimal control on the interval $[k,\ k + N]$ is given by

$$u^*_{k+j} = -R^{-1}B^T \Lambda^{-1}_{k+j+1,k+N}[M_{k+j+1,k+N}Ax_{k+j} + g_{k+j+1,k+N}]$$

$$w^*_{k+j} = \gamma^{-2}R_w^{-1}B_w^T \Lambda^{-1}_{k+j+1,k+N}[M_{k+j+1,k+N}Ax_{k+j} + g_{k+j+1,k+N}]$$

The receding horizon control is given by the first control, $j = 0$, at each interval as

$$u^*_k = -R^{-1}B^T \Lambda^{-1}_{k+1,t+N}[M_{k+1,k+N}Ax_k + g_{k+1,k+N}]$$

$$w^*_k = \gamma^{-2}R_w^{-1}B_w^T \Lambda^{-1}_{k+1,k+N}[M_{k+1,k+N}Ax_k + g_{k+1,k+N}]$$

Replace k with i as an arbitrary time point for discrete-time systems to obtain

$$u^*_i = -R^{-1}B^T \Lambda^{-1}_{i+1,i+N}[M_{i+1,i+N}Ax_i + g_{i+1,i+N}]$$

$$w^*_i = \gamma^{-2}R_w^{-1}B_w^T \Lambda^{-1}_{i+1,i+N}[M_{i+1,i+N}Ax_i + g_{i+1,i+N}].$$

In case of time-invariant systems, the simplified forms are obtained as

$$u^*_i = -R^{-1}B^T \Lambda^{-1}_1[M_1 Ax_i + g_{i+1,i+N}] \quad (3.222)$$

$$w^*_i = \gamma^{-2}R_w^{-1}B_w^T \Lambda^{-1}_1[M_1 Ax_i + g_{i+1,i+N}] \quad (3.223)$$

M_1 and $g_{i,i+N}$ can be obtained from

$$M_j = A^T \Lambda^{-1}_{j+1}M_{j+1}A + Q, \quad j = 1,\cdots,N - 1 \quad (3.224)$$

$$M_N = Q_f \quad (3.225)$$

and

$$g_{j,i+N} = -A^T \Lambda_{j+1}^{-1} g_{j+1,i+N} - Q x_j^r \qquad (3.226)$$

$$g_{i+N,i+N} = -Q_f x_{i+N}^r \qquad (3.227)$$

where

$$\Lambda_{j+1} = I + M_{j+1}(BR^{-1}B^T - \gamma^{-2}B_w R_w^{-1}B_w^T)$$

Recall through this chapter that the following condition is assumed to be satisfied:

$$R_w - \gamma^{-2}B_w^T M_i B_w > 0 , \quad i = 1, \cdots, N \qquad (3.228)$$

in order to guarantee the existence of the saddle point. Note that from (3.228), we have $M_i^{-1} > \gamma^{-2}B_w R_w^{-1}B_w^T$, from which the positive definiteness of $\Lambda_i^{-1} M_i$ is guaranteed. The positive definiteness of M_i is also guaranteed with the observability of $(A, Q^{\frac{1}{2}})$.

From (2.152) and (2.153) we have another form of the receding horizon H_∞ control:

$$u_i^* = -R^{-1}B^T P_1^{-1} A x_i \qquad (3.229)$$

$$w_i^* = \gamma^{-2}R_w^{-1}B_w^T P_1^{-1} A x_i \qquad (3.230)$$

where $\Pi = BR^{-1}B - \gamma^{-2}B_w R_w^{-1}B_w^T$,

$$P_i = A^{-1}P_{i+1}[I + A^{-1}QA^{-1}P_{i+1}]^{-1}A^{-1} + \Pi \qquad (3.231)$$

and the boundary condition $P_N = M_N^{-1} + \Pi = Q_f^{-1} + \Pi$.

We can use the following forward computation: by using the new variables $\vec{M}_j$ and $\vec{\Lambda}_j$ such that $\vec{M}_j = M_{N-j}$ and $\vec{\Lambda}_j = \Lambda_{N-j}$, (3.222) and (3.223) can be written as

$$u_i^* = -R^{-1}B^T \vec{\Lambda}_{N-1}^{-1}[\vec{M}_{N-1}A x_i + g_{i+1,i+N}] \qquad (3.232)$$

$$w_i^* = \gamma^{-2}R_w^{-1}B_w^T \vec{\Lambda}_{N-1}^{-1}[\vec{M}_{N-1}A x_i + g_{i+1,i+N}] \qquad (3.233)$$

where

$$\vec{M}_j = A^T \vec{\Lambda}_j^{-1} \vec{M}_{j-1} A + Q, \quad j = 1, \cdots, N-1$$

$$\vec{M}_0 = Q_f$$

$$\vec{\Lambda}_j = I + \vec{M}_j(BR^{-1}B^T - \gamma^{-2}B_w R_w^{-1}B_w^T)$$

3.4.2 Monotonicity of the Saddle-point Optimal Cost

In this section, terminal inequality conditions are proposed for linear discrete time-invariant systems which guarantee the monotonicity of the saddle-point value. In the next section, under the proposed terminal inequality conditions, the closed-loop stability of RHC is shown for linear discrete time-invariant systems.

Theorem 3.15. *Assume that Q_f in (3.219) satisfies the following inequality:*

$$Q_f \geq Q + H^T R H - \Gamma^T R_w \Gamma + A_{cl}^T Q_f A_{cl} \quad \text{for some } H \in \Re^{m \times n} \quad (3.234)$$

where

$$A_{cl} = A - BH + B_w \Gamma$$
$$\Gamma = \gamma^{-2} R_w^{-1} B_w^T \Lambda^{-1} Q_f A \quad (3.235)$$
$$\Lambda = I + Q_f (BR^{-1} B^T - \gamma^{-2} B_w R_w^{-1} B_w^T) \quad (3.236)$$

The saddle-point optimal cost $J^*(x_i, i, i_f)$ in (3.16) then satisfies the following relation:

$$J^*(x_\tau, \tau, \sigma + 1) \leq J^*(x_\tau, \tau, \sigma), \quad \tau \leq \sigma \quad (3.237)$$

and thus $M_{\tau,\sigma+1} \leq M_{\tau,\sigma}$.

Proof. Subtracting $J^*(x_\tau, \tau, \sigma)$ from $J^*(x_\tau, \tau, \sigma + 1)$, we can write

$$\delta J^*(x_\tau, \sigma) = \sum_{i=\tau}^{\sigma} [x_i^{1T} Q x_i^1 + u_i^{1T} R u_i^1 - \gamma^2 w_i^{1T} R_w w_i^1] + x_{\sigma+1}^{1T} Q_f x_{\sigma+1}^1 \quad (3.238)$$

$$- \sum_{i=\tau}^{\sigma-1} [x_i^{2T} Q x_i^2 + u_i^{2T} R u_i^2 - \gamma^2 w_i^{2T} R_w w_i^2] - x_\sigma^{2T} Q_f x_\sigma^2 \quad (3.239)$$

where the pair u_i^1 and w_i^1 is a saddle-point solution for $J(x_\tau, \tau, \sigma + 1)$ and the pair u_i^2 and w_i^2 is one for $J(x_\tau, \tau, \sigma)$. If we replace u_i^1 by u_i^2 and w_i^2 by w_i^1 up to $\sigma - 1$, the following inequalities are obtained by $J(u^*, w^*) \leq J(u, w^*)$:

$$\sum_{i=\tau}^{\sigma} [x_i^{1T} Q x_i^1 + u_i^{1T} R u_i^1 - \gamma^2 w_i^{1T} R_w w_i^1] + x_{\sigma+1}^{1T} Q_f x_{\sigma+1}^1$$

$$\leq \sum_{i=\tau}^{\sigma-1} [\tilde{x}_i^T Q \tilde{x}_i + u_i^{2T} R u_i^2 - \gamma^2 w_i^{1T} R_w w_i^1] + \tilde{x}_\sigma^T Q \tilde{x}_\sigma + u_\sigma^{1T} R u_\sigma^1 - \gamma^2 w_\sigma^{1T} R_w w_\sigma^1$$

$$+ x_{\sigma+1}^{1T} Q_f x_{\sigma+1}^1$$

where $u_\sigma^1 = H \tilde{x}_\sigma$, and $w_\sigma^1 = \Gamma \tilde{x}_\sigma$. By $J(u^*, w^*) \geq J(u^*, w)$, we have

$$\sum_{i=\tau}^{\sigma-1}[x_i^{2T}Qx_i^2 + u_i^{2T}Ru_i^2 - \gamma^2 w_i^{2T}R_w w_i^2] + x_\sigma^T Q_f x_\sigma$$

$$\geq \sum_{i=\tau}^{\sigma-1}[\tilde{x}_i^T Q\tilde{x}_i + u_i^{2T}Ru_i^2 - \gamma^2 w_i^{1T}R_w w_i^1] + \tilde{x}_\sigma^T Q_f \tilde{x}_\sigma$$

Note that $\tilde{x}_i$ is a trajectory associated with u_i^2 and w_i^1. We have the following inequality:

$$\delta J^*(x_\tau, \sigma) \leq \tilde{x}_\sigma^T\{Q + H^T R H - \Gamma^T R_w \Gamma + A_{cl}^T Q_f A_{cl} - Q_f\}\tilde{x}_\sigma \leq 0 \quad (3.240)$$

where the last inequality comes from (3.234).

Since $\delta J^*(x_\tau, \sigma) = x_\tau^T[M_{\tau,\sigma+1} - M_{\tau,\sigma}]x_\tau \leq 0$ for all x_τ, we have that $M_{\tau,\sigma+1} - M_{\tau,\sigma} \leq 0$. For time-invariant systems we have

$$M_{\tau+1} \leq M_\tau \qquad (3.241)$$

This completes the proof. ∎

Note that Q_f satisfying the inequality (3.234) in (3.15) should be checked for whether M_{i,i_f} generated from the boundary value Q_f satisfies $R_w - \gamma^{-2}B_w^T M_{i,i_f}B_w$. In order to obtain a feasible solution Q_f, R_w and γ can be adjusted.

Case 1: Γ in the inequality (3.234) includes Q_f, which makes it difficult to handle the inequality. We introduce the inequality without the variable Γ as follows:

$$Q + H^T R H - \Gamma^T R_w \Gamma + A_{cl}^T Q_f A_{cl}$$
$$= Q + H^T R H + \Sigma^T(B_w^T Q_f B_w - R_w)\Sigma$$
$$- (A - BH)^T Q_f B_w(B_w^T Q_f B_w - R_w)^{-1}B_w Q_f(A - BH),$$
$$\leq Q + H^T R H - (A - BH)^T(B_w^T Q_f B_w - R_w)^{-1}(A - BH) \leq Q_f \quad (3.242)$$

where $\Sigma = \Gamma + (B_w^T Q_f B_w - R)^{-1}B_w^T Q_f(A - BH)$.

Case 2:

$$Q_f \geq A^T Q_f[I + \Pi Q_f]^{-1}A + Q \qquad (3.243)$$

where $\Pi = BR^{-1}B - \gamma^{-2}B_w R_w^{-1}B_w$.

If H is replaced by an optimal gain $H = -R^{-1}B^T[I + Q_f\Pi]^{-1}Q_f A$, then by using the matrix inversion lemma in Appendix A, we can have (3.243). It is left as an exercise at the end of this chapter.

Case 3:

$$Q_f = Q + H^T R H - \Gamma^T R_w \Gamma + A_{cl}^T Q_f A_{cl} \qquad (3.244)$$

which is a special case of (3.234). Q_f has the following meaning. If the pair (A, B) is stabilizable and the system is asymptotically stable with $u_i = -Hx_i$ and $w_i = \gamma^{-1} B_\gamma^T [I + M_{i+1,\infty} \hat{Q}]^{-1} M_{i+1,\infty} A x_i$ for $\sigma \geq i \geq \tau$, then

$$\min_{u_i, i \in [\tau, \sigma-1]} \sum_{i=\tau}^{\infty} [x_i^T Q x_i + u_i^T R u_i - \gamma^2 w_i^T R_w w_i]$$

$$= \min_{u_i, i \in [\tau, \sigma-1]} \sum_{i=\tau}^{\sigma-1} [x_i^T Q x_i + u_i^T R u_i - \gamma^2 w_i^T R_w w_i] + x_\sigma^T Q_f x_\sigma \quad (3.245)$$

where Q_f can be shown to satisfy (3.244).

Case 4:

$$Q_f = Q - \Gamma^T R_w \Gamma + [A + B_w \Gamma]^T Q_f [A + B_w \Gamma] \qquad (3.246)$$

which is also a special case of (3.234). If the system matrix A is stable with $u_i = 0$ and $w_i = \gamma^{-1} R_w^{-1} B_w^T [I + M_{i+1,\infty} \hat{Q}]^{-1} M_{i+1,\infty} A x_i$ for $\sigma \geq i \geq \tau$ then, Q_f satisfies (3.246).

In the following, the nondecreasing monotonicity of the saddle-point optimal cost is studied.

Theorem 3.16. *Assume that Q_f in (3.16) satisfies the following inequality:*

$$Q_f \leq A^T Q_f [I + \Pi Q_f]^{-1} A + Q \qquad (3.247)$$

The saddle-point optimal cost $J^*(x_i, i, i_f)$ then satisfies the following relation:

$$J^*(x_\tau, \tau, \sigma + 1) \geq J^*(x_\tau, \tau, \sigma), \quad \tau \leq \sigma \qquad (3.248)$$

and thus $M_{\tau,\sigma+1} \geq M_{\tau,\sigma}$.

Proof. In a similar way to the proof of Theorem 3.15, if we replace u_i^2 by u_i^1 and w_i^1 by w_i^2 up to $\sigma - 1$, then the following inequalities are obtained by $J(u^*, w^*) \geq J(u^*, w)$:

$$J^*(x_\tau, \tau, \sigma + 1) = \sum_{i=\tau}^{\sigma} [x_i^{1T} Q x_i^1 + u_i^{1T} R u_i^1 - \gamma^2 w_i^{1T} R_w w_i^1] + x_{\sigma+1}^{1T} Q_f x_{\sigma+1}^1$$

$$\geq \sum_{i=\tau}^{\sigma-1} [\tilde{x}_i^T Q \tilde{x}_i + u_i^{1T} R u_i^1 - \gamma^2 w_i^{2T} R_w w_i^2]$$

$$+ \tilde{x}_\sigma^T Q \tilde{x}_\sigma + u_\sigma^{1T} R u_\sigma^1 - \gamma^2 w_\sigma^{2T} R_w w_\sigma^2 + \tilde{x}_{\sigma+1}^T Q_f \tilde{x}_{\sigma+1}$$

where

$$u_\sigma^1 = H\tilde{x}_\sigma$$
$$w_\sigma^1 = \Gamma\tilde{x}_\sigma$$
$$H = -R^{-1}B^T[I + Q_f\Pi]^{-1}Q_f A$$
$$\Gamma = \gamma^{-2}R_w^{-1}B_w^T\Lambda^{-1}Q_f A$$

and $\tilde{x}_i$ is the trajectory associated with x_τ, u_i^1 and w_i^2 for $i \in [\tau, \sigma]$. By $J(u^*, w^*) \leq J(u, w^*)$, we have

$$J^*(x_\tau, \tau, \sigma) = \sum_{i=\tau}^{\sigma-1}[x_i^{2T}Qx_i^2 + u_i^{2T}Ru_i^2 - \gamma^2 w_i^{2T}R_w w_i^2] + x_\sigma^{2T}Q_f x_\sigma^2$$

$$\leq \sum_{i=\tau}^{\sigma-1}[\tilde{x}_i^T Q\tilde{x}_i + u_i^{1T}Ru_i^1 - \gamma^2 w_i^{2T}R_w w_i^2] + \tilde{x}_\sigma^T Q_f \tilde{x}_\sigma$$

The difference $\delta J^*(x_\tau, \sigma)$ between $J^*(x_\tau, \tau, \sigma+1)$ and $J^*(x_\tau, \tau, \sigma)$ is represented as

$$\delta J^*(x_\tau, \sigma) \geq \tilde{x}_\sigma^T\{Q + H^T RH - \Gamma^T R_w\Gamma + A_{cl}^T Q_f A_{cl} - Q_f\}\tilde{x}_\sigma \geq 0 \quad (3.249)$$

As in the inequality (3.243), (3.249) can be changed to (3.247). The relation $M_{\sigma+1} \geq M_\sigma$ follows from $J^*(x_i, i, i_f) = x_i^T M_{i,i_f} x_i$. This completes the proof. ∎

Case 1: $Q_f = 0$

The well-known free terminal condition, i.e. $Q_f = 0$ satisfies (3.247). Thus, Theorem 3.16 includes the monotonicity of the saddle-point value of the free terminal case.

In the following theorem based on the optimality, it will be shown that when the monotonicity of the saddle-point value or the Riccati equations holds once, it holds for all subsequent times.

Theorem 3.17. *The following inequalities for the saddle-point optimal cost and the Riccati equation are satisfied:*

(1) If

$$J^*(x_{\tau'}, \tau', \sigma+1) \leq J^*(x_{\tau'}, \tau', \sigma) \quad (or \geq J^*(x_{\tau'}, \tau', \sigma)) \quad (3.250)$$

for some τ', then

$$J^*(x_{\tau''}, \tau'', \sigma+1) \leq J^*(x_{\tau''}, \tau'', \sigma) \quad (or \geq J^*(x_{\tau''}, \tau'', \sigma)) \quad (3.251)$$

where $\tau_0 \leq \tau'' \leq \tau'$.

(2) If

$$M_{\tau',\sigma+1} \leq M_{\tau',\sigma} \quad (or \geq M_{\tau',\sigma}) \tag{3.252}$$

for some τ', then,

$$M_{\tau'',\sigma+1} \leq M_{\tau'',\sigma} \quad (or \geq M_{\tau'',\sigma}) \tag{3.253}$$

where $\tau_0 \leq \tau'' \leq \tau'$.

Proof. (a) Case of $J^*(x_{\tau'},\tau',\sigma+1) \leq J^*(x_{\tau'},\tau',\sigma)$:

The pair u_i^1 and w_i^1 is a saddle-point optimal solution for $J(x_{\tau''},\tau'',\sigma+1)$ and the pair u_i^2 and w_i^2 for $J(x_{\tau''},\tau'',\sigma)$. If we replace u_i^1 by u_i^2 and w_i^2 by w_i^1 up to τ', then

$$
\begin{aligned}
J^*(x_{\tau''},\sigma+1) &= \sum_{i=\tau''}^{\tau'-1} [x_i^{1T}Qx_i^1 + u_i^{1T}Ru_i^1 - \gamma^2 w_i^{1T}R_w w_i^1] + J^*(x_{\tau'}^1,\tau',\sigma+1) \\
&\leq \sum_{i=\tau''}^{\tau'-1} [\tilde{x}_i^T Q\tilde{x}_i + u_i^{2T}Ru_i^2 - \gamma^2 w_i^{1T}R_w w_i^1] \\
&\quad + J^*(\tilde{x}_{\tau'},\tau',\sigma+1)
\end{aligned} \tag{3.254}
$$

by $J(u^*,w^*) \leq J(u,w^*)$ and

$$
\begin{aligned}
J^*(x_{\tau''},\sigma) &= \sum_{i=\tau''}^{\tau'-1} [x_i^{2T}Qx_i^2 + u_i^{2T}Ru_i^2 - \gamma^2 w_i^{2T}R_w w_i^2] + J^*(x_{\tau'}^2,\tau',\sigma) \\
&\geq \sum_{i=\tau''}^{\tau'-1} [\tilde{x}_i^T Q\tilde{x}_i + u_i^{2T}Ru_i^2 - \gamma^2 w_i^{1T}R_w w_i^1] \\
&\quad + J^*(\tilde{x}_{\tau'},\tau',\sigma)
\end{aligned} \tag{3.255}
$$

by $J(u^*,w^*) \geq J(u^*,w)$. The difference between the adjacent optimal costs can be expressed as

$$
\begin{aligned}
\delta J^*(x_{\tau''},\sigma) &= \sum_{i=\tau''}^{\tau'-1} [x_i^{1T}Qx_i^1 + u_i^{1T}Ru_i^1] + J^*(x_{\tau'}^1,\tau',\sigma+1) \\
&\quad - \sum_{i=\tau''}^{\tau'-1} [x_i^{2T}Qx_i^2 + u_i^{2T}Ru_i^2] - J^*(x_{\tau'}^2,\tau',\sigma)
\end{aligned} \tag{3.256}
$$

Substituting (3.255) and (3.255) into (3.256), we have

$$\delta J^*(x_{\tau''}, \sigma) \leq J^*(\tilde{x}_{\tau'}, \tau', \sigma + 1) - J^*(\tilde{x}_{\tau'}, \tau', \sigma)$$
$$= \delta J^*(\tilde{x}_{\tau'}, \sigma) \leq 0 \qquad (3.257)$$

Therefore,

$$\delta J^*(x_{\tau''}, \sigma) \leq \delta J^*(\tilde{x}_{\tau'}, \sigma) \leq 0$$

where $\tilde{x}_{\tau'}$ is the trajectory which consists of $x_{\tau''}$, u_i^2, and w_i^1 for $i \in [\tau'', \tau' - 1]$.

(b) Case of $J^*(x_{\tau'}, \tau', \sigma + 1) \geq J^*(x_{\tau'}, \tau', \sigma)$:
In a similar way to the case of (a), if we replace u_i^2 by u_i^1 and w_i^1 by w_i^2 up to τ', then

$$\delta J^*(x_{\tau''}, \sigma) \geq \delta J^*(x_{\tau'}, \sigma) \geq 0 \qquad (3.258)$$

The monotonicity of the Riccati equations follows from $J^*(x_i, i, i_f) = x_i^T M_{i,i_f} x_i$. This completes the proof. ∎

In the following section, stabilizing receding horizon H_∞ controls will be proposed by using the monotonicity of the saddle-point value or the Riccati equations for linear discrete time-invariant systems.

3.4.3 Stability of Receding Horizon H_∞ Control

In case of the conventional H_∞ control, the following two kinds of stability can be checked. H_∞ controls based on the infinite horizon are required to have the following properties:

1. Systems are stabilized in the case that there is no disturbance.
2. Systems are stabilized in the case that the worst-case disturbance enters the systems.

For the first case, we introduce the following result.

Theorem 3.18. *Assume that the pair (A, B) and $(A, Q^{\frac{1}{2}})$ are stabilizable and observable respectively, and that the receding horizon H_∞ control (3.222) associated with the quadratic cost $J(x_i, i, i+N)$ exists. If the following inequality holds:*

$$J^*(x_i, i, i + N + 1) \leq J^*(x_i, i, i + N) \qquad (3.259)$$

then the asymptotic stability is guaranteed in the case that there is no disturbance.

Proof. We show that the zero state is attractive. Since $J^*(x_i, i, \sigma + 1) \leq J^*(x_i, i, \sigma)$,

$$J^*(x_i, i, i+N) \tag{3.260}$$
$$= x_i^T Q x_i + u_i^{*T} R u_i^* - \gamma^2 w_i^{*T} R_w w_i^*$$
$$+ J^*(x^1(i+1; (x_i, i, u_i^*)), i+1, i+N)$$
$$\geq x_i^T Q x_i + u_i^{*T} R u_i^* + J^*(x^2(i+1; (x_i, i, u_i^*)), i+1, i+N)$$
$$\geq x_i^T Q x_i + u_i^{*T} R u_i^* + J^*(x^2(i+1; (x_i, i, u_i^*)), i+1, i+N+1) \tag{3.261}$$

where u_i^* is the optimal control at time i and x_{i+1}^2 is a state at time $i+1$ when $w_i = 0$ and the optimal control u_i^*. Therefore, $J^*(x_i, i, i+N)$ is nonincreasing and bounded below, i.e. $J^*(x_i, i, i+N) \geq 0$. $J^*(x_i, i, i+N)$ approaches some nonnegative constant c as $i \to \infty$. Hence, we have

$$x_i^T Q x_i + u_i^T R u_i \longrightarrow 0 \tag{3.262}$$

From the fact that the finite sum of the converging sequences also approaches zero, the following relation is obtained:

$$\sum_{j=i}^{i+l-1} \left[x_j^T Q x_j + u_j^T R u_j \right] \to 0, \tag{3.263}$$

leading to

$$x_i^T \left(\sum_{j=i}^{i+l-1} (A - BH)^{(j-i)T} [Q + H^T R H] (A - BH)^{j-i} \right) x_i \to 0 \tag{3.264}$$

However, since the pair $(A, Q^{\frac{1}{2}})$ is observable, $x_i \to 0$ as $i \to \infty$ independently of i_0. Therefore, the closed-loop system is asymptotically stable. This completes the proof. ∎

We suggest a sufficient condition for Theorem 3.18.

Theorem 3.19. *Assume that the pair (A, B) is stabilizable and the pair $(A, Q^{\frac{1}{2}})$ is observable. For $Q_f \geq 0$ satisfying (3.234), the system (3.216) with the receding horizon H_∞ control (3.222) is asymptotically stable for some N, $1 \leq N < \infty$.*

In the above theorem, Q must be nonzero. We can introduce another result as in a receding horizon LQ control so that Q could even be zero.

Suppose that disturbances show up. From (3.229) we have

$$u_i^* = -R^{-1} B^T P_1^{-1} A x_i \tag{3.265}$$

where

$$P_i = A^{-1} P_{i+1} [I + A^{-1} Q A^{-1} P_{i+1}]^{-1} A^{-1} + \Pi \tag{3.266}$$
$$P_N = M_N^{-1} + \Pi = Q_f^{-1} + \Pi \tag{3.267}$$

We will consider a slightly different approach. We assume that P_{i,i_f} in (2.154) is given from the beginning with a terminal constraint $P_{i_f,i_f} = P_f$ rather than P_{i_f,i_f} being obtained from (2.156).

In fact, Riccati Equation (2.154) with the boundary condition P_f can be obtained from the following problem. Consider the following system:

$$\hat{x}_{i+1} = A^{-T}\hat{x}_i + A^{-1}Q^{\frac{1}{2}}\hat{u}_i \tag{3.268}$$

where $\hat{x}_i \in \Re^n$, $\hat{u}_i \in \Re^m$, and a performance criterion

$$\hat{J}(\hat{x}_{i_0}, i_0, i_f) = \sum_{i=i_0}^{i_f-1}[\hat{x}_i^T \Pi \hat{x}_i + \hat{u}_i^T \hat{u}_i] + \hat{x}_{i_f}^T P_f \hat{x}_{i_f} \tag{3.269}$$

The optimal cost for the system (3.268) is given by $\hat{J}^*(\hat{x}_i, i, i_f) = \hat{x}_i^T P_{i,i_f}\hat{x}_i$. The optimal control $\hat{u}_i$ is

$$\hat{u}_{i,i_f} = -R^{-1}B^T P_{i+1,i_f}^{-1} A\hat{x}_i \tag{3.270}$$

From Theorem 3.16, it can be easily seen that $P_{\tau,\sigma+1} \geq P_{\tau,\sigma}$ if

$$P_f \leq A^{-1}P_f[I + A^{-T}QA^{-1}P_f]^{-1}A^{-T} + \Pi \tag{3.271}$$

Now, we are in a position to state the following result on the stabiltiy of the receding horizon H_∞ control.

Theorem 3.20. *Assume that the pair (A, B) is controllable and A is nonsingular. If the inequality (3.271) is satisfied, then the system (3.216) with the control (3.265) is asymptotically stable for $1 \leq N$.*

Proof. Consider the adjoint system of the system (3.216) with the control (3.270)

$$\hat{x}_{i+1} = [A - BR^{-1}B^T P_1^{-1}A]^{-T}\hat{x}_i \tag{3.272}$$

and the associated scalar-valued function

$$V(\hat{x}_i) = \hat{x}_i^T A^{-1}P_1 A^{-1}\hat{x}_i \tag{3.273}$$

Note that $P_1 - BR^{-1}B^T$ is nonsingular, which guarantees the nonsingularity of $A - BR^{-1}B^T P_1^{-1}A$ with a nonsingular A.

Subtracting $V(\hat{x}_{i+1})$ from $V(\hat{x}_i)$, we have

$$V(\hat{x}_i) - V(\hat{x}_{i+1}) = \hat{x}_i^T A^{-1}P_1 A^{-1}\hat{x}_i - \hat{x}_{i+1}^T A^{-1}P_1 A^{-1}\hat{x}_{i+1} \tag{3.274}$$

Recall the following relation:

$$P_0 = (A^T P_1^{-1}A + Q)^{-1} + \Pi = A^{-1}(P_1^{-1} + A^{-T}QA^{-1})^{-1}A^{-T} + \Pi$$
$$= A^{-1}\left[P_1 - P_1 A^{-T}Q^{\frac{1}{2}}(Q^{\frac{1}{2}}A^{-1}P_1 A^{-T}Q^{\frac{1}{2}} + I)^{-1}Q^{\frac{1}{2}}A^{-1}P_1\right]A^{-T} + \Pi$$
$$= A^{-1}P_1 A^{-T} + \Pi - Z \tag{3.275}$$

where

$$Z = A^{-1}P_1A^{-T}Q^{\frac{1}{2}}(Q^{\frac{1}{2}}A^{-1}P_1A^{-T}Q^{\frac{1}{2}} + I)^{-1}Q^{\frac{1}{2}}A^{-1}P_1A^{-T}$$

Replacing $\hat{x}_i$ with $[A - BR^{-1}B^TP_1^{-1}A]^T\hat{x}_{i+1}$ in (3.274) and plugging (3.275) into the second term in (3.274) yields

$$\begin{aligned}
V(\hat{x}_i) - V(\hat{x}_{i+1}) &= \hat{x}_{i+1}^T[P_1 - 2BR^{-1}B^T + BR^{-1}B^TP_1^{-1}BR^{-1}B^T]\hat{x}_{i+1} \\
&\quad - \hat{x}_{i+1}^T[P_0 - \Pi + Z]\hat{x}_{i+1} \\
&= -\hat{x}_{i+1}^T[BR^{-1}B^T - BR^{-1}B^TP_1^{-1}BR^{-1}B^T]\hat{x}_{i+1} \\
&\quad - \hat{x}_{i+1}^T[P_0 - P_1 + \gamma^{-2}B_wR_w^{-1}B_w^T + Z]\hat{x}_{i+1}
\end{aligned}$$

Since Z is positive semidefinite and $P_0 - P_1 \geq 0$, we have

$$V(\hat{x}_i) - V(\hat{x}_{i+1}) \leq -\hat{x}_{i+1}^T[BR^{-\frac{1}{2}}SR^{-\frac{1}{2}}B^T + \gamma^{-2}B_wR_w^{-1}B_w]\hat{x}_{i+1} \quad (3.276)$$

where $S = I - R^{-\frac{1}{2}}B^TP_1^{-1}BR^{-\frac{1}{2}}$.

In order to show the positive definiteness of S, we have only to prove $P_1 - BR^{-1}B^T > 0$ since

$$I - P_1^{-\frac{1}{2}}BR^{-1}B^TP_1^{-\frac{1}{2}} > 0 \Longleftrightarrow P_1 - BR^{-1}B^T > 0$$

Note that $I - AA^T > 0$ implies $I - A^TA > 0$ and vice versa for any rectangular matrix A. From the condition for the existence of the saddle point, the lower bound of P is obtained as

$$\begin{aligned}
&R_w - \gamma^{-2}B_w^TM_iB_w = R_w - \gamma^{-2}B_w^T(P_i - \Pi)^{-1}B_w > 0 \\
&\Longleftrightarrow I - \gamma^{-2}(P_i - \Pi)^{-\frac{1}{2}}B_wR_w^{-1}B_w^T(P_i - \Pi)^{-\frac{1}{2}} > 0 \\
&\Longleftrightarrow P_i - \Pi - \gamma^{-2}B_wR_w^{-1}B_w^T = P_i - BR^{-1}B^T > 0 \\
&\Longleftrightarrow P_i > BR^{-1}B^T
\end{aligned} \quad (3.277)$$

From (3.277), it can be seen that S in (3.276) is positive definite. Note that the left-hand side in (3.276) is always nonnegative. From (3.276) we have

$$V(\hat{x}(i + 1; \hat{x}_{i_0}, i_0)) - V(\hat{x}_{i_0}, i_0) \geq \hat{x}_{i_0}^T\Theta\hat{x}_{i_0}$$

where

$$\Theta \triangleq \left[\sum_{k=i_0}^{i} \Psi^{(i-i_0)T}W\Psi^{i-i_0}\right]$$

$$\Psi \triangleq A - BR^{-1}B^TP_1^{-1}A$$

$$W \triangleq BR^{-\frac{1}{2}}SR^{-\frac{1}{2}}B^T + \gamma^{-2}B_wR_w^{-1}B_w$$

If (A, B) is controllable, then the matrix Θ is positive definite. Thus, all eigenvalues of Θ are positive and the following inequality is obtained:

$$V(\hat{x}(i+1;\hat{x}_{i_0},i_0)) - V(\hat{x}_{i_0}) \geq \lambda_{\min}(\Theta)\|\hat{x}_{i_0}\| \qquad (3.278)$$

This implies that the closed-loop system (3.216) is exponentially increasing, i.e. the closed-loop system (3.216) with (3.270) is exponentially decreasing. This completes the proof. ∎

In Theorem 3.20, Q can be zero. If Q becomes zero, then P_1 can be expressed as the following closed form:

$$P_1 = \sum_{j=i+1}^{i+N} A^{j-i-1}\Pi A^{(j-i-1)T} + A^N P_f A^{TN} \qquad (3.279)$$

where A is nonsingular.

It is noted that P_f satisfying (3.271) is equivalent to Q_f satisfying (3.243) in the relation of $P_f = Q_f^{-1} + \Pi$. Replacing P_f with $Q_f^{-1} + \Pi$ in (3.271) yields the following inequality:

$$Q_f^{-1} + \Pi \leq A^{-1}[Q_f^{-1} + BR^{-1}B^T + A^{-T}QA^{-1}]^{-1}A^{-T} + \Pi$$
$$= [A^T(Q_f^{-1} + BR^{-1}B^T)^{-1}A + Q]^{-1} + \Pi$$

Finally, we have

$$Q_f \geq A^T(Q_f^{-1} + \Pi)^{-1}A + Q \qquad (3.280)$$

Therefore, if Q_f satisfies (3.280), P_f also satisfies (3.271).

Theorem 3.21. *Assume that the pair (A, B) is controllable and A is nonsingular.*

(1) If $M_{i+1} \geq M_i > 0$ for some i, then the system (3.216) with the receding horizon H_∞ control (3.222) is asymptotically stable for $1 \leq N < \infty$.
(2) For $Q_f > 0$ satisfies (3.243) for some H, then the system (3.216) with the RH H_∞ control (3.222) is asymptotically stable for $1 \leq N < \infty$.

Proof. The first part is proved as follows. $M_{i+1} \geq M_i > 0$ implies $0 < M_{i+1}^{-1} \leq M_i^{-1}$, from which we have $0 < P_{i+1} \leq P_i$ satisfying the inequality (3.271). Thus, the control (3.265) is equivalent to the control (3.222). The second part is proved as follows: inequalities $K_{i+1} \geq K_i > 0$ are satisfied for K_i generated from $Q_f > 0$ satisfying (3.234) for some H. Thus, the second result can be seen from the first one. This completes the proof. ∎

It is noted that (3.280) is equivalent to (3.247), as mentioned before.

3.4.4 Additional Properties

Now, we will show that the stabilizing receding horizon controllers guarantee the H_∞ norm bound of the closed-loop system.

Theorem 3.22. *Under the assumptions given in Theorem 3.18, the H_∞ norm bound of the closed-loop system (3.216) with (3.222) is guaranteed.*

Proof. Consider the difference of the optimal cost between the time i and $i + 1$:

$$J^*(i + 1, i + N + 1) - J^*(i, i + N)$$

$$= \sum_{j=i+1}^{i+N} \left[x_j^T Q x_j + u_j^T R u_j - \gamma^2 w_j^T R_w w_j \right] + x_{i+N+1}^T Q_f x_{i+N+1}$$

$$- \sum_{j=i}^{i+N-1} \left[x_j^T Q x_j + u_j^T R u_j - \gamma^2 w_j^T R_w w_j \right] - x_{i+N}^T Q_f x_{i+N} \qquad (3.281)$$

Note that the optimal control and the worst-case disturbance on the horizon are time-invariant with respect to the moving horizon.

Applying the state feedback control $u_{i+N} = H x_{i+N}$ at time $i + N$ yields the following inequality:

$$J^*(i + 1, i + N + 1) - J^*(i, i + N) \leq -x_i^T Q x_i - u_i^T R u_i + \gamma^{-2} w_i^T R_w w_i$$

$$+ \begin{bmatrix} w_{i+N} \\ x_{i+N} \end{bmatrix}^T \Pi \begin{bmatrix} w_{i+N} \\ x_{i+N} \end{bmatrix} \qquad (3.282)$$

where

$$\Pi \triangleq \begin{bmatrix} -\gamma^2 R_w + B_w^T Q_f B_w & B_w^T Q_f (A + BH) \\ (A + BH)^T Q_f B_w & (A + BH)^T Q_f (A + BH) - Q_f + Q + H^T R H \end{bmatrix}$$

From the cost monotonicity condition, Π is guaranteed to be positive semidefinite. The proof is left as an exercise. Taking the summation on both sides of (3.282) from $i = 0$ to ∞ and using the positiveness of Π, we have

$$J^*(0, N) - J^*(\infty, \infty + N) = \sum_{i=0}^{\infty} [J^*(i, i + N) - J^*(i + 1, i + N + 1)]$$

$$\geq \sum_{i=0}^{\infty} [x_i^T Q x_i + u_i^T R u_i - \gamma^2 w_i^T R_w w_i]$$

From the assumption $x_0 = 0$, $J^*(0, N) = 0$. The saddle-point optimal cost is guaranteed to be nonnegative, i.e. $J^*(\infty, \infty + N) \geq 0$. Therefore, it is guaranteed that

$$\sum_{i=0}^{\infty}[x_i^T Q x_i + u_i^T R u_i - \gamma^2 w_i^T R_w w_i] \leq 0$$

which implies that

$$\frac{\sum_{i=0}^{\infty}[x_i^T Q x_i + u_i^T R u_i]}{\sum_{i=0}^{\infty} w_i^T R_w w_i} \leq \gamma^2$$

This completes the proof.   ∎

In the same way, under the assumptions given in Theorem 3.20, the H_∞ norm bound of the closed-loop system (3.216) with (3.222) is guaranteed with M_1 replaced by $[P_1 - \Pi]^{-1}$. The inverse matrices exist for $N \geq l_c + 1$ since $P_{i_f - i} - \Pi = [A^T P_{i_f - i - 1}^{-1} A + Q]^{-1}$.

Example 3.3

In this example, the H_∞ RHC is compared with the LQ RHC through simulation. The target model and the reference signal are the same as those of Example 3.1. except that B_w is given by

$$B_w = \begin{bmatrix} 0.016 & 0.01 & 0.008 & 0 \\ 0.002 & 0.009 & 0 & 0.0005 \end{bmatrix}^T \tag{3.283}$$

For simulation, disturbances coming into the system are generated so that they become worst on the receding horizon. γ^2 is taken as 1.5.

As can be seen in Figure 3.10, the trajectory for the H_∞ RHC is less deviated from the reference signal than that for the LQ RHC.

The MATLAB® functions used for simulation are given in Appendix F.

3.5 Receding Horizon Control via Linear Matrix Inequality Forms

3.5.1 Computation of Cost Monotonicity Condition

Receding Horizon Linear Quadratic Control

It looks difficult to find H and Q_f that satisfy the cost monotonicity condition (3.73). However, this can be easily computed using LMI.

Pre- and post-multiplying on both sides of (3.73) by Q_f^{-1}, we obtain

$$X \geq XQX + XH^T RHX + (AX - BHX)^T X^{-1}(AX - BHX) \tag{3.284}$$

where $X = Q_f^{-1}$. Using Schur's complement, the inequality (3.284) is converted into the following:

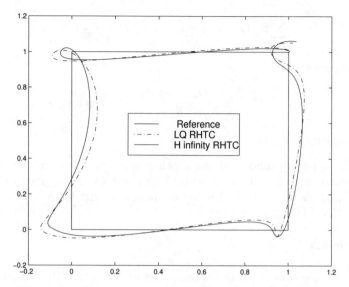

Fig. 3.10. Comparison between LQ RHTC and H_∞ RHTC

$$X - XQX - Y^T RY - (AX - BY)^T X^{-1}(AX - BY) \geq 0$$

$$\begin{bmatrix} X - XQX - Y^T RY & (AX - BY)^T \\ AX - BY & X \end{bmatrix} \geq 0 \qquad (3.285)$$

where $Y = HX$. Partitioning the left side of (3.285) into two parts, we have

$$\begin{bmatrix} X & (AX - BY)^T \\ AX - BY & X \end{bmatrix} - \begin{bmatrix} XQX - Y^T RY & 0 \\ 0 & 0 \end{bmatrix} \geq 0 \qquad (3.286)$$

In order to use Schur's complement, the second block matrix is decomposed as

$$\begin{bmatrix} X & (AX - BY)^T \\ AX - BY & X \end{bmatrix} - \begin{bmatrix} Q^{\frac{1}{2}}X & 0 \\ R^{\frac{1}{2}}Y & 0 \end{bmatrix}^T \begin{bmatrix} I & 0 \\ 0 & I \end{bmatrix}^{-1} \begin{bmatrix} Q^{\frac{1}{2}}X & 0 \\ R^{\frac{1}{2}}Y & 0 \end{bmatrix} \geq 0 \qquad (3.287)$$

Finally, we can obtain the LMI form as

$$\begin{bmatrix} X & (AX - BY)^T & (Q^{\frac{1}{2}}X)^T & (R^{\frac{1}{2}}Y)^T \\ AX - BY & X & 0 & 0 \\ Q^{\frac{1}{2}}X & 0 & I & 0 \\ R^{\frac{1}{2}}Y & 0 & 0 & I \end{bmatrix} \geq 0 \qquad (3.288)$$

Once X and Y are found, Q_f and $H = YX^{-1}$ can be known.

Example 3.4

For the following systems and the performance criterion:

$$x_{k+1} = \begin{bmatrix} 0.6831 & 0.0353 \\ 0.0928 & 0.6124 \end{bmatrix} x_k + \begin{bmatrix} 0.6085 & 0.0158 \end{bmatrix} u_k \quad (3.289)$$

$$J(x_k, k, k+N) = \sum_{i=0}^{N-1} \left[x_{k+i}^T x_{k+i} + 3u_{k+i}^2 \right] + x_{k+N}^T Q_f x_{k+N} \quad (3.290)$$

The MATLAB® code for finding Q_f satisfying the LMI (3.288) is given in Appendix F. By using this MATLAB® program, we have one possible final weighting matrix for the cost monotonicity

$$Q_f = \begin{bmatrix} 0.4205 & -0.0136 \\ -0.0136 & 0.4289 \end{bmatrix} \quad (3.291)$$

Similar to (3.73), the cost monotonicity condition (3.85) can be represented as an LMI form. First, in order to obtain an LMI form, the inequality (3.85) is converted into the following:

$$Q_f - A^T Q_f [I + BR^{-1}B^T Q_f]^{-1} A - Q \geq 0 \quad (3.292)$$

$$\begin{bmatrix} Q_f - Q & A^T \\ A & Q_f^{-1} + BR^{-1}B^T \end{bmatrix} \geq 0 \quad (3.293)$$

Pre- and post-multiplying on both sides of (3.293) by some positive definite matrices, we obtain

$$\begin{bmatrix} Q_f^{-1} & 0 \\ 0 & I \end{bmatrix}^T \begin{bmatrix} Q_f - Q & A^T \\ A & Q_f^{-1} + BR^{-1}B^T \end{bmatrix} \begin{bmatrix} Q_f^{-1} & 0 \\ 0 & I \end{bmatrix} \geq 0 \quad (3.294)$$

$$\begin{bmatrix} X - XQX & XA^T \\ AX & X + BR^{-1}B^T \end{bmatrix} \geq 0 \quad (3.295)$$

where $Q_f^{-1} = X$

Partition the left side of (3.295) into two parts, we have

$$\begin{bmatrix} X & XA^T \\ AX & X + BR^{-1}B^T \end{bmatrix} - \begin{bmatrix} XQX & 0 \\ 0 & 0 \end{bmatrix} \geq 0 \quad (3.296)$$

In order to use Schur's complement, the second block matrix is decomposed as

$$\begin{bmatrix} X & (AX + BY)^T \\ AX + BY & X \end{bmatrix} - \begin{bmatrix} Q^{\frac{1}{2}}X & 0 \\ 0 & 0 \end{bmatrix}^T \begin{bmatrix} I & 0 \\ 0 & I \end{bmatrix}^{-1} \begin{bmatrix} Q^{\frac{1}{2}}X & 0 \\ 0 & 0 \end{bmatrix} \geq 0 \quad (3.297)$$

Finally, we can obtain the LMI form as

$$\begin{bmatrix} X & XA^T & (Q^{\frac{1}{2}}X)^T & 0 \\ AX & X + BR^{-1}B^T & 0 & 0 \\ Q^{\frac{1}{2}}X & 0 & I & 0 \\ 0 & 0 & 0 & I \end{bmatrix} \geq 0 \quad (3.298)$$

Once X is obtained, Q_f is given by X^{-1}.

The cost monotonicity condition (3.98) in Theorem 3.4 can be easily obtained by changing the direction of the inequality of (3.298):

$$\begin{bmatrix} X & XA^T & (Q^{\frac{1}{2}}X)^T & 0 \\ AX & X + BR^{-1}B^T & 0 & 0 \\ Q^{\frac{1}{2}}X & 0 & I & 0 \\ 0 & 0 & 0 & I \end{bmatrix} \leq 0 \qquad (3.299)$$

In the following section, stabilizing receding horizon controls will be obtained by LMIs.

Receding Horizon H_∞ Control

The cost monotonicity condition (3.234) can be written

$$\begin{bmatrix} \Gamma \\ I \end{bmatrix}^T \begin{bmatrix} R_w - B_w^T Q_f B_w & B_w^T Q_f(A - BH) \\ (A - BH)^T Q_f B_w & \Phi \end{bmatrix} \begin{bmatrix} \Gamma \\ I \end{bmatrix} \geq 0 \qquad (3.300)$$

where

$$\Phi = Q_f - Q - H^T R H - (A - BH)^T Q_f(A - BH) \qquad (3.301)$$

From (3.300), it can be seen that we have only to find Q_f such that

$$\begin{bmatrix} R_w - B_w^T Q_f B_w & B_w^T Q_f(A - BH) \\ (A - BH)^T Q_f B_w & \Phi \end{bmatrix} \geq 0 \qquad (3.302)$$

where we have

$$\begin{bmatrix} R_w & 0 \\ 0 & Q_f - Q - H^T R H \end{bmatrix} - \begin{bmatrix} B_w^T \\ (A - BH)^T \end{bmatrix} Q_f \begin{bmatrix} B_w^T \\ (A - BH)^T \end{bmatrix}^T \geq 0 \quad (3.303)$$

By using Schur's complement, we can obtain the following matrix inequality:

$$\begin{bmatrix} R_w & 0 & B_w^T \\ 0 & Q_f - Q - H^T R H & (A - BH)^T \\ B_w & (A - BH) & Q_f^{-1} \end{bmatrix} \geq 0 \qquad (3.304)$$

Multiplying both sides of (3.304) by the matrix diag$\{I, Q_f^{-1}, I\}$ yields

$$\begin{bmatrix} R_w & 0 & B_w^T \\ 0 & X - XQX - XH^T R H X & X(A - BH)^T \\ B_w & (A - BH)X & X \end{bmatrix} \geq 0 \qquad (3.305)$$

where $Q_f^{-1} = X$. Since the matrix in (3.305) is decomposed as

$$\begin{bmatrix} R_w & 0 & B_w^T \\ 0 & X & (AX-BY)^T \\ B_w & AX-BY & X \end{bmatrix} - \begin{bmatrix} 0 & 0 \\ XQ^{\frac{1}{2}} & YR^{\frac{1}{2}} \\ 0 & 0 \end{bmatrix} \begin{bmatrix} 0 & 0 \\ XQ^{\frac{1}{2}} & YR^{\frac{1}{2}} \\ 0 & 0 \end{bmatrix}^T$$

we have

$$\begin{bmatrix} R_w & 0 & B_w^T & 0 & 0 \\ 0 & X & (AX-BY)^T & XQ^{\frac{1}{2}} & YR^{\frac{1}{2}} \\ B_w & AX-BY & X & 0 & 0 \\ 0 & Q^{\frac{1}{2}}X & 0 & I & 0 \\ 0 & R^{\frac{1}{2}}Y^T & 0 & 0 & I \end{bmatrix} \geq 0 \qquad (3.306)$$

where $Y = HX$.

3.5.2 Receding Horizon Linear Quadratic Control via Batch and Linear Matrix Inequality Forms

In the previous section, the receding horizon LQ control was obtained analytically in a closed form, and thus it can be easily computed. Here, how to achieve the receding horizon LQ control via an LMI is discussed, which will be utilized later in constrained systems.

Free Terminal Cost

The state equation in (3.3) can be written as

$$X_k = Fx_k + HU_k \qquad (3.307)$$

$$U_k = \begin{bmatrix} u_k \\ u_{k+1} \\ \vdots \\ u_{k+N-1} \end{bmatrix}, \quad X_k = \begin{bmatrix} x_k \\ x_{k+1} \\ \vdots \\ x_{k+N-1} \end{bmatrix}, \quad F = \begin{bmatrix} I \\ A \\ \vdots \\ A^{N-1} \end{bmatrix} \qquad (3.308)$$

$$H = \begin{bmatrix} 0 & 0 & 0 & \cdots & 0 \\ B & 0 & 0 & \cdots & 0 \\ AB & B & 0 & \cdots & 0 \\ \vdots & \vdots & \ddots & \vdots & \vdots \\ A^{N-2}B & A^{N-3}B & \cdots & B & 0 \end{bmatrix} \qquad (3.309)$$

The terminal state is given by

$$x_{k+N} = A^N x_k + \bar{B}U_k \qquad (3.310)$$

where

$$\bar{B} = \begin{bmatrix} A^{N-1}B & A^{N-2}B & \cdots & B \end{bmatrix} \qquad (3.311)$$

Let us define

$$\bar{Q}_N = \text{diag}\overbrace{\{Q,\cdots,Q\}}^{N}, \quad \bar{R}_N = \text{diag}\overbrace{\{R,\cdots,R\}}^{N} \tag{3.312}$$

Then, the cost function (3.22) can be rewritten by

$$J(x_k, U_k) = [X_k - X_k^r]^T \bar{Q}_N [X_k - X_k^r] + U_k^T \bar{R}_N U_k \\ + (x_{k+N} - x_{k+N}^r)^T Q_f (x_{k+N} - x_{k+N}^r)$$

where

$$X_k^r = \begin{bmatrix} x_k^r \\ x_{k+1}^r \\ \vdots \\ x_{k+N-1}^r \end{bmatrix}$$

From (3.307) and (3.310), the above can be represented by

$$\begin{aligned} J(x_k, U_k) &= [Fx_k + HU_k - X_k^r]^T \bar{Q}_N [Fx_k + HU_k - X_k^r] + U_k^T \bar{R}_N U_k \\ &\quad + [A^N x_k + \bar{B}U_k - x_{k+N}^r]^T Q_f [A^N x_k + \bar{B}U_k - x_{k+N}^r] \\ &= U_k^T [H^T \bar{Q}_N H + \bar{R}_N] U_k + 2[Fx_k - X_k^r]^T \bar{Q}_N H U_k \\ &\quad + [Fx_k - X_k^r]^T \bar{Q}_N [Fx_k - X_k^r] \\ &\quad + [A^N x_k + \bar{B}U_k - x_{k+N}^r]^T Q_f [A^N x_k + \bar{B}U_k - x_{k+N}^r] \\ &= U_k^T W U_k + w^T U_k + [Fx_k - X_k^r]^T \bar{Q}_N [Fx_k - X_k^r] \\ &\quad + [A^N x_k + \bar{B}U_k - x_{k+N}^r]^T Q_f [A^N x_k + \bar{B}U_k - x_{k+N}^r] \end{aligned} \tag{3.313}$$

where $W = H^T \bar{Q}_N H + \bar{R}_N$ and $w = 2H^T \bar{Q}_N^T [Fx_k - X_k^r]$. The optimal input can be obtained by taking $\frac{\partial J(x_k, U_k)}{\partial U_k}$. Thus we have

$$\begin{aligned} U_k &= -[W + \bar{B}^T Q_f \bar{B}]^{-1} [w + \bar{B}^T Q_f (A^N x_k - x_{k+N}^r)] \\ &= -[W + \bar{B}^T Q_f \bar{B}]^{-1} [H^T \bar{Q}_N (Fx_k - X_k^r) \\ &\quad + \bar{B}^T Q_f (A^N x_k - x_{k+N}^r)] \end{aligned} \tag{3.314}$$

The RHC can be obtained as

$$u_k = \begin{bmatrix} 1, 0, \cdots, 0 \end{bmatrix} U_k^* \tag{3.315}$$

In order to obtain an LMI form, we decompose the cost function (3.313) into two parts

$$J(x_k, U_k) = J_1(x_k, U_k) + J_2(x_k, U_k)$$

where

$$J_1(x_k, U_k) = U_k^T W U_k + w^T U_k + [Fx_k - X_k^r]^T \bar{Q}_N [Fx_k - X_k^r]$$
$$J_2(x_k, U_k) = (A^N x_k + \bar{B}U_k - x_{k+N}^r)^T Q_f (A^N x_k + \bar{B}U_k - x_{k+N}^r)$$

Assume that

$$U_k^T W U_k + w^T U_k + [F x_k - X_k^r]^T \bar{Q}_N [F x_k - X_k^r] \le \gamma_1 \quad (3.316)$$
$$(A^N x_k + \bar{B} U_k - x_{k+N}^r)^T Q_f (A^N x_k + \bar{B} U_k - x_{k+N}^r) \le \gamma_2 \quad (3.317)$$

Note that

$$J(x_k, U_k) \le \gamma_1 + \gamma_2 \quad (3.318)$$

From Schur's complement, (3.316) and (3.317) are equivalent to

$$\begin{bmatrix} \gamma_1 - w^T U_k - [F x_k - X_k^r]^T \bar{Q}_N [F x_k - X_k^r] & U_k^T \\ U_k & W^{-1} \end{bmatrix} \ge 0 \quad (3.319)$$

and

$$\begin{bmatrix} \gamma_2 & [A^N x_k + \bar{B} U_k - x_{k+N}^r]^T \\ [A^N x_k + \bar{B} U_k - x_{k+N}^r] & Q_f^{-1} \end{bmatrix} \ge 0 \quad (3.320)$$

respectively. Finally, the optimal solution U_k^* can be obtained by an LMI problem as follows:

$$\min_{U_k} \quad \gamma_1 + \gamma_2 \quad \text{subject to} \quad (3.319) \text{ and } (3.320)$$

Therefore, the RHC in a batch form is obtained by

$$u_k = \begin{bmatrix} 1, 0, \cdots, 0 \end{bmatrix} U_k^* \quad (3.321)$$

Terminal Equality Constraint

The optimal control (3.314) can be rewritten by

$$
\begin{aligned}
U_k &= - \left[\begin{bmatrix} H \\ \bar{B} \end{bmatrix}^T \begin{bmatrix} \bar{Q}_N & 0 \\ 0 & Q_f \end{bmatrix} \begin{bmatrix} H \\ \bar{B} \end{bmatrix} + \bar{R}_N \right]^{-1} \begin{bmatrix} H \\ \bar{B} \end{bmatrix}^T \begin{bmatrix} \bar{Q}_N & 0 \\ 0 & Q_f \end{bmatrix} \\
&\quad \times \left[\begin{bmatrix} F \\ A^N \end{bmatrix} x_k - \begin{bmatrix} X_k^r \\ x_{k+N}^r \end{bmatrix} \right] \\
&= -\bar{R}_N^{-1} \left[\begin{bmatrix} H \\ \bar{B} \end{bmatrix}^T \begin{bmatrix} \bar{Q}_N & 0 \\ 0 & Q_f \end{bmatrix} \begin{bmatrix} H \\ \bar{B} \end{bmatrix} \bar{R}_N^{-1} + I \right]^{-1} \begin{bmatrix} H \\ \bar{B} \end{bmatrix}^T \begin{bmatrix} \bar{Q}_N & 0 \\ 0 & Q_f \end{bmatrix} \\
&\quad \times \left[\begin{bmatrix} F \\ A^N \end{bmatrix} x_k - \begin{bmatrix} X_k^r \\ x_{k+N}^r \end{bmatrix} \right] \quad (3.322)
\end{aligned}
$$

We define

$$\bar{H} = \begin{bmatrix} H \\ \bar{B} \end{bmatrix} \quad \bar{F} = \begin{bmatrix} F \\ A^N \end{bmatrix} \quad \bar{X}_k^r = \begin{bmatrix} X_k^r \\ x_{k+N}^r \end{bmatrix} \quad (3.323)$$

Then, using the formula $(I + AB)^{-1} A = A(I + BA)^{-1}$, we have

$$U_k = -\bar{R}_N^{-1}\bar{H}^T[\hat{Q}_N\bar{H}\bar{R}_N^{-1}\bar{H}^T + I]^{-1}\hat{Q}_N[\bar{F} - I]\begin{bmatrix} x_k \\ \bar{X}_k^r \end{bmatrix}$$

$$= -\bar{R}_N^{-1}\bar{H}^T[\tilde{Q}_{N2}\bar{H}\bar{R}_N^{-1}\bar{H}^T + \tilde{Q}_{N1}^{-1}]^{-1}\tilde{Q}_{N2}[\bar{F} - I]\begin{bmatrix} x_k \\ \bar{X}_k^r \end{bmatrix} \quad (3.324)$$

where

$$\hat{Q}_N = \begin{bmatrix} \bar{Q}_N & 0 \\ 0 & Q_f \end{bmatrix} = \begin{bmatrix} I & 0 \\ 0 & Q_f \end{bmatrix}\begin{bmatrix} \bar{Q}_N & 0 \\ 0 & I \end{bmatrix} = \tilde{Q}_{N1}\tilde{Q}_{N2} \quad (3.325)$$

For terminal equality constraint, we take $Q_f = \infty I$ ($Q_f^{-1} = 0$). So U_k is given as (3.324) with $\tilde{Q}_{N1}^{-1}$ replaced by $\begin{bmatrix} I & 0 \\ 0 & 0 \end{bmatrix}$.

We introduce an LMI-based solution. In a fixed terminal case, (3.317) is not used. Instead, the condition $A^N x_k + \bar{B}U_k = x_{k+N}^r$ should be met. Thus, we need an equality condition together with an LMI. In order to remove the equality representation, we parameterize U_k in terms of known variables according to Theorem A.3. We can set U_k as

$$U_k = -\bar{B}^{-1}(A^N x_k - x_{k+N}^r) + M\hat{U}_k \quad (3.326)$$

where $\bar{B}^{-1}$ is the right inverse of $\bar{B}$ and columns of M are orthogonal to each other, spanning the null space of $\bar{B}$.

From (3.316) we have

$$\begin{aligned}
J(x_k, U_k) &= U_k^T W U_k + w^T U_k + [Fx_k - X_k^r]^T \bar{Q}_N[Fx_k - X_k^r] \\
&= (-\bar{B}^{-1}(A^N x_k - x_{k+N}^r) + M\hat{U}_k)^T W(-\bar{B}^{-1}(A^N x_k - x_{k+N}^r) \\
&\quad + M\hat{U}_k) + w^T(-\bar{B}^{-1}(A^N x_k - x_{k+N}^r) + M\hat{U}_k) + [Fx_k - X_k^r]^T \bar{Q}_N \\
&\quad \times [Fx_k - X_k^r] \\
&= \hat{U}_k^T \mathcal{V}_1 \hat{U}_k + \mathcal{V}_2 \hat{U}_k + \mathcal{V}_3
\end{aligned}$$

where

$$\begin{aligned}
\mathcal{V}_1 &= M^T W M \\
\mathcal{V}_2 &= -2(A^N x_k - x_{k+N}^r)^T \bar{B}^{-T} W M + w^T M \\
\mathcal{V}_3 &= (A^N x_k - x_{k+N}^r)^T \bar{B}^{-T} W B^{-1}(A^N x_k - x_{k+N}^r) \\
&\quad + [Fx_k - X_k^r]^T \bar{Q}_N[Fx_k - X_k^r] - w^T \bar{B}^{-1}(A^N x_k - x_{k+N}^r)
\end{aligned}$$

The optimal input can be obtained by taking $\frac{\partial J(x_k, \hat{U}_k)}{\partial \hat{U}_k}$. Thus we have

$$\hat{U}_k = -\mathcal{V}_1^{-1}\mathcal{V}_2^T$$

The RHC in a batch form can be obtained as in (3.315). The optimal control for the fixed terminal case can be obtained from the following inequality:

$$J(x_k, \hat{U}_k) = \hat{U}_k^T \mathcal{V}_1 \hat{U}_k + \mathcal{V}_2 \hat{U}_k + \mathcal{V}_3 \leq \gamma_1$$

which can be transformed into the following LMI:

$$\min \gamma_1$$

$$\begin{bmatrix} \gamma_1 - \mathcal{V}_2 \hat{U}_k - \mathcal{V}_3 & -\hat{U}_k^T \mathcal{V}_1^{\frac{1}{2}} \\ -\mathcal{V}_1^{\frac{1}{2}} \hat{U}_k & I \end{bmatrix} \geq 0$$

where $\hat{U}_k$ is obtained. U_k is computed from this according to (3.326). What remains to do is just to pick up the first one among U_k as in (3.321).

GPC for the CARIMA model (3.194) can be obtained in a batch form similar to that presented above. From the state-space model (3.200), we have

$$y_{k+j} = \bar{C}\bar{A}^j x_k + \sum_{i=0}^{j-1} \bar{C}\bar{A}^{j-i-1}\bar{B}\triangle u_{k+i} \tag{3.327}$$

The performance index (3.204) can be represented by

$$J = \left[Y_k^r - V x_k - W\triangle U_k\right]^T \bar{Q}\left[Y_k^r - V x_k - W\triangle U_k\right] + \triangle U_k^T \bar{R}\triangle U_k$$
$$+ \left[Y_{k+N_c}^r - V_f x_k - W_f\triangle U_k\right]^T \bar{Q}_f \left[Y_{k+N_c}^r - V_f x_k - W_f\triangle U_k\right] \tag{3.328}$$

where

$$Y_k^r = \begin{bmatrix} y_{k+1}^r \\ \vdots \\ y_{k+N_c}^r \end{bmatrix}, \quad V = \begin{bmatrix} \bar{C}\bar{A} \\ \vdots \\ \bar{C}\bar{A}^{N_c} \end{bmatrix}, \quad \triangle U_k = \begin{bmatrix} \triangle u_k \\ \vdots \\ \triangle u_{k+N_c-1} \end{bmatrix}$$

$$Y_{k+N_c}^r = \begin{bmatrix} y_{k+N_c+1}^r \\ \vdots \\ y_{k+N_p}^r \end{bmatrix}, \quad V_f = \begin{bmatrix} \bar{C}\bar{A}^{N_c+1} \\ \vdots \\ \bar{C}\bar{A}^{N_p} \end{bmatrix}, \quad W = \begin{bmatrix} \bar{C}\bar{B} & \cdots & 0 \\ \vdots & \ddots & \vdots \\ \bar{C}\bar{A}^{N_c-1}\bar{B} & \cdots & \bar{C}\bar{B} \end{bmatrix}$$

$$W_f = \begin{bmatrix} \bar{C}\bar{A}^{N_c}\bar{B} & \cdots & \bar{C}\bar{A}\bar{B} \\ \vdots & \ddots & \vdots \\ \bar{C}\bar{A}^{N_p-1}\bar{B} & \cdots & \bar{C}\bar{A}^{N_p-N_c}\bar{B} \end{bmatrix}, \quad \bar{R} = [\mathrm{diag}(\overbrace{r\ r\ \cdots\ r}^{N_c})]$$

$$\bar{Q}_f = [\mathrm{diag}(\overbrace{q_f\ q_f\ \cdots\ q_f}^{N_p-N_c})], \quad \bar{Q} = [\mathrm{diag}(\overbrace{q\ q\ \cdots\ q}^{N_c})].$$

Using

$$\frac{\partial J}{\partial \triangle U_k} = 0$$

we can obtain

$$\triangle U_k = \left[W^T \bar{Q}W + W_f^T \bar{Q}_f W_f + \bar{R}\right]^{-1}\left\{W^T \bar{Q}\left[Y_k^r - V x_k\right]\right.$$

$$\left. + W_f^T \bar{Q}_f \left[Y_{k+N_c}^r - V_f x_k\right]\right\}$$

Therefore, $\triangle u_k$ is given by

$$
\triangle u_k = \begin{bmatrix} I & 0 & \cdots & 0 \end{bmatrix} \left[W^T \bar{Q} W + W_f^T \bar{Q}_f W_f + \bar{R} \right]^{-1} \left\{ W^T \bar{Q} \left[Y_k^r - V x_k \right] \right.
$$

$$
\left. + W_f^T \bar{Q}_f \left[Y_{k+N_c}^r - V_f x_k \right] \right\}
\tag{3.329}
$$

3.5.3 Receding Horizon H_∞ Control via Batch and Linear Matrix Inequality Forms

In the previous section, the receding horizon H_∞ control was obtained analytically in a closed form and thus it can be easily computed. Here, how to achieve the receding horizon H_∞ control via LMI is discussed.

The state equation (3.9) can be represented by

$$
X_k = F x_k + H U_k + H_w W_k
\tag{3.330}
$$

where H_w is given by

$$
H_w = \begin{bmatrix}
0 & 0 & 0 & \cdots & 0 \\
G & 0 & 0 & \cdots & 0 \\
AG & G & 0 & \cdots & 0 \\
\vdots & \vdots & \ddots & \vdots & \vdots \\
A^{N-2}G & A^{N-3}G & \cdots & G & 0
\end{bmatrix}
\tag{3.331}
$$

and U_k, F, X_k, and H are defined in (3.308) and (3.309).

The H_∞ performance criterion can be written in terms of the augmented matrix as

$$
\begin{aligned}
J(x_k, U_k, W_k) &= [F x_k + H U_k + H_w W_k - X_k^r]^T \bar{Q}_N [F x_k + H U_k + H_w W_k \\
&\quad - X_k^r] + [A^N x_k + \bar{B} U_k + \bar{G} W_k - x_{k+N}^r]^T Q_f [A^N x_k + \bar{B} U_k \\
&\quad + \bar{G} W_k - x_{k+N}^r] + U_k^T \bar{R}_N U_k - \gamma^2 W_k^T W_k
\end{aligned}
$$

Representing $J(x_k, U_k, W_k)$ in quadratic form with respect to W_k yields the following equation:

$$
\begin{aligned}
J(x_k, U_k, W_k) &= W_k^T \mathcal{V}_1 W_k + 2 W_k^T \mathcal{V}_2 + [F x_k + H U_k - X_k^r]^T \bar{Q}_N [F x_k + H U_k \\
&\quad - X_k^r] + U_k^T \bar{R}_N U_k + [A^N x_k + \bar{B} U_k - x_{k+N}^r]^T Q_f [A^N x_k + \bar{B} U_k \\
&\quad - x_{k+N}^r] \\
&= [\mathcal{V}_1 W_k + \mathcal{V}_2]^T \mathcal{V}_1^{-1} [\mathcal{V}_1 W_k + \mathcal{V}_2] - \mathcal{V}_2^T \mathcal{V}_1^{-1} \mathcal{V}_2 + U_k^T \bar{R}_N U_k \\
&\quad + [F x_k + H U_k - X_k^r]^T \bar{Q}_N [F x_k + H U_k - X_k^r] \\
&\quad + [A^N x_k + \bar{B} U_k - x_{k+N}^r]^T Q_f [A^N x_k + \bar{B} U_k - x_{k+N}^r] \\
&= [\mathcal{V}_1 W_k + \mathcal{V}_2]^T \mathcal{V}_1^{-1} [\mathcal{V}_1 W_k + \mathcal{V}_2] + U_k^T \mathcal{P}_1 U_k + 2 U_k^T \mathcal{P}_2 \\
&\quad + \mathcal{P}_3
\end{aligned}
\tag{3.332}
$$

where

$$\mathcal{V}_1 \overset{\triangle}{=} -\gamma^2 I + \bar{G}^T Q_f \bar{G} + H_w^T \bar{Q}_N H_w \tag{3.333}$$

$$\mathcal{V}_2 \overset{\triangle}{=} H_w^T \bar{Q}_N^T [F x_k + H U_k - X_k^r] + \bar{G}^T Q_f^T [A^N x_k + \bar{B} U_k - x_{k+N}^r] \tag{3.334}$$

$$\mathcal{P}_1 \overset{\triangle}{=} -(H_w^T \bar{Q}_N^T H + \bar{G}^T Q_f^T \bar{B})^T \mathcal{V}_1^{-1} (H_w^T \bar{Q}_N^T H + \bar{G}^T Q_f^T \bar{B})$$
$$+ H^T \bar{Q}_N H + \bar{R}_N + \bar{B}^T Q_f \bar{B} \tag{3.335}$$

$$\mathcal{P}_2 \overset{\triangle}{=} -(H_w^T \bar{Q}_N^T H + \bar{G}^T Q_f^T \bar{B})^T \mathcal{V}_1^{-1} (H_w^T \bar{Q}_N^T (F x_k - X_k^r)$$
$$+ \bar{G}^T Q_f^T (A^N x_k - x_{k+N}^r)) + H^T \bar{Q}_N F x_k + \bar{B}^T Q_f A^N x_k \tag{3.336}$$

and $\mathcal{P}_3$ is a constant that is independent of U_k and W_k.

In order that the solution to the saddle point exists, $\mathcal{V}_1$ must be negative. Thus, we have

$$-\gamma^2 I + \bar{G}^T Q_f \bar{G} + H_w^T \bar{Q}_N H_w < 0$$

In order to maximize (3.332) with respect to W_k, we have only to maximize

$$[\mathcal{V}_1 W_k + \mathcal{V}_2]^T \mathcal{V}_1^{-1} [\mathcal{V}_1 W_k + \mathcal{V}_2] \tag{3.337}$$

to obtain

$$W_k = -\mathcal{V}_1^{-1} \mathcal{V}_2 \tag{3.338}$$

If we put (3.338) into (3.332), (3.332) can be represented by

$$J(x_k, U_k, W_k) = U_k^T \mathcal{P}_1 U_k + 2 U_k^T \mathcal{P}_2 + \mathcal{P}_3 \tag{3.339}$$

Then the optimal input can be obtained by taking $\frac{\partial J(x_k, U_k, W_k)}{\partial U_k}$. Thus we have

$$U_k = -\mathcal{P}_1^{-1} \mathcal{P}_2$$

Now we can introduce an LMI form for the receding horizon H_∞ control. In order to maximize (3.332) with respect to W_k, we have only to minimize

$$-[\mathcal{V}_1 W_k + \mathcal{V}_2]^T \mathcal{V}_1^{-1} [\mathcal{V}_1 W_k + \mathcal{V}_2] \tag{3.340}$$

Then we try to minimize (3.339). It follows finally that we have the following LMI:

$$\min_{U_k, W_k} \quad r_1 + r_2 \tag{3.341}$$

$$\begin{bmatrix} r_1 - \mathcal{P}_2^T U_k & U_k^T \\ U_k & \mathcal{P}_1^{-1} \end{bmatrix} \geq 0$$

$$\begin{bmatrix} r_2 & (\mathcal{V}_1 W_k + \mathcal{V}_2)^T \\ (\mathcal{V}_1 W_k + \mathcal{V}_2) & -\mathcal{V}_1 \end{bmatrix} \geq 0$$

The stabilizing RH H_∞ control can be obtained by solving the semidefinite program (3.306) and (3.341) where $Q_f = X^{-1}$. What remains to do is just to pick up the first one among U_k as in (3.321).

An LMI representation in this section would be useful for constrained systems.

3.6 References

In order to explain the receding horizon concept, the predictor form and the reference predictive form are introduced first in Section 3.2.1 of this chapter.

The primitive form of the RH control was given in [Kle70] [Kle74], where only input energy with fixed terminal constraint is concerned without the explicit receding horizon concept. The general form of the RHC was first given with receding horizon concepts in [KP77a], where state weighting is considered. The RHTC presented in Section 3.3.1 is similar to that in [KB89].

With a terminal equality constraint which corresponds to the infinite terminal weighting matrix, the closed-loop stability of the RHC was first proved in a primitive form [Kle70] and in a general form [KP77a]. There are also some other results in [Kle74] [KP78] [AM80] [NS97].

The terminal equality constraint in Theorem 3.1 is a well-known result.

Since the terminal equality constraint is somewhat strong, finite terminal weighting matrices for the free terminal cost have been investigated in [Yaz84] [BGP85] [KB89] [PBG88] [BGW90] [DC93] [NP97] [LKC98]. The monotone property of the Ricatti equation is used for the stability [KP77a]. Later, the monotone property of the optimal cost was introduced not only for linear, but also for nonlinear systems. At first, the cost monotonicity condition was used for the terminal equality constraint [KRC92] [SC94][RM93][KBM96] [LKL99]. The cost monotonicity condition for free terminal cost in Theorem 3.2 is first given in [LKC98]. The general proof of Theorem 3.2 is a discrete version of [KK00]. The inequality (3.84) is a special case of (3.73) and is partly studied in [KB89] [BGW90]. The terminal equality constraint comes historically before the free terminal cost. The inequality between the terminal weighting matrix and the steady-state Riccati solution in Proposition 3.3 appeared first in this book.

The opposite direction of the cost monotonicity in Theorem 3.4 is first introduced for discrete systems in this book. It is shown in [BGW90] that once the monotonicity of the Riccati equation holds at a certain point it holds for all subsequent times as in Theorem 3.5.

The stability of RHCs in Theorems 3.6 and 3.7 is first introduced in [LKC98] and the general proofs of these theorems in this book are discrete versions of [KK00].

The stability of the RHC in the case of the terminal equality constraint in Theorem 3.7 is derived by using Theorems 3.1 and 3.6.

A stabilizing control in Theorem 3.9 is first introduced in [LKC98] without a proof, and thus a proof is included in this book by using Lyapunov theory.

The observability in Theorems 3.6 and 3.9 can be weakened with detectability, similar to that in [KK00].

The results on Theorems 3.10 and 3.11 appear first in this book and are extensions of [KP77a]. For time-invariant systems, the controllability in Theorems 3.10 and 3.11 can be weakened with stabilizability, as shown in [RM93] and [KP77a]. The closed-loop stability of the RHC via FARE appears in [PBG88, BGW90]

The lower bound of the horizon size stabilizing the system in Theorem 3.12 appeared in [JHK04].

The RH LQ control with a prescribed degree of stability appeared in [KP77b] for continuous-time systems. In Section 3.3.5 of this book, slight modifications are made to obtain it for discrete-time systems.

The upper and lower bounds of the performance criteria in Theorems 3.13 and 3.14 are discrete versions of the result [KBK83] for continuous-time systems.

It was shown in [KBK83] that the RH LQ control stabilizes the system for a sufficiently large horizon size irrespective of the final weighting matrix.

The RH H_∞ control presented in Section 3.4.1 is a discrete version of the work by [KYK01]. The cost monotonicity condition of the RH H_∞ control in Theorems 3.15, 3.16, and 3.17 is a discrete version of the work by [KYK01]. The stability of the RH H_∞ control in Theorems 3.18 and 3.19 also appeared in [KYK01]. The free terminal cost in the above theorems was proposed in [LG94] [LKL99]. The relation between the free terminal cost and the monotonicity of the saddle point value was fully discussed in [KYK01].

The RH H_∞ control without requiring the observability of $(A, Q^{\frac{1}{2}})$, as in Theorems 3.20 and 3.21, is first discussed in this book in parallel with the RH LQ control.

The guaranteed H_∞ norm of the H_∞ RHC in Theorem 3.22 is first given in this book for discrete-time systems by a modification of the result on continuous-time systems in [KYK01].

In [LKC98], how to obtain the receding horizon control and a final weighting matrix satisfying the cost monotonicity condition was discussed by using LMIs. Sections 3.5.1 and 3.5.2 are mostly based on [LKC98].

The RHLQC with the equality constraint and the cost monotonicity condition for the H_∞ RHC in an LMI form appear first in Sections 3.5.2 and 3.5.3 of this book respectively.

3.7 Problems

3.1. Referring to Problem 2.6, make simulations for three kinds of planning based on Table 1.1. α, γ, β, $\bar{u}$ are set to 0.8, 1.3, 10, and 1 respectively. For long-term planning, use $N = 100$. For periodic and short-term planning, use $N = 5$ and a simulation time of 100.

3.2. Derive a cost monotonicity condition for the following performance criterion for the system (3.1):

$$J(x_{i_0}, u_\cdot) = \sum_{i=i_0}^{i_f-1} \begin{bmatrix} x_i \\ u_i \end{bmatrix}^T \begin{bmatrix} Q & S \\ S^T & R \end{bmatrix} \begin{bmatrix} x_i \\ u_i \end{bmatrix} + x_{i_f}^T Q_f x_{i_f}$$

3.3. (1) If Q_f satisfies a cost monotonicity condition, show that the RHC with this Q_f can be an infinite horizon optimal control (2.107) with some nonnegative symmetric Q and some positive definite R.

(2) Verify that the RHC with the equality constraint has the property that it is an infinite horizon optimal control (2.107) associated with some nonnegative symmetric Q and some positive definite R.

3.4. Consider the cost monotonicity condition (3.73).

(1) Show that the condition (3.73) can be represented as

$$Q_f \geq \min_{H} \left\{ Q + H^T R H + (A - BH)^T Q_f (A - BH) \right\} \quad (3.342)$$

(2) Choose H so that the right side of (3.342) is minimized.

3.5. Consider a discrete-time system as

$$x_{i+1} = \begin{bmatrix} 0 & 0 \\ 1 & 0 \end{bmatrix} x_i + \begin{bmatrix} 1 \\ 0 \end{bmatrix} u_i \quad (3.343)$$

(1) Find an RHC for the following performance criterion:

$$x_{k+1|k}^T \begin{bmatrix} 1 & 2 \\ 2 & 6 \end{bmatrix} x_{k+1|k} + u_{k|k}^2 \quad (3.344)$$

where the horizon size is 1. Check the stability.

(2) Find an RHC for the following performance criterion:

$$\sum_{j=0}^{1} \{ x_{k+j|k}^T \begin{bmatrix} 1 & 2 \\ 2 & 6 \end{bmatrix} x_{k+j|k} + u_{k+j|k}^2 \} + x_{k+2|k}^T \begin{bmatrix} 1 & 0 \\ 0 & 0 \end{bmatrix} x_{k+2|k} \quad (3.345)$$

where the horizon size is 2. Check the stability.

(3) In the problem (b), introduce the final weighting matrix as

$$\sum_{j=0}^{1} \{ x_{k+j|k}^T \begin{bmatrix} 1 & 2 \\ 2 & 6 \end{bmatrix} x_{k+j|k} + u_{k+j|k}^2 \} + x_{k+2|k}^T Q_f x_{k+2|k} \quad (3.346)$$

and find Q_f such that the system is stabilized.

3.6. Suppose that Q_f is positive definite and the system matrix A is nonsingular.

(1) Prove that the solution to Riccati Equation (3.49) is positive definite.

(2) Let $V(x_i) = x_i^T A^{-1} (K_1^{-1} + BR^{-1}B^T) A^{-T} x_i$, where K_1 is obtained from the Riccati equation starting from $K_N = Q_f$, then show that the system can be stabilized. (Hint: use Lasalle's theorem and the fact that if A is Hurwitz, then so is A^T.)

Remark: in the above problem, the observability of $(A, Q^{\frac{1}{2}})$ is not required.

3.7. Prove the stability of the RHC (3.49) by using Lyapunov theory.

(1) Show that K_1 defined in (3.47) satisfies

$$K_1 \geq (A - BL_1)^T K_1 (A - BL_1) + L_1^T R L_1 + Q \qquad (3.347)$$

starting from $K_N = Q_f$ satisfying (3.73) and $L_1 = [R + B^T K_1 B]^{-1} B^T K_1 A$.

(2) Show that $x_i^T K_1 x_i$ in (3.347) can be a Lyapunov function. Additionally, show the stability of the RHC under assumptions that (A, B) and $(A, Q^{\frac{1}{2}})$ are stabilizable and observable respectively.

3.8. Consider the FARE (3.122). Suppose that (A, B) is stabilizable, $\bar{Q} \geq 0$, and $(A, \bar{Q}^{\frac{1}{2}})$ is observable. If $K_{i_0+2} - 2K_{i_0+1} + K_{i_0} \leq 0$ for some i_0, then the system with the RHC (3.55) is stable for any $N \geq i_0$.

3.9. * Denote the control horizon and the prediction horizon as N_c and N_p respectively. This book introduces various RHC design methods in the case of $N = N_c = N_p$. When we use different control and prediction horizons ($N_c \neq N_p$):

(1) discuss the effect on the computational burden.
(2) discuss the effect on the optimal performance.

3.10. In this chapter, $\|A\|_{\rho,\epsilon}$ is introduced.

(1) Take an example that does not satisfy the following inequality

$$\rho(AB) \leq \rho(A)\rho(B)$$

where $\rho(A)$ is the spectral radius.
(2) Show that there always exists a matrix norm $\|A\|_{\rho,\epsilon}$ such that

$$\rho(A) \leq \|A\|_{\rho,\epsilon} \leq \rho(A) + \epsilon \qquad (3.348)$$

for any $\epsilon > 0$.
(3) Disprove that $\rho(A) \leq 1$ implies $\|A\|_2 \leq 1$

3.11. Let K_i be the solution to the difference Riccati equation (2.45) and L_i its corresponding state feedback gain (2.57). K and L are the steady-state values of K_i and L_i.

(1) Show that

$$L_{i+1} - L = -R_{o,i+1}^{-1} B^T \triangle K_{i+1} A_c \qquad (3.349)$$

$$A_{c,i+1} = A - BL_{i+1} = (I - BR_{o,i+1}^{-1} B^T \triangle K_{k+1}) A_c \qquad (3.350)$$

where

$$R_{o,i+1} \overset{\triangle}{=} R + B^T K_i B, \quad \triangle K_i \overset{\triangle}{=} K_i - K, \quad A_c \overset{\triangle}{=} A - BL$$

(2) Show that

$$\triangle K_i = A_c^T [\triangle K_{i+1} - \triangle K_{i+1} B R_{o,i+1}^{-1} B^T \triangle K_{i+1}] A_c \qquad (3.351)$$

3.12. Suppose that the pair (A, B) and $(A, Q^{\frac{1}{2}})$ are controllable and observable respectively.

(1) Show that the closed-loop system can be written as

$$x_{i+1} = G_i x_i + BR^{-1}B^T \hat{K}^e_{i+1,N} A x_i \qquad (3.352)$$

with

$$G_i = A - BR^{-1}B^T \hat{K}_{i+1,\infty} A \qquad (3.353)$$

$$\hat{K}_{i+1,i+N} = [K^{-1}_{i+1,i+N} + BR^{-1}B^T]^{-1} \qquad (3.354)$$

$$\hat{K}^e_{i+1,N} = \hat{K}_{i+1,\infty} - \hat{K}_{i+1,i+N} \qquad (3.355)$$

where $K_{i+1,i+N}$ is given in (3.47) and $K_{i+1,\infty}$ is the steady-state solution of (3.47).

(2) Prove that, for all x,

$$\lim_{N \to \infty} \frac{|BR^{-1}B^T \hat{K}^e_{i+1,N} A x|}{|x|} = 0 \qquad (3.356)$$

(3) Show that there exists a finite horizon size N such that the RHC (3.56) stabilizes the closed-loop system.

Hint. Use the following fact: suppose that $x_{i+1} = f(x_i)$ is asymptotically stable and $g(x_i, i)$ satisfies the equality $\lim_{i \to \infty} \frac{g(x_i,i)}{x_i} = 0$. Then, $x_{i+1} = f(x_i) + g(x_i, i)$ is also stable.

3.13. A state-space model is given as

$$x_{i+1} = \begin{bmatrix} 2 & 1 \\ 3 & 4 \end{bmatrix} x_i + \begin{bmatrix} 2 \\ 3 \end{bmatrix} u_i \qquad (3.357)$$

where

$$Q = \begin{bmatrix} 2 & 0 \\ 0 & 2 \end{bmatrix}, \quad R = 2 \qquad (3.358)$$

(1) According to the formula (3.149), find a lower bound of the horizon size N that guarantees the stability irrespective of the final weighting matrix Q_f.

(2) Calculate a minimum horizon size stabilizing the closed-loop systems by direct computation of the Riccati equation and closed-loop poles.

3.14. MAC used the following model:

$$y_k = \sum_{i=0}^{n-1} h_i u_{k-i}$$

(1) Obtain a state-space model.

(2) Obtain an RHC with the following performance:

$$J = \sum_{j=0}^{N-1} \{q[y_{k+j|k} - y_{k+j|k}^r]^2 + ru_{k+j|k}^2\}$$

3.15. DMC used the following model:

$$y_k = \sum_{i=0}^{n-1} g_i \triangle u_{k-i}$$

where $\triangle u_k = u_k - u_{k-1}$.

(1) Obtain a state-space model.

(2) Obtain an RHC with the following performance:

$$J = \sum_{j=0}^{N-1} \{q[y_{k+j|k} - y_{k+j|k}^r]^2 + r[\triangle u_{k+j|k}]^2\}.$$

3.16. Consider the CARIMA model (3.194). Find an optimal solution for the performance criterion (3.204).

3.17. (1) Show that

$$Q_f - Q + H^T R H - \Gamma^T R_w \Gamma + (A - BH + B_w \Gamma)^T Q_f (A - BH + B_w \Gamma)$$
$$\geq Q_f - Q + H^T R H + (A - BH)(Q_f^{-1} - B_w R_w^{-1} B_w^T)^{-1}(A - BH) \quad (3.359)$$

holds irrespective of Γ.

(2) Find out Γ such that the equality holds in (3.359).

(3) Show that

$$Q_f - Q + H^T R H + (A - BH)(Q_f^{-1} - B_w R_w^{-1} B_w^T)^{-1}(A - BH) \geq 0$$

can be represented in the following LMI form:

$$\begin{bmatrix} X & (AX - BY)^T & (Q^{\frac{1}{2}}X)^T & (R^{\frac{1}{2}}Y)^T \\ AX - BY & X - B_w R_w^{-1} B_w^T & 0 & 0 \\ Q^{\frac{1}{2}}X & 0 & I & 0 \\ R^{\frac{1}{2}}Y & 0 & 0 & I \end{bmatrix} \geq 0 \quad (3.360)$$

where $X = Q_f^{-1}$ and $Y = HQ_f^{-1}$.

3.18. Consider the cost monotonicity condition (3.234) in the RH H_∞ control.

(1) Show that (3.234) is equivalent to the following performance criterion:

$$\max_w [(x^T Q x + u^T R u - r^2 w^T w) - x^T Q_f x] \leq 0 \quad (3.361)$$

(2) Show that if (3.234) holds, then the following inequality is satisfied:

$$\begin{bmatrix} -\gamma^2 I + B_w^T Q_f B_w & B_1^T Q_f (A + B_2 H) \\ (A + B_2 H)^T Q_f B_1 & (A + BH)^T Q_f (A + BH) - Q_f + Q + H^T H \end{bmatrix} \leq 0$$

3.19. If H is replaced by an optimal gain $H = -R^{-1} B^T [I + Q_f \Pi]^{-1} Q_f A$, then show that we can have (3.243) by using the matrix inversion lemma.

3.20. As shown in Figure 3.11, suppose that there exists an input uncertainty $\triangle$ described by

$$\widetilde{x}_{k+1} = \widetilde{A}\widetilde{x}_k + \widetilde{B}\widetilde{u}_k$$
$$\widetilde{y}_k = \widetilde{C}\widetilde{x}_k$$

where the feedback interconnection is given by

$$\widetilde{u}_k = u_k^{RHC}$$
$$u_k = -\widetilde{y}_k$$

The input $\widetilde{y}_k$ and output $\widetilde{u}_k$ of the uncertainty $\triangle$ satisfy

$$\mathcal{V}(\widetilde{x}_{k+1}) - \mathcal{V}(\widetilde{x}_k) \leq \widetilde{y}_k^T \widetilde{u}_k - \rho \widetilde{u}_k^T \widetilde{u}_k$$

where $\mathcal{V}(x_k)$ is some nonnegative function (this is called the dissipative property) and ρ is a constant. If ρ is greater than $\frac{1}{4}$ and the H_∞ RHC (3.222) is adopted, show that the H_∞ norm bound of the closed-loop system with this input uncertainty is still guaranteed.

Hint: use the cost monotonicity condition.

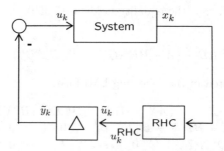

Fig. 3.11. Feedback Interconnection of Problem 3.20

3.21. The state equation (3.1) can be transformed into

$$X_{k+j} = F_j x_{k+j} + H_j U_{k+j}$$
$$x_{k+N} = A^{N-j} x_{k+j} + \bar{B}_j U_{k+j}$$
$$\bar{X}_{k+j} = \bar{F}_j x_{k+j} + \bar{H}_j U_{k+j}$$

where $0 \leq j \leq N - 1$.

$$U_{k+j} = \begin{bmatrix} u_{k+j} \\ u_{k+j+1} \\ \vdots \\ u_{k+N-1} \end{bmatrix}, \quad X_{k+j} = \begin{bmatrix} x_{k+j} \\ x_{k+j+1} \\ \vdots \\ x_{k+N-1} \end{bmatrix}$$

$$F_j = \begin{bmatrix} I \\ A \\ \vdots \\ A^{N-1-j} \end{bmatrix} = \begin{bmatrix} I \\ F_{j+1}A \end{bmatrix}$$

$$H_j = \begin{bmatrix} 0 & 0 & 0 & \cdots & 0 \\ B & 0 & 0 & \cdots & 0 \\ AB & B & 0 & \cdots & 0 \\ \vdots & \vdots & & \ddots & \vdots & \vdots \\ A^{N-2-j}B & A^{N-3-j}B & \cdots & B & 0 \end{bmatrix} = \begin{bmatrix} 0 & 0 \\ F_{j+1}B & H_{j+1} \end{bmatrix}$$

$$\bar{B}_j = \begin{bmatrix} A^{N-1-j}B & A^{N-2-j}B & \cdots & AB & B \end{bmatrix}$$

$$\bar{X}_{k+j} = \begin{bmatrix} X_{k+j} \\ x_{k+N} \end{bmatrix}, \quad \bar{F}_j = \begin{bmatrix} F_j \\ A^{N-j} \end{bmatrix}, \quad \bar{H}_j = \begin{bmatrix} H_j \\ \bar{B}_j \end{bmatrix}$$

(1) We define

$$K_j = \bar{F}_j^T \hat{Q}_j \bar{F}_j - \bar{F}_j^T \hat{Q}_j \bar{H}_j (\bar{H}_j^T \hat{Q}_j \bar{H}_j + \bar{R}_j)^{-1} \bar{H}_j^T \hat{Q}_j \bar{F}_j$$

where

$$\hat{Q}_j = \mathrm{diag}\{\overbrace{Q, \cdots, Q}^{N-j+1} Q_f\}, \quad \bar{R}_j = \mathrm{diag}\{\overbrace{R, \cdots, R}^{N-j+1}\}.$$

Then, show that the optimal control (3.314) can be rewritten by

$$U_{k+j} = -[\bar{H}_j^T \hat{Q}_j \bar{H}_j + \bar{R}_j]^{-1} \bar{H}_j^T \hat{Q}_j \bar{F}_j x_{k+j} \qquad (3.362)$$

$$= \begin{bmatrix} -[R + B^T K_{j+1}B]^{-1} B^T K_{j+1}A x_{k+j} \\ -[\bar{R}_{j+1} + \bar{H}_{j+1}^T \hat{Q}_{j+1} \bar{H}_{j+1}]^{-1} \bar{H}_{j+1}^T \hat{Q}_{j+1} \bar{F}_{j+1} x_{k+j+1} \end{bmatrix}$$

$$= \begin{bmatrix} u_{k+j} \\ U_{k+j+1} \end{bmatrix}$$

(2) Show that the above-defined K_j satisfies (3.47), i.e. the recursive solution can be obtained from a batch form of solution.

3.22. Consider the GPC (3.329) for the CARIMA model (3.194).

(a) Using (3.329), obtain the GPC $\triangle u_k$ when $Q_f = \infty I$.

(b) Show that the above GPC is asymtotically stable.

3.23. In Section 2.5, the optimal control U_k on the finite horizon was obtained from the LMI approach. Derive an LMI for the control gain H of $U_k = Hx_k$, not the control U_k itself.

4

Receding Horizon Filters

4.1 Introduction

In the previous chapters we have used state feedback under the assumption that all the state variables are available. However, this assumption may not hold in practice, since the state information may not be measurable or it may cost a lot to measure all state variables. In this case we have to use output feedback controls that utilize measured inputs and outputs. A state can be estimated from measured inputs and outputs. Thus, output feedback controls can be obtained by replacing the real state in state feedback controls by the estimated state. In this chapter we shall introduce state observers, or filters that estimate the state from the measured inputs and outputs.

In the control area, filters can be used for the output feedback control, and in the signal processing area, filters can be used for the separation of real signals from noisy signals. Therefore, the results in this chapter will be useful not only for control, but also for signal processing. In Chapter 1, control models, control objectives, control structures, and control performance criteria were discussed for control designs. In this chapter, signal models, filter objectives, filter structures, and filter performance criteria are introduced, as in Figure 4.1, for filter designs.

Signal Models

Models, often called signal models, are mathematical description of physical systems under consideration. Models are constructed from the measured data about physical systems. The more *a priori* information we can incorporate into the model, the better model we can obtain. Models fall into the several classes by the type of underlying mathematical equations, i.e. linear or nonlinear, state space or I/O, deterministic or stochastic, time-invariant or varying. Undesirable elements such as disturbances, noises, and uncertainties of dynamics can be included in the model. In this chapter we focus on linear time-invariant systems that are represented by state-space models with deterministic disturbances or stochastic noises.

Filter Objectives

There can be several objectives of filter designs. Actually, the stability and the performance even under the above undesirable elements are considered important. Since it is difficult to obtain a filter under general frameworks, the filter can be designed first for simpler models, i.e. nominal systems and then for general systems, i.e. uncertain systems. Thus, the goals can be categorized into some simpler intermediate ones, such as:

- Nominal stability. The estimation error approaches zero for signal models without uncertainties.
- Nominal performance. The filter satisfies the performance criteria for signal models without uncertainties.
- Robust stability. The estimation error approaches zero for signal models with uncertainties.
- Robust performance. The filter satisfies the performance criteria for signal models with uncertainties.

Filter Structure

Filters are of various forms. Filters can process the given information linearly or nonlinearly. Linear filters are easily implemented and have less computation load than nonlinear filters.

Filters can be divided into FIR filters and IIR filters on the basis of the duration of the impulse response. Recursive IIR filters such as the conventional Kalman filters are especially popular in control areas, but they may have some undesirable problems, such as divergences. Nonrecursive FIR filters are popular in signal processing areas and may have some good properties, such as the guaranteed stability.

Filters can be independent of the initial state information. There are some cases in which initial state information is not used: the initial state information is unknown, as it often is in the real world; the measurement of the initial state is avoided due to its cost; or the initial state can be ignored because we can use certain approaches that do not require the initial state information. Kalman filters are implemented under the assumption that the information of the initial state is available. Thus, filter coefficients or gains may be dependent on the initial state, which does not make sense.

Unbiasedness has been considered to be a good property in the literature for a long time. Conventionally, this is checked after a filter is designed. However, the unbiased property can be built in to a filter during the design phase.

Performance Criterion

The performance criteria can take many forms. An optimal filter is obtained by minimizing or mini-maximizing a certain performance criterion. In the stochastic case, the performance criterion such as a minimum variance can be cast as a minimization problem. In the deterministic case, the performance criteria such as H_∞ can be posed mathematically as a mini-maximization problem. A least squares operation for the minimization performance criterion can be used for both deterministic and stochastic systems.

In this chapter we introduce either IIR or FIR filters which correspond to receding horizon controls in Chapter 3. In particular, we introduce a dual IIR filter whose gain is dual to the receding horizon control. Then we will introduce optimal FIR filters which use the receding horizon concept for the measured data. FIR filters are implemented each time with the finite recent measurement information.

For FIR filters, linearity, unbiasedness, FIR structure, and the independence of the initial state information are built in during the design phase. In the signal processing area, the FIR filter has been widely used for unmodelled signals due to its many good properties, such as guaranteed stability, linear phase(zero error), robustness to temporary parameter changes and round-off error, etc. Good properties of FIR filters for the state estimation will be investigated in this chapter.

The sections with an asterisk are provided for a general system matrix A. Complex notation is required, so that this section may be skipped for a first reading.

The organization of this chapter is as follows. In Section 4.2, dual filters to receding horizon LQ controls are introduced with inherent properties. In Section 4.3, minimum variance FIR filters are first given for the nonsingular matrix A for simplicity and then for the general (possibly singular) matrix A. Both batch and recursive forms are provided. In Section 4.4, dual filters to receding horizon H_∞ controls are introduced with inherent properties. In Section 4.5, minimax FIR filters, such as L_2-E FIR filters and H_∞ filters, are introduced.

4.2 Dual Infinite Impulse Response Filter Based on Minimum Criterion

Consider a linear discrete-time state-space model:

$$x_{i+1} = Ax_i + Bu_i + Gw_i \tag{4.1}$$
$$y_i = Cx_i + v_i \tag{4.2}$$

where $x_i \in \Re^n$ is the state, and $u_i \in \Re^l$ and $y_i \in \Re^q$ are the input and measurement respectively. At the initial time i_0 of the system, the state x_{i_0} is a random variable with a mean $\bar{x}_{i_0}$ and a covariance P_{i_0}. The system noise $w_i \in \Re^p$ and the measurement noise $v_i \in \Re^q$ are zero-mean white Gaussian and mutually uncorrelated. The covariances of w_i and v_i are denoted by Q_w

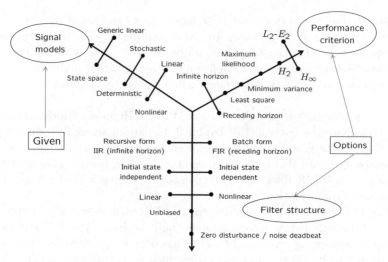

Fig. 4.1. Filter components

and R_v respectively, which are assumed to be positive definite matrices. These noises are uncorrelated with the initial state x_{i_0}. (A, C) of the system (4.1) and (4.2) is assumed to be observable, so that the stable observer can be constructed.

Before we introduce a filter for (4.1) and (4.2), we will compare the LQ control with the Kalman filter and investigate structural differences between them. The LQ control (2.48) with a zero reference signal and a terminal time $i_0 + N$ is given by

$$u_i^* = -[R + B^T K_{i+1} B]^{-1} B^T K_{i+1} A x_i \qquad (4.3)$$

where Riccati Equation (2.45) is given by

$$K_i = A^T [I + K_{i+1} B R^{-1} B^T]^{-1} K_{i+1} A + Q \qquad (4.4)$$

with

$$K_{i_0+N} = Q_f \qquad (4.5)$$

Meanwhile, the receding horizon control (3.56) is given by

$$\begin{aligned} u_i^* &= -R^{-1} B^T [I + K_1 B R^{-1} B^T]^{-1} K_1 A x_i \\ &= -[R + B^T K_1 B]^{-1} B^T K_1 A x_i \end{aligned} \qquad (4.6)$$

where K_1 is computed from (4.4) with the terminal condition $K_N = Q_f$. K_i in the control gain matrix in (4.3) is time-varying and K_1 in the gain matrix (4.6) is fixed. That is, K_1 is said to be frozen at time 1 calculating from time N in backward time.

The standard Kalman filter (2.179) can be written as

$$\hat{x}_{i+1|i} = A\hat{x}_{i|i-1} + AP_iC^T(CP_iC^T + R_v)^{-1}(y_i - C\hat{x}_{i|i-1}) \qquad (4.7)$$

where

$$P_{i+1} = AP_iA^T - AP_iC^T(R_v + CP_iC^T)^{-1}CP_iA^T + GQ_wG^T \qquad (4.8)$$
$$= A[I + P_iC^TR_v^{-1}C]^{-1}P_iA^T + GQ_wG^T$$

with the given initial condition P_{i_0}.

The Kalman filter is said to be dual to the LQC in the sense that two Riccati equations, (4.4) and (4.8), become the same each other if some variables are changed as

$$
\begin{aligned}
A &\longleftrightarrow A^T \\
B &\longleftrightarrow C^T \\
Q &\longleftrightarrow GQ_wG^T \\
R &\longleftrightarrow R_v \\
K_{i+1} &\longleftrightarrow P_i
\end{aligned}
\qquad (4.9)
$$

Note that indexes of K_{i+1} and P_i are based for backward and forward computations respectively. We can see that the filter gain in (4.7) and the control gain in (4.3) are similar to each other, in the sense that the transpose of the control gain (4.3) replaced by (4.9) is the same as the filter gain in (4.7). In that view, the corresponding closed-loop system matrices of the LQC and the Kalman filter are also similar. It is noted that properties of the Kalman filter can be obtained from properties of the LQC, since they are structurally dual.

Likewise, we can introduce the dual filter to the receding horizon LQ control as follows:

$$\hat{x}_{i+1|i} = A\hat{x}_{i|i-1} + AP_NC^T(R_v + CP_NC^T)^{-1}(y_i - C\hat{x}_{i|i-1}) + Bu_i \qquad (4.10)$$

where P_N is obtained from (4.8) starting from P_1. It turns out that the new filter (4.10) is just the Kalman filter with a frozen gain from (4.7). This filter will be called the dual filter to the RHLQC or the Kalman filter with frozen gains throughout this book. The final weighting matrix Q_f and K_1 in the RHLQC correspond to the initial covariance P_1 and P_N in the dual filter as

$$Q_f \longleftrightarrow P_1 \qquad (4.11)$$
$$K_1 \longleftrightarrow P_N \qquad (4.12)$$

Since some important properties of the receding horizon controls are obtained in Chapter 3, we can utilize those results for the investigation of the stability and additional properties of the dual filter.

Theorem 4.1. *Suppose that* (A, C) *is observable. The dual filter to the RHLQC is stabilized under the following condition:*

$$P_1 \geq AP_1A^T + GQ_wG^T - AP_1C^T(R_v + CP_1C^T)^{-1}CP_1A^T \qquad (4.13)$$

In other words, the following matrix is Hurwitz:

$$A - AP_N C^T (R_v + CP_N C^T)^{-1} C$$

Proof. The closed-loop system matrix of the filter is as follows:

$$A - AP_N C^T (CP_N C^T + R_v)^{-1} C = A(I - P_N C^T (CP_N C^T + R_v)^{-1} C) \quad (4.14)$$

where the transpose of the right side is

$$(I - C^T (CP_N C^T + R_v)^{-1} CP_N) A^T \quad (4.15)$$

The original matrix (4.14) is stable if and only if its transpose (4.15) is stable. Compare (4.15) with the closed-loop system matrix of the control

$$(I - B(B^T K_1 B + R)^{-1} B^T K_1) A \quad (4.16)$$

which is obtained with the changes listed in (4.9) and in this case $K_i = P_{N+1-i}$. Therefore, the above result follows from Theorem 3.6. This completes the proof.   ∎

The error covariance of the filter is bounded above by a constant matrix as in the following theorem.

Theorem 4.2. *The error covariance matrix of the dual filter to the RHLQC $\hat{x}_{i|i-1}$, i.e. P_i^r at time i, is bounded as the following inequality*

$$P_i^* \leq P_i^r \leq P_{N+1} + \Theta^{i-i_0} [P_{i_0}^r - P_{N+1}] \Theta^{(i-i_0)T}$$

where P_i^* is the covariance matrix of the optimal Kalman filter and $P_i^r = E[(\hat{x}_{i|i-1} - x_i)(\hat{x}_{i|i-1} - x_i)^T]$.

Proof. From the state-space model and the dual filter to the RHLQC, error dynamics are represented as

$$e_{i+1} = (A - LC)e_i - Gw_i + Lv_i \quad (4.17)$$

where $L = AP_N C^T (R_v + CP_N C^T)^{-1}$.

From the error dynamics we can obtain the propagating error covariance

$$P_{i+1}^r = (A - LC)P_i^r (A - LC) + GQ_w G^T + LR_v L^T \quad (4.18)$$

If we use the following results:

$$AP_N C^T (R_v + CP_N C^T)^{-1} CP_N A^T = LCP_N A^T$$
$$= AP_N C^T L^T = L(R_v + CP_N C^T)L^T$$

then we can easily have

$$P_{N+1} = (A - LC)P_N(A - LC)^T + GQ_wG^T + LR_vL^T \qquad (4.19)$$

Define

$$T_i \triangleq P_i^r - P_{N+1} \qquad (4.20)$$

Subtracting (4.19) from (4.18) yields

$$T_{i+1} = (A - LC)[T_i + P_{N+1} - P_N](A - LC)^T \qquad (4.21)$$
$$\leq (A - LC)T_i(A - LC)^T \qquad (4.22)$$

Therefore, we can obtain T_i by evaluating recursively, and finally we have

$$T_i \leq \Theta^{i-i_0}T_{i_0}\Theta^{(i-i_0)T}$$

where $\Theta = A - LC$. Thus, we have

$$P_i^r \leq P_{N+1} + \Theta^{i-i_0}[P_{i_0}^r - P_{N+1}]\Theta^{(i-i_0)T}$$

from which follows the result. This completes the proof.   ∎

It is noted that, as in the receding horizon control, the steady-state filter has the following bounds:

$$P_\infty^r \leq P_{N+1} \qquad (4.23)$$

since $\Theta^{i-i_0} \to 0$ as i goes to ∞.

4.3 Optimal Finite Impulse Response Filters Based on Minimum Criterion

4.3.1 Linear Unbiased Finite Impulse Response Filters

The concept of FIR filters was given in Section 1.3 and is also depicted in Figure 4.2.

The FIR filter can be represented by

$$\hat{x}_{k|k-1} = \sum_{i=k-N}^{k-1} H_{k-i}y_i + \sum_{i=k-N}^{k-1} L_{k-i}u_i \qquad (4.24)$$

for discrete-time systems, where N is a filter horizon. The IIR filter has a similar form to (4.24), with $k - N$ replaced by the initial time i_0. For IIR and FIR types, the initial state means x_{i_0} and x_{k-N} respectively. The FIR filter (4.24) does not have an initial state term and the filter gain must be independent of the initial state information.

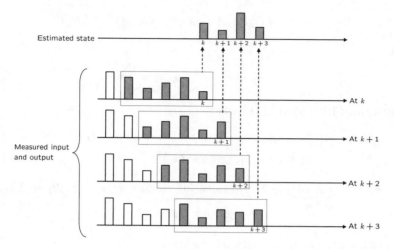

Fig. 4.2. Receding horizon filter

It is noted that a standard Kalman filter (4.7) has an initial state term and the filter gain depends on the initial state information, such as

$$\hat{x}_{k|k-1} = M_{k-i_0}x_{i_0} + \sum_{i=i_0}^{k-1} H_{k-i}y_i + \sum_{i=i_0}^{k-1} L_{k-i}u_i \qquad (4.25)$$

As can be seen in (4.24), FIR filters make use of finite measurements of inputs and outputs on the most recent time interval $[k - N, k]$, called the receding horizon or horizon.

Since filters need to be unbiased as a basic requirement, it is desirable that the linear FIR filter (4.24) must be unbiased. The unbiased condition for the FIR filter (4.24) can be

$$E[\hat{x}_{k|k-1}] = E[x_k] \qquad (4.26)$$

for any x_{k-N} and any u_i on $k - N \le i \le k - 1$.

If there exist no noises on the horizon $[k - N, k - 1]$, $\hat{x}_{k|k-1}$ and x_k become deterministic values and $\hat{x}_{k|k-1} = x_k$. This is a deadbeat property, as seen in Figure 4.3. Thus, the constraint (4.26) can be called the deadbeat condition.

Among linear FIR filters with the unbiased condition, optimal filters will be obtained to minimize the estimation error variance in the next section. These filters are called minimum variance FIR (MVF) filters.

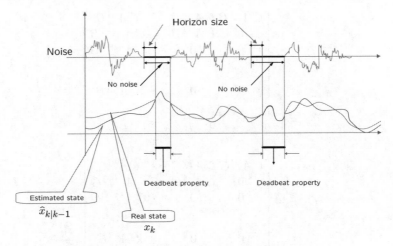

Fig. 4.3. Deadbeat property without noise

4.3.2 Minimum Variance Finite Impulse Response Filters with Nonsingular A

Batch Form

The system (4.1) and (4.2) will be represented in a batch form on the most recent time interval $[k-N, k]$, called the horizon. We assume that the system matrix A is nonsingular. The case of a general matrix A will be discussed in the next section. On the horizon $[k-N, k]$, the finite number of measurements is expressed in terms of the state x_k at the current time k as follows:

$$Y_{k-1} = \bar{C}_N x_k + \bar{B}_N U_{k-1} + \bar{G}_N W_{k-1} + V_{k-1} \qquad (4.27)$$

where

$$Y_{k-1} \triangleq [y_{k-N}^T \; y_{k-N+1}^T \; \cdots \; y_{k-1}^T]^T \qquad (4.28)$$

$$U_{k-1} \triangleq [u_{k-N}^T \; u_{k-N+1}^T \; \cdots \; u_{k-1}^T]^T \qquad (4.29)$$

$$W_{k-1} \triangleq [w_{k-N}^T \; w_{k-N+1}^T \; \cdots \; w_{k-1}^T]^T \qquad (4.30)$$

$$V_{k-1} \triangleq [v_{k-N}^T \; v_{k-N+1}^T \; \cdots \; v_{k-1}^T]^T$$

and $\bar{C}_N$, $\bar{B}_N$, $\bar{G}_N$ are obtained from

$$\bar{C}_i \triangleq \begin{bmatrix} CA^{-i} \\ CA^{-i+1} \\ CA^{-i+2} \\ \vdots \\ CA^{-1} \end{bmatrix} = \begin{bmatrix} \bar{C}_{i-1} \\ C \end{bmatrix} A^{-1} \qquad (4.31)$$

$$\bar{B}_i \triangleq - \begin{bmatrix} CA^{-1}B & CA^{-2}B & \cdots & CA^{-i}B \\ 0 & CA^{-1}B & \cdots & CA^{-i+1}B \\ 0 & 0 & \cdots & CA^{-i+2}B \\ \vdots & \vdots & \vdots & \vdots \\ 0 & 0 & \cdots & CA^{-1}B \end{bmatrix}$$

$$= \begin{bmatrix} \bar{B}_{i-1} & -\bar{C}_{i-1}A^{-1}B \\ 0 & -CA^{-1}B \end{bmatrix} \tag{4.32}$$

$$\bar{G}_i \triangleq - \begin{bmatrix} CA^{-1}G & CA^{-2}G & \cdots & CA^{-i}G \\ 0 & CA^{-1}G & \cdots & CA^{-i+1}G \\ 0 & 0 & \cdots & CA^{-i+2}G \\ \vdots & \vdots & \vdots & \vdots \\ 0 & 0 & \cdots & CA^{-1}G \end{bmatrix}$$

$$= \begin{bmatrix} \bar{G}_{i-1} & -\bar{C}_{i-1}A^{-1}G \\ 0 & -CA^{-1}G \end{bmatrix} \tag{4.33}$$

$$1 \leq i \leq N$$

The noise term $\bar{G}_N W_{k-1} + V_{k-1}$ in (4.27) can be shown to be zero-mean with covariance Ξ_N given by

$$
\begin{aligned}
\Xi_i & \triangleq \bar{G}_i [\mathrm{diag}(\overbrace{Q_w \ Q_w \ \cdots \ Q_w}^{i})]\bar{G}_i^T + [\mathrm{diag}(\overbrace{R_v \ R_v \ \cdots \ R_v}^{i})] \\
& = \begin{bmatrix} \bar{G}_{i-1}[\mathrm{diag}(\overbrace{Q_w \ Q_w \ \cdots \ Q_w}^{i-1})]\bar{G}_{i-1}^T + [\mathrm{diag}(\overbrace{R_v \ R_v \ \cdots \ R_v}^{i-1})] & 0 \\ 0 & R_v \end{bmatrix} \\
& + \begin{bmatrix} \bar{C}_{i-1} \\ C \end{bmatrix} A^{-1}GQ_wG^TA^{-T}\begin{bmatrix} \bar{C}_{i-1} \\ C \end{bmatrix}^T \\
& = \begin{bmatrix} \Xi_{i-1} & 0 \\ 0 & R_v \end{bmatrix} + \begin{bmatrix} \bar{C}_{i-1} \\ C \end{bmatrix} A^{-1}GQ_wG^TA^{-T}\begin{bmatrix} \bar{C}_{i-1} \\ C \end{bmatrix}^T \tag{4.34}
\end{aligned}
$$

An FIR filter with a batch form for the current state x_k can be expressed as a linear function of the finite measurements Y_{k-1} (4.28) and inputs U_{k-1} (4.29) on the horizon $[k-N, k]$ as follows:

$$\hat{x}_{k|k-1} = HY_{k-1} + LU_{k-1} \tag{4.35}$$

where

$$H \triangleq \begin{bmatrix} H_N & H_{N-1} & \cdots & H_1 \end{bmatrix}$$

$$L \triangleq \begin{bmatrix} L_N & L_{N-1} & \cdots & L_1 \end{bmatrix}$$

and matrices H and L will be chosen to minimize a given performance criterion later.

Equation (4.35) can be written as

$$\hat{x}_{k|k-1} = H(\bar{C}_N x_k + \bar{B}_N U_{k-1} + \bar{G}_N W_{k-1} + V_{k-1}) + LU_{k-1} \qquad (4.36)$$

Taking the expectation on both sides of (4.36), the following relations are obtained:

$$E[\hat{x}_{k|k-1}] = H\bar{C}_N E[x_k] + (H\bar{B}_N + L)U_{k-1}$$

To satisfy the unbiased condition, $E[\hat{x}_{k|k-1}] = E[x_k]$, irrespective of the input, the following constraint must be met:

$$H\bar{C}_N = I, \ HB_N = -L \qquad (4.37)$$

Substituting (4.37) into (4.36) yields

$$\hat{x}_{k|k-1} = x_k + H\bar{G}_N W_{k-1} + HV_{k-1} \qquad (4.38)$$

Thus, the estimation error can be represented as

$$e_k \overset{\triangle}{=} \hat{x}_{k|k-1} - x_k = H\bar{G}_N W_{k-1} + HV_{k-1} \qquad (4.39)$$

The objective now is to obtain the optimal gain matrix H_B, subject to the unbiasedness constraint or the deadbeat constraint (4.37), in such a way that the estimation error e_k of the estimate $\hat{x}_{k|k-1}$ has minimum variance as follows:

$$
\begin{aligned}
H_B &= \arg\min_H E[e_k^T e_k] = \arg\min_H E \ \mathrm{tr}[e_k e_k^T] \\
&= \arg\min_H \ \mathrm{tr}[H\bar{G}_N Q_N G_N^T H^T + HR_N H^T]
\end{aligned}
\qquad (4.40)
$$

where

$$Q_N = [\mathrm{diag}(\overbrace{Q_w \ Q_w \ \cdots \ Q_w}^{N})] \quad \text{and} \quad R_N = [\mathrm{diag}(\overbrace{R_v \ R_v \ \cdots \ R_v}^{N})] (4.41)$$

Before obtaining the solution to the optimization problem (4.40), we introduce a useful result on a constraint optimization.

Lemma 4.3. *Suppose that the following general trace optimization problem is given:*

$$\min_H tr\left[(HA - B)C(HA - B)^T + HDH^T\right] \qquad (4.42)$$

$$\text{subject to}$$

$$HE = F \qquad (4.43)$$

where $C = C^T > 0$, $D = D^T > 0$, $\mathrm{tr}(M)$ is the sum of the main diagonal of M, and A, B, C, D, E, and F are constant matrices and have appropriate *dimensions. The solution to the optimization problem (4.42) and (4.43) is as follows:*

$$H = \begin{bmatrix} F & B \end{bmatrix} \begin{bmatrix} (E^T \Pi^{-1} E)^{-1} E^T \Pi^{-1} \\ C A^T \Pi^{-1} (I - E(E^T \Pi^{-1} E)^{-1} E^T \Pi^{-1}) \end{bmatrix} \qquad (4.44)$$

where $\Pi \triangleq A C A^T + D$. H has an alternative representation as

$$H = \begin{bmatrix} F & B \end{bmatrix} \begin{bmatrix} W_{1,1} & W_{1,2} \\ W_{1,2}^T & W_{2,2} \end{bmatrix}^{-1} \begin{bmatrix} E^T \\ A^T \end{bmatrix} D^{-1} \qquad (4.45)$$

where

$$W_{1,1} \triangleq E^T D^{-1} E \qquad (4.46)$$

$$W_{1,2} \triangleq E^T D^{-1} A \qquad (4.47)$$

$$W_{2,2} \triangleq A^T D^{-1} A + C^{-1} \qquad (4.48)$$

Proof. For convenience, partition the matrix H in (4.43) as

$$H^T = \begin{bmatrix} h_1 & h_2 & \cdots & h_n \end{bmatrix} \qquad (4.49)$$

From $HE = F$, the sth unbiasedness constraint can be written as

$$E^T h_s = f_s, \quad 1 \leq s \leq n \qquad (4.50)$$

where f_s is the sth column vector of F. In terms of the partitioned vector h_s, the cost function (4.42) is represented as

$$\sum_{s=1}^{n} \left[(h_s^T A - b_s^T) C (h_s^T A - b_s^T)^T + h_s^T D h_s \right] \qquad (4.51)$$

where b_s is the sth column vector of B. It can be seen that the sth constraint (4.50) and sth term in the summation (4.51) are dependent only on h_s, not h_p, $p \neq s$. Thus, the optimization problem (4.43) is reduced to n independent optimization problems

$$\min_{h_s} (h_s^T A - b_s^T) C (h_s^T A - b_s^T)^T + h_s^T D h_s \qquad (4.52)$$

subject to

$$E^T h_s = f_s \qquad (4.53)$$

for $1 \leq s \leq n$. Obtaining the solutions to each optimization problem (4.53) and putting them together, we can finally obtain the solution to (4.42) and (4.43).

In order to solve the optimization problem, we establish the following cost function:

$$\phi_s \overset{\triangle}{=} (h_s^T A - b_s^T) C (h_s^T A - b_s^T)^T + h_s^T D h_s + \lambda_s^T (E^T h_s - f_s)$$
$$= h_s^T (A C A^T + D) h_s - b_s^T C A^T h_s - h_s^T A C b_s + b_s^T C b_s$$
$$+ \lambda_s^T (E^T h_s - f_s) \tag{4.54}$$

where λ_s is the sth vector of a Lagrange multiplier, which is associated with the sth unbiased constraint. Under the constraint (4.53), h_s will be chosen to optimize (4.52) with respect to h_s and λ_s for $s = 1, 2, \cdots, n$.

In order to minimize ϕ_s, two necessary conditions

$$\frac{\partial \phi_s}{\partial h_s} = 0 \quad \text{and} \quad \frac{\partial \phi_s}{\partial \lambda_s} = 0$$

are necessary, which give

$$h_s = (A C A^T + D)^{-1} (A C b_s - \frac{1}{2} E \lambda_s) = \Pi^{-1} (A C b_s - \frac{1}{2} E \lambda_s) \tag{4.55}$$

where $\Pi = A C A^T + D$ and the inverse of Π is guaranteed since $C > 0$ and $D > 0$. Pre-multiplying (4.55) by E^T, we have

$$E^T h_i = E^T \Pi^{-1} (A C b_s - \frac{1}{2} E \lambda_s) = f_s \tag{4.56}$$

where the second equality comes from (4.50). From (4.56), we can obtain

$$\lambda_s = 2 (E^T \Pi^{-1} E)^{-1} [E^T \Pi^{-1} A C b_s - f_s]$$

from which h_s is represented as

$$h_s^T = [f_s^T - b_s^T C A^T \Pi^{-1} E] (E^T \Pi^{-1} E)^{-1} E^T \Pi^{-1} + b_s^T C A^T \Pi^{-1}$$

by using (4.55). H is reconstructed from h_s as follows:

$$H = B C A^T \Pi^{-1} [I - E (E^T \Pi^{-1} E)^{-1} E^T \Pi^{-1}] + F (E^T \Pi^{-1} E)^{-1} E^T \Pi^{-1}$$
$$= [F \ B] \begin{bmatrix} (E^T \Pi^{-1} E)^{-1} E^T \Pi^{-1} \\ C A^T \Pi^{-1} (I - E (E^T \Pi^{-1} E)^{-1} E^T \Pi^{-1}) \end{bmatrix} \tag{4.57}$$

Now, we shall derive a more compact form by using variables (4.46), (4.47), and (4.48). It is first noted that the nonsingularity of $W_{1,1}$ is guaranteed to exist under the assumption that E is of a full rank. The first row block of the second block matrix in (4.57) can be represented using the defined variables (4.46)–(4.48) as

$$(E^T \Pi^{-1} E)^{-1} E^T \Pi^{-1}$$
$$= \left(E_N^T D^{-1} E - E^T D^{-1} A (A^T D^{-1} A + C^{-1})^{-1} A^T C^{-1} E \right)^{-1}$$
$$\times \left(E^T D^{-1} - E^T D^{-1} A (A^T D^{-1} A + C^{-1})^{-1} A^T D^{-1} \right)$$
$$= \left(W_{1,1} - W_{1,2} W_{2,2}^{-1} W_{1,2}^T \right)^{-1} (E^T D^{-1} - W_{1,2} W_{2,2}^{-1} A^T D^{-1}) \tag{4.58}$$

It is observed from (4.46)–(4.48) that

$$CA^T \Pi^{-1} = CA^T(D^{-1}ACA^T + I)^{-1}D^{-1}$$
$$= (A^T D^{-1}A + C^{-1})^{-1}A^T D^{-1} = W_{2,2}A^T D^{-1}$$

and

$$CA^T \Pi^{-1}E = W_{2,2}A^T D^{-1}E = W_{2,2}^{-1}W_{1,2}^T$$

from which we obtain

$$CA^T \Pi^{-1}(I - E(E^T \Pi^{-1}E)^{-1}E^T \Pi^{-1})$$
$$= W_{2,2}^{-1}A^T D^{-1}$$
$$- W_{2,2}^{-1}W_{1,2}^T(W_{1,1} - W_{1,2}W_{2,2}^{-1}W_{1,2}^T)^{-1}(E^T D^{-1} - W_{1,2}W_{2,2}^{-1}AD^{-1})$$
$$= (W_{2,2} - W_{1,2}^T W_{1,1}^{-1}W_{1,2})^{-1}W_{1,2}^T W_{1,1}^{-1}ED^{-1}$$
$$+ (W_{2,2} - W_{1,2}^T W_{1,1}^{-1}W_{1,2})^{-1}A^T D^{-1} \tag{4.59}$$

where the second equality can be derived from

$$W_{2,2}^{-1}W_{1,2}^T(W_{1,1} - W_{1,2}W_{2,2}^{-1}W_{1,2}^T)^{-1}$$
$$= W_{2,2}^{-1}W_{1,2}^T(I - W_{1,1}^{-1}W_{1,2}W_{2,2}^{-1}W_{1,2}^T)^{-1}W_{1,1}^{-1}$$
$$= W_{2,2}^{-1}(I - W_{1,2}^T W_{1,1}^{-1}W_{1,2}W_{2,2}^{-1}W_{1,2}^T)^{-1}W_{1,2}^T$$
$$= (W_{2,2} - W_{1,2}^T W_{1,1}^{-1}W_{1,2})^{-1}W_{1,2}^T W_{1,1}^{-1}$$

and

$$W_{2,2}^{-1} - W_{2,2}^{-1}W_{1,2}^T(W_{1,1} - W_{1,2}W_{2,2}^{-1}W_{1,2}^T)^{-1}W_{1,2}W_{2,2}^{-1}$$
$$= (W_{2,2} - W_{1,2}^T W_{1,1}^{-1}W_{1,2})^{-1}$$

Substituting (4.58) and (4.59) into (4.57) yields

$$\begin{bmatrix} (E^T \Pi^{-1}E)^{-1}E^T \Pi^{-1} \\ CA^T \Pi^{-1}(I - E(E^T \Pi^{-1}E)^{-1}E^T \Pi^{-1}) \end{bmatrix}$$
$$= \begin{bmatrix} (W_{1,1} - W_{1,2}W_{2,2}^{-1}W_{1,2}^T)^{-1} & -Z \\ Z^T & (W_{2,2} - W_{1,2}^T W_{1,1}^{-1}W_{1,2})^{-1} \end{bmatrix} \begin{bmatrix} E^T D^{-1} \\ A^T D^{-1} \end{bmatrix}$$
$$= \begin{bmatrix} W_{1,1} & W_{1,2} \\ W_{1,2}^T & W_{2,2} \end{bmatrix}^{-1} \begin{bmatrix} E^T D^{-1} \\ A^T D^{-1} \end{bmatrix} \tag{4.60}$$

where $Z \triangleq (W_{1,1} - W_{1,2}W_{2,2}^{-1}W_{1,2}^T)^{-1}W_{1,2}W_{2,2}^{-1}$. Therefore, the gain matrix H in (4.57) can be written as (4.45). This completes the proof. ∎

By using the result of Lemma 4.3, the solution to the optimization problem (4.40) can be obtained according to the following correspondence:

$$
\begin{array}{ccc}
A & \longleftarrow & \bar{G}_N \\
B & \longleftarrow & O \\
C & \longleftarrow & Q_N \\
D & \longleftarrow & R_N \\
E & \longleftarrow & \bar{C}_N \\
F & \longleftarrow & I
\end{array}
\tag{4.61}
$$

In the following theorem, the optimal filter gain H_B is represented in an explicit form.

Theorem 4.4.   When $\{A, C\}$ is observable and $N \geq n$, the MVF filter $\hat{x}_{k|k-1}$ with a batch form on the horizon $[k - N, k]$ is given as follows:

$$
\hat{x}_{k|k-1} = H_B(Y_{k-1} - \bar{B}_N U_{k-1})
\tag{4.62}
$$

with the optimal gain matrix H_B determined by

$$
H_B = (\bar{C}_N^T \Xi_N^{-1} \bar{C}_N)^{-1} \bar{C}_N^T \Xi_N^{-1}
\tag{4.63}
$$

where Y_{k-1}, U_{k-1}, $\bar{C}_N$, $\bar{B}_N$, and Ξ_N are given by (4.28), (4.29), (4.31), (4.32), and (4.34) respectively.   ∎

From Theorem 4.4, it can be known that the MVF filter $\hat{x}_{k|k-1}$ (4.62) processes the finite measurements and inputs on the horizon $[k - N, k]$ linearly and has the properties of unbiasedness and minimum variance *by design*. Note that the optimal gain matrix H_B (4.63) requires computation only on the interval $[0, N]$ once and is time-invariant for all horizons. This means that the MVF filter is time-invariant. It is a general rule of thumb that, due to the FIR structure, the MVF filter may also be robust against temporary modelling uncertainties or round-off errors, as seen in Example 4.1.

It is true that even though the MVF filter $\hat{x}_{k|k-1}$ (4.62) is designed with nonzero Q and R, it has the deadbeat property, i.e. $\hat{x}_{k|k-1} = x_k$, when applied to the following noise-free observable systems:

$$
x_{i+1} = Ax_i + Bu_i
\tag{4.64}
$$
$$
y_i = Cx_i
\tag{4.65}
$$

If there is no noise on the horizon $[k - N, k - 1]$, then $\hat{x}_{k|k-1}$ and x_k become deterministic. Since $\hat{x}_{k|k-1}$ is unbiased, $\hat{x}_{k|k-1}$ must be equal to x_k. Since MVF filter (4.62) is obtained with nonzero Q_w and R_v, it may be less sensitive to noises or disturbances than existing deterministic deadbeat observers that are obtained from (4.64) and (4.65), as seen in Example 4.2. The deadbeat property indicates finite convergence time, and thus fast tracking ability for noise-free systems. Therefore, an MVF filter may have a faster tracking ability than an IIR filter even with noises.

An MVF filter can be used in many problems, such as fault detection and diagnosis of various systems, manoeuvre detection and target tracking of flying objects, and model-based signal processing.

Recursive Form

In this section, the MVF filter $\hat{x}_{k|k-1}$ (4.62) with a batch form is represented in an iterative form for computational advantage in the case of nonsingular A.

First, the MVF filter $\hat{x}_{k|k-1}$ (4.62) will be represented in an iterative form. Define

$$\Omega_i \triangleq \bar{C}_i^T \Xi_i^{-1} \bar{C}_i \tag{4.66}$$

Then it can be represented in the following discrete Riccati Equation using [Lew86a]:

$$\Omega_{i+1} = \bar{C}_{i+1}^T \Xi_{i+1}^{-1} \bar{C}_{i+1}$$

$$= A^{-T} \begin{bmatrix} \bar{C}_i \\ C \end{bmatrix}^T \left(\Delta_i + \begin{bmatrix} \bar{C}_i \\ C \end{bmatrix} A^{-1} G Q_w G^T A^{-T} \begin{bmatrix} \bar{C}_i \\ C \end{bmatrix}^T \right)^{-1} \begin{bmatrix} \bar{C}_i \\ C \end{bmatrix} A^{-1}$$

$$= A^{-T} \begin{bmatrix} \bar{C}_i \\ C \end{bmatrix}^T \left(\Delta_i^{-1} - \Delta_i^{-1} \begin{bmatrix} \bar{C}_i \\ C \end{bmatrix} A^{-1} G \times \left\{ I + G^T A^{-T} \begin{bmatrix} \bar{C}_i \\ C \end{bmatrix}^T \Delta_i^{-1} \right.$$

$$\times \left. \begin{bmatrix} \bar{C}_i \\ C \end{bmatrix} A^{-1} G \right\} G^T A^{-T} \begin{bmatrix} \bar{C}_i \\ C \end{bmatrix}^T \Delta_i^{-1} \right)^{-1} \begin{bmatrix} \bar{C}_i \\ C \end{bmatrix} A^{-1}$$

$$= \left(I - A^{-T} \begin{bmatrix} \bar{C}_i \\ C \end{bmatrix}^T \Delta_i^{-1} \begin{bmatrix} \bar{C}_i \\ C \end{bmatrix} A^{-1} G \times \left\{ I + G^T A^{-T} \begin{bmatrix} \bar{C}_i \\ C \end{bmatrix}^T \Delta_i^{-1} \right. \right.$$

$$\times \left. \left. \begin{bmatrix} \bar{C}_i \\ C \end{bmatrix} A^{-1} G \right\} G^T \right) A^{-T} \begin{bmatrix} \bar{C}_i \\ C \end{bmatrix}^T \Delta_i^{-1} \begin{bmatrix} \bar{C}_i \\ C \end{bmatrix} A^{-1}$$

$$= \left(I - A^{-T}(\Omega_i + C^T R_v^{-1} C) A^{-1} G \left\{ I + G^T A^{-T}(\Omega_i + C^T R_v^{-1} C) \right. \right.$$

$$\times \left. \left. A^{-1} G \right\} G^T \right) A^{-T} \begin{bmatrix} \bar{C}_i \\ C \end{bmatrix}^T \Delta_i^{-1} \begin{bmatrix} \bar{C}_i \\ C \end{bmatrix} A^{-1}$$

$$= \left(I + A^{-T}(\Omega_i + C^T R_v^{-1} C) A^{-1} G Q_w G^T \right)^{-1}$$

$$\times A^{-T} \begin{bmatrix} \bar{C}_i \\ C \end{bmatrix}^T \Delta_i^{-1} \begin{bmatrix} \bar{C}_i \\ C \end{bmatrix} A^{-1} \tag{4.67}$$

Finally, we have

$$\Omega_{i+1} = [I + A^{-T}(\Omega_i + C^T R_v^{-1} C) A^{-1} G Q_w G^T]^{-1}$$
$$\times A^{-T}(\Omega_i + C^T R_v^{-1} C) A^{-1} \tag{4.68}$$

Since Ω_1 in the batch form (4.66) can be expressed as

$$\Omega_1 = \bar{C}_1^T \Xi_1^{-1} \bar{C}_1 = (CA^{-1})^T (CA^{-1} G Q_w G^T A^{-T} C^T + R_v)^{-1} CA^{-1}$$
$$= [I + A^{-T} C^T R_v^{-1} CA^{-1} G Q_w G^T]^{-1} A^{-T} C^T R_v^{-1} CA^{-1}$$

$\Omega_0 = 0$ should be satisfied to obtain the above Ω_1 in Riccati Equation (4.68).

Let

$$\check{x}_{k-N+i} \triangleq \bar{C}_i^T \Xi_i^{-1}(Y_{k-N+i-1} - \bar{B}_i U_{k-N+i-1}) \tag{4.69}$$

where

$$Y_{k-N+i} \triangleq [y_{k-N}^T \ \ y_{k-N+1}^T \ \ \cdots \ \ y_{k-N+i}^T]^T$$
$$U_{k-N+i} \triangleq [u_{k-N}^T \ \ u_{k-N+1}^T \ \ \cdots \ \ u_{k-N+i}^T]^T$$

Then, on the interval $0 \leq i \leq N - 1$, the subsidiary estimate $\check{x}_{k-N+i+1}$ at time $k - N + i + 1$ is obtained from the definition (4.69) as follows:

$$
\begin{aligned}
&\check{x}_{k-N+i+1} \\
&= \bar{C}_{i+1}^T \Xi_{i+1}^{-1}(Y_{k-N+i} - \bar{B}_{i+1} U_{k-N+i}) \\
&= \bar{C}_{i+1}^T \Xi_{i+1}^{-1}\left(\begin{bmatrix} Y_{k-N+i-1} - \bar{B}_i U_{k-N+i-1} \\ y_{k-N+i} \end{bmatrix} + \begin{bmatrix} \bar{C}_i \\ C \end{bmatrix} A^{-1} B u_{k-N+i} \right)
\end{aligned}
\tag{4.70}
$$

In (4.70), the matrix $\bar{C}_{i+1}^T \Xi_{i+1}^{-1}$ can be written as

$$
\begin{aligned}
&\bar{C}_{i+1}^T \Xi_{i+1}^{-1} \\
&= A^{-T} \begin{bmatrix} \bar{C}_i \\ C \end{bmatrix}^T \left(\Delta_i + \begin{bmatrix} \bar{C}_i \\ C \end{bmatrix} A^{-1} G Q_w G^T A^{-T} \begin{bmatrix} \bar{C}_i \\ C \end{bmatrix}^T \right)^{-1} \\
&= A^{-T} \begin{bmatrix} \bar{C}_i \\ C \end{bmatrix}^T \left(\Delta_i^{-1} - \Delta_i^{-1} \begin{bmatrix} \bar{C}_i \\ C \end{bmatrix} A^{-1} G \right. \\
&\quad \times \left. \left\{ Q_w^{-1} + G^T A^{-T} \begin{bmatrix} \bar{C}_i \\ C \end{bmatrix}^T \Delta_i^{-1} \begin{bmatrix} \bar{C}_i \\ C \end{bmatrix} A^{-1} G \right\}^{-1} G^T A^{-T} \begin{bmatrix} \bar{C}_i \\ C \end{bmatrix}^T \Delta_i^{-1} \right) \\
&= \left(I - A^{-T} \begin{bmatrix} \bar{C}_i \\ C \end{bmatrix}^T \Delta_i^{-1} \begin{bmatrix} \bar{C}_i \\ C \end{bmatrix} A^{-1} G \right. \\
&\quad \times \left. \left\{ Q_w^{-1} + G^T A^{-T} \begin{bmatrix} \bar{C}_i \\ C \end{bmatrix}^T \Delta_i^{-1} \begin{bmatrix} \bar{C}_i \\ C \end{bmatrix} A^{-1} G \right\}^{-1} G^T \right) A^{-T} \begin{bmatrix} \bar{C}_i \\ C \end{bmatrix}^T \Delta_i^{-1} \\
&= \left(I - A^{-T}(\Omega_i + C^T R_v^{-1} C) A^{-1} G \right. \\
&\quad \times \left. \left\{ Q_w^{-1} + G^T A^{-T}(\Omega_i + C^T R_v^{-1} C) A^{-1} G \right\}^{-1} G^T \right) A^{-T} \begin{bmatrix} \bar{C}_i \\ C \end{bmatrix}^T \Delta_i^{-1} \\
&= [I + A^{-T}(\Omega_i + C^T R_v^{-1} C) A^{-1} G Q_w G^T]^{-1} A^{-T} \begin{bmatrix} \bar{C}_i \\ C \end{bmatrix}^T \Delta_i^{-1} \tag{4.71}
\end{aligned}
$$

where Δ_i is defined as

$$\Delta_i \triangleq \begin{bmatrix} \Xi_i & 0 \\ 0 & R_v \end{bmatrix}$$

Substituting (4.71) into (4.70) gives

$$\check{x}_{k-N+i+1}$$
$$= [I + A^{-T}(\Omega_i + C^T R_v^{-1} C)A^{-1} G Q_w G^T]^{-1} A^{-T} \begin{bmatrix} \bar{C}_i \\ C \end{bmatrix}^T \Delta_i^{-1}$$
$$\times \left(\begin{bmatrix} Y_{k-N+i-1} - \bar{B}_i U_{k-N+i-1} \\ y_{k-N+i} \end{bmatrix} + \begin{bmatrix} \bar{C}_i \\ C \end{bmatrix} A^{-1} B u_{k-N+i} \right)$$
$$= [I + A^{-T}(\Omega_i + C^T R_v^{-1} C)A^{-1} G Q_w G^T]^{-1} A^{-T}$$
$$\times [\bar{C}_i^T \Xi_i^{-1}(Y_{k-N+i-1} - \bar{B}_i U_{k-N+i-1}) + C^T R_v^{-1} y_{k-N+i}$$
$$+ (\Omega_i + C^T R_v^{-1} C)A^{-1} B u_{k-N+i}]$$

and thus the subsidiary estimate $\check{x}_{k-N+i}$ becomes

$$\check{x}_{k-N+i+1} = [I + A^{-T}(\Omega_i + C^T R_v^{-1} C)A^{-1} G Q_w G^T]^{-1} A^{-T}[\check{x}_{k-N+i}$$
$$+ C^T R_v^{-1} y_{k-N+i} + (\Omega_i + C^T R_v^{-1} C)A^{-1} B u_{k-N+i}] \qquad (4.72)$$

$\check{x}_{k-N+1}$ in a batch form (4.70) can be expressed as

$$\check{x}_{k-N+1} = \bar{C}_1^T \Xi_1^{-1}(y_{k-N} - \bar{B}_1 U_{k-N})$$
$$= (CA^{-1})^T (CA^{-1} G Q_w G^T A C^T + R)^{-1}(y_{k-N} + CA^{-1} B u_{k-N})$$
$$= [I + A^{-T} C^T R_v^{-1} CA^{-1} G Q_w G^T]^{-1} A^{-T}$$
$$\times [C^T R_v^{-1} y_{k-N+i} + C^T R_v^{-1} CA^{-1} B u_{k-N+i}] \qquad (4.73)$$

where $\check{x}_{k-N} = 0$ should be satisfied to obtain the same $\check{x}_{k-N+1}$ in Riccati Equation (4.72).

Therefore, from (4.69) and (4.72), the MVF filter $\hat{x}_{k|k-1}$ with an iterative form is depicted in Figure 4.4 and given in the following theorem.

Theorem 4.5.   *Assume that $\{A, C\}$ is observable and $N \geq n$. Then the MVF filter $\hat{x}_{k|k-1}$ with an iterative form is given on the horizon $[k - N, k]$ as follows:*

$$\hat{x}_{k|k-1} = \Omega_N^{-1} \check{x}_k \qquad (4.74)$$

where Ω_N and $\check{x}_k$ are obtained from (4.67) and (4.72) respectively.   ∎

The recursive filter (4.74) is depicted in Figure 4.4.

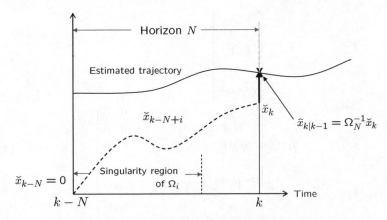

Fig. 4.4. Obtaining the state from recursive equations

4.3.3* Minimum Variance Finite Impulse Response Filters with General A

Batch Form

If it is assumed that the system matrix is nonsingular, then the derivation of the FIR filter becomes somewhat easy. Thus, many results are based on these assumptions. These restrictions may prevent an FIR filter from being applied to many applications. For example, in the case of high-order systems, the system matrix may be a sparse matrix so that there is much chance to have the singularity in the system matrix. Thus it is desirable to derive the FIR filter for general cases. In this section, the MVF filter is derived based on the general systems that do not require the inverse of the system matrix.

The system (4.1) and (4.2) will be represented in a batch form on the most recent time interval $[k - N, k]$, called the horizon. On the horizon $[k - N, k]$, the finite number of measurements is expressed in terms of the state x_{k-N} at the initial time $k - N$ on the horizon as follows:

$$Y_{k-1} = \tilde{C}_N x_{k-N} + \tilde{B}_N U_{k-1} + \tilde{G}_N W_{k-1} + V_{k-1} \qquad (4.75)$$

where

$$Y_{k-1} \triangleq [y_{k-N}^T \ y_{k-N+1}^T \ \cdots \ y_{k-1}^T]^T \qquad (4.76)$$

$$U_{k-1} \triangleq [u_{k-N}^T \ u_{k-N+1}^T \ \cdots \ u_{k-1}^T]^T \qquad (4.77)$$

$$W_{k-1} \triangleq [w_{k-N}^T \ w_{k-N+1}^T \ \cdots \ w_{k-1}^T]^T$$

$$V_{k-1} \triangleq [v_{k-N}^T \ v_{k-N+1}^T \ \cdots \ v_{k-1}^T]^T$$

and $\tilde{C}_N$, $\tilde{B}_N$, and $\tilde{G}_N$ are obtained from

$$\tilde{C}_i \triangleq \begin{bmatrix} C \\ CA \\ CA^2 \\ \vdots \\ CA^{i-1} \end{bmatrix} \tag{4.78}$$

$$\tilde{B}_i \triangleq \begin{bmatrix} 0 & 0 & \cdots & 0 & 0 \\ CB & 0 & \cdots & 0 & 0 \\ CAB & CB & \cdots & 0 & 0 \\ \vdots & \vdots & \vdots & \vdots & \vdots \\ CA^{i-2}B & CA^{i-3}B & \cdots & CB & 0 \end{bmatrix} \tag{4.79}$$

$$\tilde{G}_i \triangleq \begin{bmatrix} 0 & 0 & \cdots & 0 & 0 \\ CG & 0 & \cdots & 0 & 0 \\ CAG & CG & \cdots & 0 & 0 \\ \vdots & \vdots & \vdots & \vdots & \vdots \\ CA^{i-2}G & CA^{i-3}G & \cdots & CG & 0 \end{bmatrix} \tag{4.80}$$

The noise term $\tilde{G}_N W_{k-1} + V_{k-1}$ in (4.75) can be shown to be zero-mean with covariance Π_N given by

$$\Pi_N = \tilde{G}_N Q_N \tilde{G}_N^T + R_N \tag{4.81}$$

where Q_N and R_N are given in (4.41). The current state x_k can be represented using the initial state x_{k-N} on the horizon as

$$x_k = A^N x_{k-N} + \begin{bmatrix} A^{N-1}G & A^{N-2}G & \cdots & G \end{bmatrix} W_{k-1}$$
$$+ \begin{bmatrix} A^{N-1}B & A^{N-2}B & \cdots & B \end{bmatrix} U_{k-1} \tag{4.82}$$

Augmenting (4.75) and (4.82) yields the following linear model:

$$\begin{bmatrix} Y_{k-1} \\ 0 \end{bmatrix} = \begin{bmatrix} \tilde{C}_N & 0 \\ A^N & -I \end{bmatrix} \begin{bmatrix} x_{k-N} \\ x_k \end{bmatrix} + \begin{bmatrix} \tilde{B}_N & \\ A^{N-1}B & \cdots & B \end{bmatrix} U_{k-1}$$
$$+ \begin{bmatrix} \tilde{G}_N & \\ A^{N-1}G & \cdots & G \end{bmatrix} W_{k-1} + \begin{bmatrix} V_{k-1} \\ 0 \end{bmatrix} \tag{4.83}$$

Using Y_{k-1} (4.76) and U_{k-1} (4.77), the FIR filter is represented as

$$\hat{x}_{k|k-1} = HY_{k-1} + LU_{k-1} \tag{4.84}$$

By using Equation (4.83), the FIR filter (4.35) can be rewritten as

$$\hat{x}_{k|k-1} = \begin{bmatrix} H & -I \end{bmatrix} \begin{bmatrix} Y_{k-1} \\ 0 \end{bmatrix} + LU_{k-1},$$

$$= \begin{bmatrix} H & -I \end{bmatrix} \begin{bmatrix} \tilde{C}_N & 0 \\ A^N & -I \end{bmatrix} \begin{bmatrix} x_{k-N} \\ x_k \end{bmatrix}$$

$$+ \begin{bmatrix} H & -I \end{bmatrix} \begin{bmatrix} \tilde{B}_N \\ A^{N-1}B \ \cdots \ B \end{bmatrix} U_{k-1}$$

$$+ \begin{bmatrix} H & -I \end{bmatrix} \begin{bmatrix} \tilde{G}_N \\ A^{N-1}G \ \cdots \ G \end{bmatrix} W_{k-1} + \begin{bmatrix} H & -I \end{bmatrix} \begin{bmatrix} V_{k-1} \\ 0 \end{bmatrix}$$

$$+ LU_{k-1}, \tag{4.85}$$

and taking the expectation on both sides of (4.85) yields the following equation:

$$E[\hat{x}_{k|k-1}] = (H\tilde{C}_N - A^N)E[x_{k-N}] + E[x_k] + \begin{bmatrix} H & -I \end{bmatrix}$$

$$\times \begin{bmatrix} \tilde{B}_N \\ A^{N-1}B \ \cdots \ B \end{bmatrix} U_{k-1} + LU_{k-1}$$

To satisfy the unbiased condition, i.e. $E[\hat{x}_{k|k-1}] = E[x_k]$, irrespective of the initial state and the input, the following constraint is required:

$$H\tilde{C}_N = A^N \tag{4.86}$$

$$L = -\begin{bmatrix} H & -I \end{bmatrix} \begin{bmatrix} \tilde{B}_N \\ A^{N-1}B \ \cdots \ B \end{bmatrix} \tag{4.87}$$

Substituting (4.86) and (4.87) into (4.85) yields the somewhat simplified equation as

$$\hat{x}_{k|k-1} = x_k + H\tilde{G}_N W_{k-1} - \begin{bmatrix} A^{N-1}G \ \cdots \ G \end{bmatrix} W_{k-1} + HV_{k-1}$$

$$e_k \triangleq \hat{x}_{k|k-1} - x_k = H\tilde{G}_N W_{k-1} - \begin{bmatrix} A^{N-1}G \ \cdots \ G \end{bmatrix} W_{k-1} + HV_{k-1}$$

Note that the error between the real state and the estimated state depends only on H, not L, so that we have only to find H optimizing the give performance criterion.

The objective now is to obtain the optimal gain matrix H_B, subject to the unbiasedness constraints (4.86) and (4.87), in such a way that the estimation error e_k of the estimate $\hat{x}_{k|k-1}$ has minimum variance as follows:

$$H_B = \arg\min_H E[e_k^T e_k] = \arg\min_H E \ \text{tr}[e_k e_k^T]$$

$$= \arg\min_H \text{tr}[\Phi Q_N \Phi^T + H R_N H^T] \tag{4.88}$$

where $\Phi = H\tilde{G}_N - \begin{bmatrix} A^{N-1}G \ \cdots \ G \end{bmatrix}$.

By using the result of Lemma 4.3, the solution to the optimization problem (4.88) can be obtained according to the following correspondence:

$$
\begin{aligned}
A &\longleftarrow \tilde{G}_N \\
B &\longleftarrow \begin{bmatrix} A^{N-1}G \cdots G \end{bmatrix} \\
C &\longleftarrow Q_N \\
D &\longleftarrow R_N \\
E &\longleftarrow \tilde{C}_N \\
F &\longleftarrow A^N \\
W_{1,1} &\longleftarrow \tilde{C}_N^T R_N^{-1} \tilde{C}_N \\
W_{1,2} &\longleftarrow \tilde{C}_N^T R_N^{-1} \tilde{G}_N \\
W_{2,2} &\longleftarrow \tilde{G}_N^T R_N^{-1} \tilde{G}_N + Q_N^{-1}
\end{aligned}
\tag{4.89}
$$

The filter gain L can be obtained from (4.87). In the following theorem, we summarize what we have done so far.

Theorem 4.6.   *When $\{A, C\}$ is observable and $N \geq n$, for a general A, the MVF filter $\hat{x}_{k|k-1}$ with a batch form on the horizon $[k - N, k]$ is given as*

$$
\begin{aligned}
\hat{x}_{k|k-1} = {}& \begin{bmatrix} A^N\ A^{N-1}G\ A^{N-2}G \cdots AG\ G \end{bmatrix} \begin{bmatrix} W_{1,1} & W_{1,2} \\ W_{1,2}^T & W_{2,2} \end{bmatrix}^{-1} \\
& \times \begin{bmatrix} \tilde{C}_N^T \\ \tilde{G}_N^T \end{bmatrix} R_N^{-1} \left(Y_{k-1} - \tilde{B}_N U_{k-1} \right) \\
& + \begin{bmatrix} A^{N-1}B\ A^{N-2}B\ A^{N-3}B \cdots AB\ B \end{bmatrix} U_{k-1}
\end{aligned}
\tag{4.90}
$$

∎

As can be seen in (4.90), the inverse of the system matrix A does not appear in the filter coefficients. The batch form proposed in this section requires the inversion computation of matrices $\varXi_N$ and $\tilde{C}_N^T \varXi_N^{-1} \tilde{C}_N$. The dimension of these matrices becomes large as the horizon length N increases. To avoid the inversion computation of large-dimensional matrices, the iterative form is proposed in the next section.

Recursive Form

Now we introduce how to obtain a recursive form without using the inverse of A. An system without inputs is concerned for simple derivation. Before deriving the recursive form of the MVF filter for general systems, we shall introduce a batch form of the Kalman filter.

Theorem 4.7. *On the horizon $[k - N\ k]$, the Kalman filter can be represented in a batch form as*

$$
\begin{aligned}
\hat{x}_{k|k-1} = {}& \begin{bmatrix} A^N\ A^{N-1}G\ A^{N-2}G \cdots AG\ G \end{bmatrix} \begin{bmatrix} W_{1,1} + P_{k-N}^{-1} & W_{1,2} \\ W_{1,2}^T & W_{2,2} \end{bmatrix}^{-1} \\
& \times \left(\begin{bmatrix} P_{k-N}^{-1} \\ 0 \end{bmatrix} \hat{x}_{k-N} + \begin{bmatrix} \tilde{C}_N^T \\ \tilde{G}_N^T \end{bmatrix} R_N^{-1} Y_k \right)
\end{aligned}
\tag{4.91}
$$

where $k - N$ is the initial time and the covariance and the estimated value for the initial state on the horizon are given as P_{k-N} and $\hat{x}_{k-N}$ respectively. The solution to Riccati equation for the Kalman filter is given as

$$P_k = \begin{bmatrix} A^N & A^{N-1}G & A^{N-2}G & \cdots & AG & G \end{bmatrix} \begin{bmatrix} W_{1,1} + P_{k-N}^{-1} & W_{1,2} \\ W_{1,2}^T & W_{2,2} \end{bmatrix}^{-1}$$

$$\times \begin{bmatrix} A^N & A^{N-1}G & A^{N-2}G & \cdots & AG & G \end{bmatrix}^T \tag{4.92}$$

Proof. In time-invariant systems we can always assume, without a loss of generality, that the initial time is 0. Therefore, for simple notations, 0 and N will be used instead of $k - N$ and N. In other words, we will focus on obtaining $\hat{x}_{N|N-1}$ using the information on $\hat{x}_{0|-1}$, P_0, and Y_{N-1} where

$$Y_i \triangleq [y_0^T \ y_1^T \ \cdots \ y_{i-1}^T]^T \tag{4.93}$$

for $0 \le i \le N - 1$.

Define variables L_i, M_i, and N_i as

$$L_i \triangleq \begin{bmatrix} A^i & A^{i-1}G & A^{i-2}G & \cdots & AG & G \end{bmatrix} \tag{4.94}$$

$$M_i \triangleq \begin{bmatrix} \tilde{C}_i^T R_i^{-1} \tilde{C}_i + P_0^{-1} & \tilde{C}_i^T R_i^{-1} \tilde{G}_i^o \\ \tilde{G}_i^{oT} R_i^{-1} \tilde{C}_i & \tilde{G}_i^{oT} R_i^{-1} \tilde{G}_i^o + Q_i^{-1} \end{bmatrix} \tag{4.95}$$

$$N_i \triangleq \begin{bmatrix} M_{i-1} & 0 \\ 0 & Q_w^{-1} \end{bmatrix} \tag{4.96}$$

$$T_i \triangleq \begin{bmatrix} P_0^{-1} \\ 0 \end{bmatrix} \hat{x}_{k-i} + \begin{bmatrix} \tilde{C}_i^T \\ \tilde{G}_i^T \end{bmatrix} R_i^{-1} Y_i \tag{4.97}$$

where $\tilde{C}_i$ and $\tilde{G}_i$ are defined in (4.78)–(4.80) and $\tilde{G}_i^o$ is the matrix which is obtained by removing the last zero column block from $\tilde{G}_i$, i.e. $\tilde{G}_i = \begin{bmatrix} \tilde{G}_i^o & 0 \end{bmatrix}$.

Using the defined variables (4.94)–(4.97), we can represent $\hat{x}_N$ of (4.91) and P_N of (4.92) as

$$\hat{x}_N = L_N N_N^{-1} T_N \tag{4.98}$$

$$P_N = L_N N_N^{-1} L_N^T \tag{4.99}$$

First, we will try to derive the batch form of covariance matrix (4.92). Recall that Riccati equation of the Kalman filter is

$$P_{i+1} = AP_i A^T + GQ_w G^T - AP_i C^T (R_v + CP_i C^T)^{-1} CP_i A^T \tag{4.100}$$

which is used to solve for the estimation error covariance matrix.

Given an initial covariance matrix P_0, we can calculate P_1 from (4.100) as follows:

$$P_1 = AP_0 A^T + GQ_w G^T - AP_0 C^T (R_v + CP_0 C^T)^{-1} CP_0 A^T \tag{4.101}$$

$$= A(P_0^{-1} + C^T R_v^{-1} C)^{-1} A^T + GQ_w G^T \tag{4.102}$$

$$= \begin{bmatrix} A & G \end{bmatrix} \begin{bmatrix} P_0^{-1} + C^T R_v^{-1} C & 0 \\ 0 & Q_w^{-1} \end{bmatrix}^{-1} \begin{bmatrix} A & G \end{bmatrix}^T \tag{4.103}$$

(4.103) can be written in terms of L_i and N_i as

$$P_1 = L_1 N_1^{-1} L_1^T \tag{4.104}$$

By an induction method, P_{i+1} will be calculated from P_i of the batch form, i.e. $P_i = L_i N_i^{-1} L_i^T$. Before proceeding, we introduce the following relation:

$$N_i + L_i^T C^T R_v^{-1} C L_i = M_i \tag{4.105}$$

which is proved as follows. If S_i is defined as

$$S_i \triangleq \begin{bmatrix} A^{i-1}G & A^{i-2}G & \cdots & AG & G \end{bmatrix} \tag{4.106}$$

then L_i can be represented as

$$L_i = \begin{bmatrix} A^i & S_i \end{bmatrix} \tag{4.107}$$

from which we have

$$L_i^T C^T R_v^{-1} C L_i = \begin{bmatrix} A^{iT} C^T R_v^{-1} C A^i & A^{iT} C^T R_v^{-1} C S_i \\ S_i^T C^T R_v^{-1} C A^i & S_i^T C^T R_v^{-1} C S_i \end{bmatrix} \tag{4.108}$$

The four block elements in M_i can be expressed recursively:

$$\tilde{C}_i^T R_i^{-1} \tilde{C}_i = \tilde{C}_{i-1}^T R_{i-1}^{-1} \tilde{C}_{i-1} + A^{Ti} C^T R_v^{-1} C A^i \tag{4.109}$$

$$\tilde{C}_i^T R_i^{-1} \tilde{G}_i^o = \begin{bmatrix} \tilde{C}_{i-1}^T R_{i-1}^{-1} \tilde{G}_{i-1}^o | 0 \end{bmatrix} + A^{Ti} C^T R_v^{-1} C S_i \tag{4.110}$$

$$\tilde{G}_i^{oT} R_i^{-1} \tilde{G}_i^o = \begin{bmatrix} \tilde{G}_{i-1}^{oT} R_{i-1}^{-1} \tilde{G}_{i-1}^o & 0 \\ 0 & 0 \end{bmatrix} + S_i^T C^T R_v^{-1} C S_i \tag{4.111}$$

Using (4.108) and (4.109)–(4.111), we have

$$N_i + L_i^T C^T R_v^{-1} C L_i \tag{4.112}$$

$$= \begin{bmatrix} \tilde{C}_{i-1}^T R_{i-1}^{-1} \tilde{C}_{i-1} & \begin{bmatrix} \tilde{C}_{i-1}^T R_{i-1}^{-1} \tilde{G}_{i-1} | 0 \end{bmatrix} \\ \begin{bmatrix} \tilde{C}_{i-1}^T R_{i-1}^{-1} \tilde{G}_{i-1} | 0 \end{bmatrix}^T & \begin{bmatrix} \tilde{G}_{i-1}^{oT} R_{i-1}^{-1} \tilde{G}_{i-1}^o + Q_{i-1}^{-1} & 0 \\ 0 & Q_w^{-1} \end{bmatrix} \end{bmatrix}$$

$$+ \begin{bmatrix} A^{iT} C^T R_v^{-1} C A^i & A^{iT} C^T R_v^{-1} C S_i \\ S_i^T C^T R_v^{-1} C A_i & S_i^T C^T R_v^{-1} C S_i \end{bmatrix} \tag{4.113}$$

$$= \begin{bmatrix} \tilde{C}_i^T R_i^{-1} \tilde{C}_i & \tilde{C}_i^T R_i^{-1} \tilde{G}_i \\ \tilde{G}_i^T R_i^{-1} \tilde{C}_i & \tilde{G}_i^T R_i^{-1} \tilde{G}_i + Q_i^{-1} \end{bmatrix} = M_i \tag{4.114}$$

Starting from (4.100) and using the relation (4.105), we obtain the P_{i+1} of the batch form

$$P_{i+1} = AL_i \left\{ N_i^{-1} - N_i^{-1} L_i^T C^T (R_v + C L_i N_i^{-1} L_i^T C^T)^{-1} C L_i N_i^{-1} \right\} L_i^T A$$

$$+ GQ_w G^T,$$

$$= AL_i (N_i + L_i^T C^T R_v^{-1} C L_i)^{-1} L_i^T A + GQ_w G^T$$

$$= AL_i M_i^{-1} L_i^T A + GQ_w G^T = L_{i+1} N_{i+1}^{-1} L_{i+1}^T \tag{4.115}$$

Now, using the solution of Riccati equation, i.e. the covariance matrix represented in the batch form, we obtain a batch form of the Kalman filter. The dynamic equation for the Kalman filter is written as follows:

$$\hat{x}_{i+1|i} = A\hat{x}_{i|i-1} + AP_iC^T(R_v + CP_iC^T)^{-1}(y_i - C\hat{x}_{i|i-1})$$
$$= (A - AP_iC^T(R_v + CP_iC^T)^{-1}C)\hat{x}_{i|i-1}$$
$$+ AP_iC^T(R_v + CP_iC^T)^{-1}y_i \qquad (4.116)$$

where P_i is given by (4.92). To obtain a batch form of the Kalman filter, we shall also use an induction method.

First, by using (4.100), $\hat{x}_{1|0}$ will be obtained from $\hat{x}_{0|-1}$ and P_0:

$$\hat{x}_{1|0} = (A - AP_0C^T(R_v + CP_0C^T)^{-1}C)\hat{x}_{0|-1} + AP_0C^T(R_v + CP_0C^T)^{-1}y_0,$$
$$= A(I + P_0C^TR_v^{-1}C)^{-1}\hat{x}_{0|-1} + AP_0C^T(R_v + CP_0C^T)^{-1}y_0$$
$$= A(P_0^{-1} + C^TR_v^{-1}C)^{-1}P_0^{-1}\hat{x}_{0|-1} + AP_0C^T(I + R_v^{-1}CP_0C^T)^{-1}R_v^{-1}y_0,$$
$$= A(P_0^{-1} + C^TR_v^{-1}C)^{-1}(\hat{x}_{0|-1} + C^TR_v^{-1}y_0)$$

In terms of L_i, N_i, and T_i, $\hat{x}_{1|0}$ can be written as

$$\hat{x}_{1|0} = L_1N_1^{-1}T_1 \qquad (4.117)$$

By an induction method, $\hat{x}_{i+1|i}$ will be calculated from $\hat{x}_{i|i-1}$ of the batch form, i.e. $\hat{x}_{i|i-1} = L_iN_i^{-1}T_i$.

First, observe that the first and second terms on the right-hand side of the Kalman filter reduce to

$$\left\{A - AP_iC^T(R_v + CP_iC^T)^{-1}C\right\}\hat{x}_{i|i-1}$$
$$= AL_i\left\{N_i^{-1} - N_i^{-1}L_i^TC^T(R_v + CL_iN_i^{-1}L_i^TC^T)^{-1}CL_iN_i^{-1}\right\}T_i$$
$$= AL_i\left\{N_i + L_i^TC^TR_v^{-1}CL_i\right\}^{-1}T_i = AL_iM_i^{-1}T_i \qquad (4.118)$$

and

$$AP_iC^T(R_v + CP_iC^T)^{-1}y_i$$
$$= AL_iN_i^{-1}L_i^TC^T(R_v + CL_iN_i^{-1}L_i^TC^T)^{-1}y_i$$
$$= AL_i(I + N_i^{-1}L_i^TC^TR_v^{-1}CL_i)^{-1}N_i^{-1}L_i^TC^TR_v^{-1}y_i$$
$$= AL_i(N_i + L_i^TC^TR_v^{-1}CL_i)^{-1}L_i^TC^TR_v^{-1}y_i$$
$$= AL_iM_i^{-1}L_i^TC^TR_v^{-1}y_i \qquad (4.119)$$

respectively. Substituting (4.118) and (4.119) into the Kalman filter yields

$$\hat{x}_{i+1|i} = AL_i M_i^{-1} T_i + AL_i M_i^{-1} L_i^T C^T R_v^{-1} y_i$$
$$= AL_i M_i^{-1}(T_i + L_i^T C^T R_v^{-1} y_i)$$
$$= AL_i M_i^{-1}\left\{\begin{bmatrix} P_0^{-1} \\ 0 \end{bmatrix}\hat{x}_{0|-1} + \begin{bmatrix} \tilde{C}_i^T \\ \tilde{G}_i^T \end{bmatrix} R_i^{-1} Y_i + L_i^T C^T R_v^{-1} y_i\right\} \quad (4.120)$$

Combining the second and third terms in brackets in (4.120) yields

$$\hat{x}_{i+1|i} = \begin{bmatrix} AL_i & 0 \end{bmatrix}\begin{bmatrix} M_i^{-1} & 0 \\ 0 & 0 \end{bmatrix}\left\{\begin{bmatrix} P_0^{-1} \\ 0 \end{bmatrix}\hat{x}_{0|-1} + \begin{bmatrix} \tilde{C}_{i+1}^T \\ [\tilde{G}_{i+1}^o|0]^T \end{bmatrix} R_{i+1}^{-1} Y_{i+1}\right\}$$
$$= \begin{bmatrix} AL_i & G \end{bmatrix}\begin{bmatrix} M_i^{-1} & 0 \\ 0 & Q_w \end{bmatrix}\left\{\begin{bmatrix} P_0^{-1} \\ 0 \end{bmatrix}\hat{x}_{0|-1} + \begin{bmatrix} \tilde{C}_{i+1}^T \\ \tilde{G}_{i+1}^T \end{bmatrix} R_{i+1}^{-1} Y_{i+1}\right\}$$
$$= L_{i+1} N_{i+1}^{-1} T_{i+1}$$

where G and Q_w in the first and second matrix blocks on the right-hand side of the second equality have no effect on the equation. This completes the proof. ∎

In the following theorem, we will show that the FIR filter can be obtained from the appropriate initial covariance and mean with the Kalman filter.

Theorem 4.8. *The FIR filter (4.35) can be obtained by replacing $\hat{x}_{k-N}$ and P_{k-N} in the Kalman filter (4.91) by $(\tilde{C}_N^T \Pi_N^{-1} \tilde{C}_N)^{-1}$ and $P_{k-N}\tilde{C}_N^T \Pi_N^{-1} Y_k$ respectively.*

Proof. Using (4.89), we can write P_{k-N} and $P_{k-N}^{-1}\hat{x}_{k-N}$ as

$$P_{k-N} = (W_{1,1} - W_{1,2}W_{2,2}^{-1}W_{1,2}^T)^{-1} \quad (4.121)$$
$$P_{k-N}^{-1}\hat{x}_{k-N} = -W_{1,2}W_{2,2}^{-1}\tilde{G}^T R_N^{-1} Y_k \quad (4.122)$$

Replacing $\hat{x}_{k-N}$ with $P_{k-N}\tilde{C}_N^T \Pi_N^{-1} Y_k$ and using (4.122) yields

$$\begin{bmatrix} P_{k-N}^{-1} \\ 0 \end{bmatrix}\hat{x}_{k-N} + \begin{bmatrix} \tilde{C}_N^T \\ \tilde{G}_N^T \end{bmatrix} R_N^{-1} Y_k = \begin{bmatrix} 2\tilde{C}_N^T R_N^{-1} Y_k - W_{1,2}W_{2,2}^{-1}\tilde{G}_N^T R_N^{-1} Y_k \\ \tilde{G}_N^T R_N^{-1} Y_k \end{bmatrix}$$
$$= \begin{bmatrix} 2I & -W_{1,2}W_{2,2}^{-1} \\ 0 & I \end{bmatrix}\begin{bmatrix} \tilde{C}_N^T \\ \tilde{G}_N^T \end{bmatrix} R_N^{-1} Y_k$$

By the following relation:

$$\begin{bmatrix} W_{1,1} + P_{k-N}^{-1} & W_{1,2} \\ W_{1,2}^T & W_{2,2} \end{bmatrix}^{-1}\begin{bmatrix} 2I & -W_{1,2}W_{2,2}^{-1} \\ 0 & I \end{bmatrix}$$
$$= \left\{\begin{bmatrix} \frac{1}{2}I & \frac{1}{2}W_{1,2}W_{2,2}^{-1} \\ 0 & I \end{bmatrix}\begin{bmatrix} W_{1,1} + P_{k-N}^{-1} & W_{1,2} \\ W_{1,2}^T & W_{2,2} \end{bmatrix}\right\}^{-1} = \begin{bmatrix} W_{1,1} & W_{1,2} \\ W_{1,2}^T & W_{2,2} \end{bmatrix}^{-1} \quad (4.123)$$

it can be shown that (4.91) with the prescribed P_{k-N} and $\hat{x}_{k-N}$ is equivalent to (4.90). This completes the proof. ∎

In the following lemma, $\hat{x}_{k-N}$ and P_{k-N} in Theorem 4.8 are shown to be computed recursively.

Lemma 4.9. $\hat{x}_{k-N}$ and P_{k-N} in Theorem 4.8 can be computed from

$$\hat{x}_{k-N} = (C^T R_v^{-1} C + \hat{P}_N)^{-1}(C^T R_v^{-1} y_{k-N} + \hat{\omega}_N) \qquad (4.124)$$
$$P_{k-N} = (C^T R_v^{-1} C + \hat{P}_N)^{-1} \qquad (4.125)$$

where $\hat{P}_N$ and $\hat{w}_N$ can be obtained from the following recursive equations:

$$\hat{P}_{i+1} = A^T C^T R_v^{-1} C A + A^T \hat{P}_i A$$
$$- A^T(C^T R_v^{-1} C + \hat{P}_i)G\left\{Q_w^{-1} + G^T A^T(C^T R_v^{-1} C + \hat{P}_i)AG\right\}^{-1}$$
$$\times G^T(C^T R_v^{-1} C + \hat{P}_i)A \qquad (4.126)$$

and

$$\hat{\omega}_{i+1} = A^T C^T R_v y_{k-i} + A\hat{\omega}_i$$
$$- A^T(C^T R_v^{-1} C + \hat{P}_i)G\left\{Q_w^{-1} + G^T A^T(C^T R_v^{-1} C + \hat{P}_i)AG\right\}^{-1}$$
$$\times G^T(C^T R_v^{-1} y_{k-i} + \hat{\omega}_i) \qquad (4.127)$$

with $\hat{P}_1 = 0$, $\hat{w}_1 = 0$, and $1 \leq i \leq N - 1$.

Proof. Define $\hat{C}_i$, $\hat{N}_i$, $\hat{\Pi}_i$, $\hat{P}_i$, and $\hat{\omega}_i$ as

$$\hat{C}_i \triangleq \begin{bmatrix} CA \\ \vdots \\ CA^{i-3} \\ CA^{i-2} \\ CA^{i-1} \end{bmatrix} = \begin{bmatrix} C \\ \hat{C}_{i-1} \end{bmatrix} A \qquad (4.128)$$

$$\hat{N}_i \triangleq \begin{bmatrix} CG & 0 & \cdots & 0 \\ CAG & CG & \cdots & 0 \\ \vdots & \vdots & \vdots & \vdots \\ CA^{i-2}G & CA^{i-3}G & \cdots & CG \end{bmatrix} \qquad (4.129)$$

$$\hat{\Pi}_i \triangleq \hat{N}_i Q_{i-1}\hat{N}_i^T + R_{i-1} \qquad (4.130)$$

$$\hat{P}_i \triangleq \hat{C}_i^T \hat{\Pi}_i^{-1}\hat{C}_i \qquad (4.131)$$

$$\hat{w}_i \triangleq \hat{C}_i^T \hat{\Pi}_i^{-1}Y_{k-i+1,k-1} \qquad (4.132)$$

$$Y_{k-i+1,k-1} \triangleq [y_{k-i+1}^T \ y_{k-i+2}^T \ \cdots \ y_{k-1}^T]^T \qquad (4.133)$$

Using (4.128)–(4.131), P_{k-N} and $\hat{x}_{k-N}$ can be represented as

$$P_{k-N} = (\tilde{C}_N^T \Pi_N^{-1} \tilde{C}_N)^{-1} = (C^T R_v^{-1} C + \hat{P}_N)^{-1} \qquad (4.134)$$

$$\hat{x}_{k-N} = P_{k-N}(C^T R_v^{-1} y_{k-N} + \hat{C}_N^T \hat{\Pi}_N^{-1} Y_{k-N+1,k-1}) \qquad (4.135)$$

In order to obtain $\hat{x}_{k-N}$ and P_{k-N}, we have only to know $\hat{C}_N^T \hat{\Pi}_N^{-1} Y_{k-N+1,k-1}$ and $\hat{P}_N$, which are calculated recursively.

First, a useful equality is introduced for obtaining the recursive form:

$$\hat{\Pi}_{i+1}^{-1} = (\hat{N}_{i+1} Q_i \hat{N}_{i+1}^T + R_i)^{-1}$$

$$= \left(\begin{bmatrix} R_v & 0 \\ 0 & \hat{\Pi}_i \end{bmatrix} + \begin{bmatrix} C \\ \hat{C}_i \end{bmatrix} G Q_w G^T \begin{bmatrix} C \\ \hat{C}_i \end{bmatrix}^T \right)^{-1}$$

$$= \Delta_i^{-1} - \Delta_i^{-1} \begin{bmatrix} C \\ \hat{C}_i \end{bmatrix} G \left\{ Q_w^{-1} + G^T A^T \begin{bmatrix} C \\ \hat{C}_i \end{bmatrix}^T \Delta_i^{-1} \begin{bmatrix} C \\ \hat{C}_i \end{bmatrix} A G \right\}^{-1}$$

$$G^T \begin{bmatrix} C \\ \hat{C}_i \end{bmatrix} \Delta_i^{-1} \qquad (4.136)$$

where

$$\Delta_i = \begin{bmatrix} R_v & 0 \\ 0 & \hat{\Pi}_i \end{bmatrix} \qquad (4.137)$$

Pre- and post-multiplying on both sides of (4.136) by

$$\begin{bmatrix} C \\ \hat{C}_i \end{bmatrix} A \qquad (4.138)$$

we have

$$\hat{P}_{i+1} = A^T C^T R_v^{-1} C A + A^T \hat{P}_i A$$

$$- A^T (C^T R_v^{-1} C + \hat{P}_i) G \left\{ Q_w^{-1} + G^T A^T (C^T R_v^{-1} C + \hat{P}_i) A G \right\}^{-1}$$

$$\times G^T (C^T R_v^{-1} C + \hat{P}_i) A \qquad (4.139)$$

Pre-multiplying on both sides of (4.136) by (4.138), we obtain another recursive equation:

$$\hat{\omega}_{i+1} = A^T C^T R_v y_{k-i} + A \hat{\omega}_i$$

$$- A^T (C^T R_v^{-1} C + \hat{P}_i) G \left\{ Q_w^{-1} + G^T A^T (C^T R_v^{-1} C + \hat{P}_i) A G \right\}^{-1}$$

$$\times G^T (C^T R_v^{-1} y_{k-i} + \hat{\omega}_i) \qquad (4.140)$$

After obtaining $\hat{P}_N$ from (4.139), we can calculate P_{k-N} from (4.134). $\hat{x}_{k-N}$ in (4.135) can be obtained from P_{k-N} and $\hat{\omega}_N$ that comes from (4.140). This completes the proof. ∎

In the following theorem, we summarize what we have done so far.

Theorem 4.10. *For general A, the MVF filter is given on the horizon $[k - N, k]$ as follows:*

$$\hat{x}_{k|k-1} = \beta_k \tag{4.141}$$

where β_i is obtained from the following recursive equations:

$$\begin{aligned}
\beta_{k-N+i+1} &= A\beta_{k-N+i} + AP_iC^T(R_v + CP_iC^T)^{-1}(y_{k-N+i} - C\beta_{k-N+i}) \\
&= (A - AP_iC^T(R_v + CP_iC^T)^{-1}C)\beta_{k-N+i} \\
&\quad + AP_iC^T(R_v + CP_iC^T)^{-1}y_{k-N+i} \tag{4.142} \\
P_{i+1} &= AP_iA^T + GQ_wG^T - AP_iC^T(R_v + CP_iC^T)^{-1}CP_iA^T \tag{4.143}
\end{aligned}$$

Here, $P_0 = (C^TR_v^{-1}C + \hat{P}_N)^{-1}$ and $\beta_{k-N} = P_0(C^TR_v^{-1}y_{k-N} + \hat{\omega}_N)$. $\hat{P}_N$ and $\hat{\omega}_N$ are obtained from (4.139) and (4.140) respectively. ∎

Note that (4.142) and (4.143) are in a form of the Kalman filter with a special initial condition. The recursive filter in Theorem 4.10 is depicted in Figure 4.5.

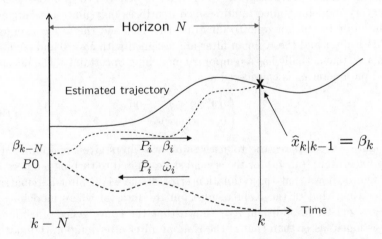

Fig. 4.5. Backward and forward estimations for general matrix A

4.3.4 Numerical Examples for Minimum Variance Finite Impulse Response Filters

Example 4.1: Robustness of Minimum Variance Finite Impulse Response Filters

To demonstrate the validity of the MVF filter, numerical examples on the discretized model of an F-404 engine [EWMR94] are presented via simulation studies. The corresponding dynamic model is written as

$$x_{k+1} = \begin{bmatrix} 0.9305 + \delta_k & 0 & 0.1107 \\ 0.0077 & 0.9802 + \delta_k & -0.0173 \\ 0.0142 & 0 & 0.8953 + 0.1\delta_k \end{bmatrix} x_k + \begin{bmatrix} 1 \\ 1 \\ 1 \end{bmatrix} w_k$$

$$y_k = \begin{bmatrix} 1 + 0.1\delta_k & 0 & 0 \\ 0 & 1 + 0.1\delta_k & 0 \end{bmatrix} x_k + v_k \tag{4.144}$$

where δ_k is an uncertain model parameter. The system noise covariance Q_w is 0.02 and the measurement noise covariance R_v is 0.02. The horizon length is taken as $N = 10$. We perform simulation studies for the system (4.144) with a temporary modelling uncertainty.

As mentioned previously, the MVF filter is believed to be robust against temporary modelling uncertainties since it utilizes only finite measurements on the most recent horizon. To illustrate this fact and the fast convergence, the MVF filter and the Kalman filter are designed with $\delta_k = 0$ and compared when a system actually has a temporary modelling uncertainty. The uncertain model parameter δ_k is considered as

$$\delta_k = \begin{cases} 0.1, \ 50 \leq k \leq 100 \\ 0, \quad \text{otherwise} \end{cases} \tag{4.145}$$

Figure 4.6 compares the robustness of two filters given temporary modelling uncertainty (4.145) for the second state related to turbine temperature. This figure shows that the estimation error of the MVF filter is remarkably smaller than that of the Kalman filter on the interval where modelling uncertainty exists. In addition, it is shown that the convergence of estimation error is much faster than that of the Kalman filter after temporary modelling uncertainty disappears. Therefore, it can be seen that the suggested MVF filter is more robust than the Kalman filter when applied to systems with a model parameter uncertainty.

Example 4.2: Comparison with Deadbeat Infinite Impulse Response Filters

It is known that the MVF filter has the deadbeat property without external disturbances. Via numerical simulation, we compare the MVF filter with

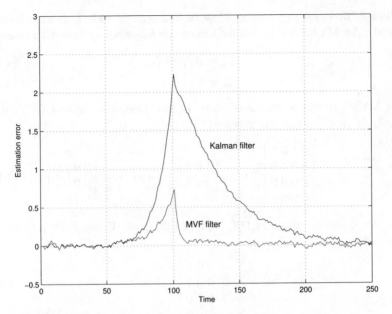

Fig. 4.6. Estimation errors of MVF filter and Kalman filter

the conventional deadbeat IIR filter, which is known to be sensitive to external disturbance.

From the following simple model:

$$x_{i+1} = Ax_i \tag{4.146}$$
$$y_i = Cx_i \tag{4.147}$$

the deadbeat filter of the partial state $z_i = Nx_i$ will be obtained as

$$\hat{w}_{i+1} = A_o\hat{w}_i + Ly_i \tag{4.148}$$
$$\hat{z}_i = M\hat{w}_i \tag{4.149}$$

where $\hat{z}_i$ estimates z_i exactly in a finite time irrespective of the initial state error. First, it is assumed that there exist matrices W and A_o such that

$$WA = A_oW + LC , \quad N = MW, \quad A_o^d = 0 \tag{4.150}$$

for some positive integer d. Letting $e_i = \hat{w}_i - Wx_i$, we have

$$e_{i+1} = \hat{w}_{i+1} - Wx_{i+1} = A_o\hat{w}_i + Ly_i - WAx_i$$
$$= A_o\hat{w}_i + LCx_i - WAx_i = A_o\hat{w}_i - A_oWx_i = A_oe_i \tag{4.151}$$

and the estimation error $z_i - \hat{z}_i$ is represented as $Nx_i - M\hat{w}_i = MWx_i - M\hat{w}_i = -Me_i$. According to (4.150) and (4.151), e_i should be zero in a finite time so that $\hat{z}_i$ becomes equal to z_i. How to find out W and A_o is introduced in [Kuc91]. An MVF filter is obtained from the following system with noise:

$$x_{i+1} = Ax_i + Gw_i \tag{4.152}$$
$$y_i = Cx_i + v_i \tag{4.153}$$

This MVF filter is a deadbeat filter for the noise-free system (4.146) and (4.147). If A, C, and N are given by

$$A = \begin{bmatrix} 0 & 1 & 0 & 1 \\ 0 & 0 & 1 & 0 \\ 0 & 0 & 0 & 0 \\ 0 & 0 & 0 & 1 \end{bmatrix}, \ C = \begin{bmatrix} 1 & 0 & 0 & 0 \\ 0 & 1 & 0 & 0 \end{bmatrix}, \ \text{and} \ N = \begin{bmatrix} 0 & 0 & 1 & 0 \\ 0 & 0 & 0 & 1 \end{bmatrix} \tag{4.154}$$

then the deadbeat filter (4.148) and (4.149) and W satisfying (4.150) are obtained as

$$W = \begin{bmatrix} -1 & 0 & 0 & 1 \\ 0 & 0 & 0 & 1 \\ 0 & 0 & 1 & 0 \end{bmatrix} \tag{4.155}$$

$$\hat{w}_{i+1} = \begin{bmatrix} 0 & 0 & 0 \\ 1 & 0 & 0 \\ 0 & 0 & 0 \end{bmatrix} \hat{w}_i + \begin{bmatrix} 0 & -1 \\ 1 & 0 \\ 0 & 0 \end{bmatrix} y_i \tag{4.156}$$

$$\hat{z}_i = \begin{bmatrix} 0 & 0 & 1 \\ 0 & 1 & 0 \end{bmatrix} \hat{w}_i \tag{4.157}$$

It can be easily seen that $A_o^2 = 0$, i.e. $d = 2$. We compare the deadbeat filter (4.156) and (4.157) and the MVF filter obtained from (4.152) and (4.153). For simulation, the system noise covariance Q is 0.2 and the measurement noise covariance R is 0.4. The horizon length is taken as $N = 15$. As seen in Figure 4.7, the deadbeat filter is sensitive to noises. However, the MVF filter is robust against noises.

4.4 Dual Infinite Impulse Response Filters Based on Minimax Criterion

In this section, the following linear discrete-time state-space signal model is considered:

$$x_{i+1} = Ax_i + B_w w_i + Bu_i \tag{4.158}$$
$$y_i = Cx_i + D_w w_i \tag{4.159}$$
$$z_i = C_z x_i \tag{4.160}$$

where $x_i \in \Re^n$ is the state, and $u_i \in \Re^l$, $y_i \in \Re^q$, and $w_i \in \Re^p$ are the input, measurement, and disturbance respectively.

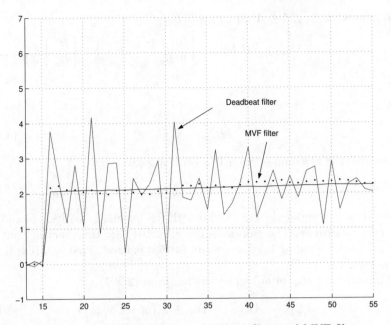

Fig. 4.7. Comparison between deadbeat filter and MVF filter

Before we introduce a filter for (4.158) and (4.159), we will compare the H_∞ control with the H_∞ filter and investigate the structural differences between them.

The H_∞ control (2.146) with a zero reference signal and terminal time i_f, is given by

$$u_i^* = -R^{-1}B^T \Lambda_{i+1,i_f}^{-1} M_{i+1,i_f} A x_i \tag{4.161}$$

where Riccati equation is given by

$$
\begin{aligned}
M_{i,i_f} &= A^T[I + M_{i+1,i_f}(BR^{-1}B^T - \gamma^{-2}B_w R_w^{-1} B_w^T)]^{-1} M_{i+1,i_f} A + Q \\
&= A^T M_{i+1,i_f} A - A^T M_{i+1,i_f}[(BR^{-1}B^T - \gamma^{-2}B_w R_w^{-1} B_w^T)M_{i+1,i_f} \\
&\quad + I]^{-1}(BR^{-1}B^T - \gamma^{-2}B_w R_w^{-1} B_w^T)M_{i+1,i_f} A + Q \tag{4.162}
\end{aligned}
$$

Use the following relation:

$$BR^{-1}B^T - \gamma^{-2}B_w R_w^{-1} B_w^T = \begin{bmatrix} B\ B_w \end{bmatrix} \begin{bmatrix} R & 0 \\ 0 & R_w - \gamma^2 \end{bmatrix}^{-1} \begin{bmatrix} B^T \\ B_w^T \end{bmatrix}$$

we have

$$M_{i,i_f} = A^T M_{i+1,i_f} A - A^T M_{i+1,i_f} \begin{bmatrix} B & B_w \end{bmatrix}$$
$$\times \left\{ \begin{bmatrix} B^T \\ B_w^T \end{bmatrix} M_{i+1,i_f} \begin{bmatrix} B & B_w \end{bmatrix} + \begin{bmatrix} R & 0 \\ 0 & R_w - \gamma^2 \end{bmatrix}^{-1} \right\}^{-1} \begin{bmatrix} B^T \\ B_w^T \end{bmatrix} M_{i+1,i_f} A$$
$$+ Q \tag{4.163}$$

with

$$M_{i_f,i_f} = Q_f \tag{4.164}$$

Meanwhile, the receding horizon H_∞ control is given by

$$u_i^* = -R^{-1} B^T \Lambda_{1,N}^{-1} M_{1,N} A x_i \tag{4.165}$$

where $M_{1,N}$ is computed from (4.163) with the terminal condition (4.164) by replacing i_f with N. The gain matrices in (4.161) are time-varying, but the gain M_1 in the receding horizon control (4.165) is fixed. That is, M_1 is frozen at the time 1.

The standard H_∞ filter can be written from (2.201) as

$$\hat{x}_{i+1|i} = A\hat{x}_{i|i-1} + Bu_i + L_i(y_i - C\hat{x}_{i|i-1}) \tag{4.166}$$
$$\hat{z}_i = C_z \hat{x}_{i|i-1} \tag{4.167}$$

where

$$L_i = AS_i \Gamma_i^{-1} C^T \tag{4.168}$$

and the matrices S_i and Γ_i are given by

$$S_{i+1} = AS_i \Gamma_i^{-1} A^T + B_w B_w^T \tag{4.169}$$
$$\Gamma_i = I + (C^T C - \gamma^{-2} C_z^T C_z) S_i \tag{4.170}$$

where $S_0 = 0$. The recursive equation (4.169) with respect to S_i can be written as

$$S_{i+1} = AS_i A^T - AS_i \begin{bmatrix} C^T & C_z^T \end{bmatrix} \tag{4.171}$$
$$\times \left\{ \begin{bmatrix} C \\ C_z \end{bmatrix} S_i \begin{bmatrix} C^T & C_z^T \end{bmatrix} + \begin{bmatrix} I & 0 \\ 0 & -\gamma^2 I \end{bmatrix}^{-1} \right\}^{-1} \begin{bmatrix} C \\ C_z \end{bmatrix} S_i A^T + B_w B_w^T \tag{4.172}$$

The H_∞ filter is said to be dual to the H_∞ control in the sense that two Riccati equations become the same as each other if some variables are changed, such as

$$\begin{matrix} A & \longleftrightarrow & A^T \\ B & \longleftrightarrow & C^T \\ B_w & \longleftrightarrow & C_z^T \\ Q & \longleftrightarrow & B_w B_w^T \\ R & \longleftrightarrow & I \\ M_i & \longleftrightarrow & S_i \end{matrix} \tag{4.173}$$

Note that indexes of M_{i+1} and Q_i are based on backward and forward computations respectively. We can see that the filter gain in (4.168) and the control gain in (4.161) are similar to each other in the sense that the transpose of the control gain replaced by (4.173) is the same as the filter gain in (4.168). In that view, the corresponding closed-loop system matrices of the H_∞ control and the H_∞ filter are also similar. It is noted that the properties of the H_∞ filter can be obtained from that of the dual to the H_∞ control, since they are structurally dual.

Likewise, we can introduce the dual filter to the receding horizon H_∞ control as follows:

$$\hat{x}_{i+1|i} = A\hat{x}_{i|i-1} + \left[AS_N[I + (C^TC - \gamma^{-2}C_z^TC_z)S_N]^{-1}C^T \right]$$
$$\times (y_i - C\hat{x}_{i|i-1}) \qquad (4.174)$$

where S_N is obtained from (4.169). It turns out that the new filter (4.174) is just the H_∞ filter with a frozen gain from (4.166). This filter will be called the dual filter to the RH H_∞ control or the H_∞ filter with frozen gains throughout this book.

By using the duality (4.173), the inequality (3.234) in Theorem 3.15 is changed to

$$K \geq B_wB_w^T + LL^T$$
$$- \Gamma_f\Gamma_f^T + (A - LC + \Gamma_fC_z)K(A - LC + \Gamma_fC_z)^T \qquad (4.175)$$

where matrices L and K correspond to H and Q_f in (3.234) and $\Gamma_f = \gamma^{-2}AK(I + (C^TC - \gamma^{-2}C_z^TC_z)K)^{-1}C_z^T$.

The stability of the dual filter to the RH H_∞ control can be obtained.

Theorem 4.11. *The dual filter to the RH H_∞ control is stabilized if there exist K and L satisfying (4.175). In other words, the following matrix is Hurwitz:*

$$A - AS_N[I + \{C^TC - \gamma^{-2}C_z^TC_z\}S_N]^{-1}C^TC$$

where S_N is the solution calculated from (4.169) with $S_0 = K$.

Proof. The closed-loop system matrix of the filter is as follows:

$$A - AS_N[I + (C^TC - \gamma^{-2}C_z^TC_z)S_N]^{-1}C^TC$$
$$= A[I - S_N[I + (C^TC - \gamma^{-2}C_z^TC_z)S_N]^{-1}C^TC] \qquad (4.176)$$

where the transpose of the right side is

$$[I - C^TC\{I + S_N(C^TC - \gamma^{-2}C_z^TC_z)\}^{-1}S_N]A^T \qquad (4.177)$$

The matrix (4.176) is stable if and only if its transpose (4.177) is stable.

Compare (4.177) with the closed-loop system matrix of the control

$$[I - BB^T\{I + M_N(BB^T - \gamma^{-2}B_w B_w^T)\}^{-1}M_N]A \tag{4.178}$$

which is obtained with the changes listed in (4.173) and in this case $S_i = M_{N+1-i}$ and $K_0 = M_N$. Therefore, the above result follows from Theorem 3.15. This completes the proof.   ∎

The error covariance of the dual filter to the receding horizon H_∞ control is bounded above by a constant matrix as in the following theorem.

Theorem 4.12. *If there exist K and L satisfying (4.175), then the dual IIR filter $\hat{x}_i$ has the following bound:*

$$\frac{\sum_{i=i_0}^{\infty} e_i^T e_i}{\sum_{i=i_0}^{\infty} w_i^T w_i} \le \gamma^2 \tag{4.179}$$

Proof. The error dynamics can be represented as

$$e_{i+1} = (A - L_N C)e_i + (B_w - L_N D_w)w_i \tag{4.180}$$

where L_N is defined in (4.168). Let the function V_i be defined by

$$V_i = e_i^T K e_i \tag{4.181}$$

where K is positive definite. e_{i_0} is assumed to be zero. Then we have

$$\begin{aligned}
J &= \sum_{i=i_0}^{\infty}\left[e_i^T e_i - \gamma^2 w_i^T w_i\right] \le \sum_{i=i_0}^{\infty}\left[e_i^T e_i - \gamma^2 w_i^T w_i\right] + V_\infty - V_{i_0} \\
&\le \sum_{i=i_0}^{\infty}\left[e_i^T e_i - \gamma^2 w_i^T w_i + e_{i+1}^T K e_{i+1} - e_i^T K e_i\right] \\
&= \begin{bmatrix} e_i \\ w_i \end{bmatrix}^T \Theta \begin{bmatrix} e_i \\ w_i \end{bmatrix} \le 0
\end{aligned} \tag{4.182}$$

where

$$\Theta =$$

$$\begin{bmatrix} (A - L_N C)^T K(A - L_N C) - K + I & (A - L_N C)^T K(B_w - L_N D_w) \\ (B_w - L_N D_w)^T K(A - L_N C) & (B_w - L_N D_w)^T K(B_w - L_N D_w) - \gamma^2 I \end{bmatrix}$$

and the negative semidefiniteness of Θ is proved in a similar way to the cost monotonicity of the RH H_∞ control in Theorem 3.22. This completes the proof.   ∎

4.5 Finite Impulse Response Filters Based on Minimax Criterion

4.5.1 Linear Unbiased Finite Impulse Response Filters

In this section, the FIR filter without *a priori* initial state information represented by

$$\hat{x}_{k|k-1} = \sum_{i=k-N}^{k-1} H_{k-i} y_i + \sum_{i=k-N}^{k-1} L_{k-i} u_i \qquad (4.183)$$

is introduced for deterministic systems (4.158) and (4.159).

The system (4.158) and (4.159) will be represented in a standard batch form on the most recent time interval $[k - N, k]$, called the horizon. On the horizon $[k - N, k]$, the finite number of measurements is expressed in terms of the state x_k at the current time k as follows:

$$Y_{k-1} = \bar{C}_N x_k + \bar{B}_N U_{k-1} + \bar{G}_N W_{k-1} + \bar{D}_N W_{k-1} \qquad (4.184)$$

where $Y_{k-1}, U_{k-1}, W_{k-1}, \bar{C}_N, \bar{B}_N,$ and $\bar{G}_N$ are given by (4.28), (4.29), (4.30), (4.31), (4.32), and (4.33) respectively and $\bar{D}_N$ is defined by

$$\bar{D}_i \triangleq \left[\mathrm{diag}(\overbrace{D\ D\ \cdots\ D}^{i}) \right] = \left[\mathrm{diag}(\bar{D}_{i-1}, D) \right] \qquad (4.185)$$

$$1 \le i \le N \qquad (4.186)$$

An FIR filter with a batch form for the current state x_k can be expressed as a linear function of the finite measurements Y_{k-1} (4.28) and inputs U_{k-1} (4.29) on the horizon $[k - N, k]$ as follows:

$$\hat{x}_{k|k-1} \triangleq HY_{k-1} + LU_{k-1} \qquad (4.187)$$

where H and L are gain matrices of a linear filter. It is noted that the filter defined in (4.187) is an FIR structure without the requirement of any *a priori* information about the horizon initial state x_{k-N}.

Substituting (4.184) into (4.187) yields

$$\begin{aligned} \hat{x}_{k|k-1} &= HY_{k-1} + LU_{k-1} \\ &= H(\bar{C}_N x_k + \bar{B}_N U_{k-1} + \bar{G}_N W_{k-1} + \bar{D}_N W_{k-1}) + LU_{k-1} \end{aligned} \qquad (4.188)$$

Among multiple H and L, we want to choose one that satisfies the following relation:

$$\hat{x}_{k|k-1} = x_k \quad \text{for any } x_{k-N} \text{ and any } U_{k-1} \qquad (4.189)$$

for a zero disturbance $w_i = 0$. For stochastic systems, the unbiasedness constraint $E[\hat{x}_{k|k-1}] = E[x_k]$ is an accepted requirement for the filter. If noises become zero on the horizon $[k-N, k-1]$, then $\hat{x}_{k|k-1}$ and x_k become deterministic values, i.e. $\hat{x}_{k|k-1} = x_k$. For the corresponding property, if disturbances become zero on the horizon $[k - N, k - 1]$, then we have

$$x_{k+1} = Ax_k + Bu_k \tag{4.190}$$
$$y_k = Cx_k \tag{4.191}$$

where we require $\hat{x}_{k|k-1} = x_k$. This is called a deadbeat condition and the model (4.190) and (4.191) is called a nominal model. It is noted that, conventionally, noises are assumed to have zero mean and thus noises fluctuate around zero. Likewise, the disturbance is assumed to fluctuate around zero. The relation (4.189) is called an unbiased condition in the deterministic sense, as seen in Figure 4.8.

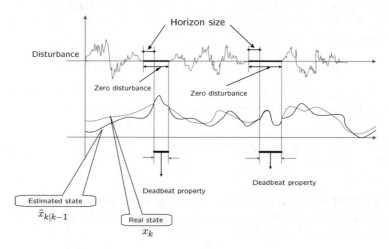

Fig. 4.8. Deadbeat property without disturbance

In the case of $w_i = 0$ we have

$$\hat{x}_{k|k-1} = H(\bar{C}_N x_k - \bar{B}_N U_{k-1}) + LU_{k-1} \tag{4.192}$$

Therefore, the following constraints are required for the unbiased condition:

$$H\bar{C}_N = I \quad \text{and} \quad L = -H\bar{B}_N \tag{4.193}$$

which will be called the unbiased constraint. The FIR filter with the unbiased constraints for the current state x_k can be expressed as

$$\hat{x}_{k|k-1} = H(Y_{k-1} - \bar{B}_N U_{k-1}) \tag{4.194}$$

and substituting (4.193) into (4.188) yields

$$e_k \overset{\triangle}{=} x_k - \hat{x}_{k|k-1} = H(\bar{G}_N + \bar{D}_N)W_{k-1} \tag{4.195}$$

4.5.2 L_2-E Finite Impulse Response Filters

The objective now is to obtain the optimal gain matrices H and L in such a way that the worst-case gain between the estimation error of the estimate $\hat{x}_{k|k-1}$ and disturbance has a minimum value as follows:

$$\min_{H,L} \max_{w_i} \frac{[x_k - \hat{x}_{k|k-1}]^T [x_k - \hat{x}_{k|k-1}]}{\sum_{i=1}^N w_{k-i}^T w_{k-i}} \qquad (4.196)$$

The numerator of the performance criterion (4.196) considers only the current estimation error. The FIR filter obtain from Equation (4.196) will be called an $L_2 - E$ FIR (LEF) filter. The H and L coming from the performance criterion (4.196) are dependent on system parameters such as A, B, C, and G, not dependent on any specific real disturbances.

By using (4.195), (4.196) can be written as

$$\min_{H,L} \max_{W_{k-1}} \frac{W_{k-1}^T (\bar{G}_N + \bar{D}_N)^T H^T H (\bar{G}_N + \bar{D}_N) W_{k-1}}{W_{k-1}^T W_{k-1}}$$

$$= \min_{H,L} \lambda_{\max} \left[(\bar{G}_N + \bar{D}_N)^T H^T H (\bar{G}_N + \bar{D}_N) \right]$$

$$= \min_{H,L} \lambda_{\max} \left[H (\bar{G}_N + \bar{D}_N)(\bar{G}_N + \bar{D}_N)^T H^T \right] \qquad (4.197)$$

for H and L satisfying (4.193).

Before obtaining the solution to the optimization problem (4.197), we introduce a useful result in the following lemma.

Lemma 4.13. *Suppose that the following general maximum eigenvalue optimization problem is given:*

$$\min_{H} \lambda_{\max} \left[(HA - B)C(HA - B)^T \right] \qquad (4.198)$$

$$\text{subject to}$$

$$HE = F \qquad (4.199)$$

where $C = C^T > 0$, $D = D^T > 0$, and A, B, C, D, E, and F are constant matrices and have appropriate dimensions. The solution to the optimization problem (4.198) and (4.199) is given by

$$H = \left[BCA^T + (F - BCA^T \Pi^{-1} E)(E^T \Pi^{-1} E)^{-1} E^T \right] \Pi^{-1} \qquad (4.200)$$

where $\Pi = ACA^T$.

Proof. If there exists an H_B such that

$$(H_B A - B)C(H_B A - B)^T \leq (HA - B)C(HA - B)^T \qquad (4.201)$$

where $HE = H_B E = F$ and H is arbitrary, we have

$$\alpha^T[(H_B A - B)C(H_B A - B)^T]\alpha = \lambda_{max}[(H_B A - B)C(H_B A - B)^T]\alpha^T\alpha$$
$$\leq \alpha^T(HA - B)C(HA - B)^T\alpha$$
$$= \lambda_{max}[(HA - B)C(HA - B)^T]\alpha^T\alpha \quad (4.202)$$

for the vector α that is an eigenvector of $(H_B A - B)C(H_B A - B)^T$ corresponding to $\lambda_{\max}(H_B A - B)C(H_B A - B)^T$. Since α is nonzero, the following inequality is obtained from (4.202):

$$\lambda_{max}[(H_B A - B)C(H_B A - B)^T] \leq \lambda_{max}[(HA - B)C(HA - B)^T] \quad (4.203)$$

which means that H_B is the solution to the optimization problem (4.198) and (4.199).

Thus, we try to find out H_B such that the inequality (4.201) is satisfied. In order to make the problem tractable, we change the inequality (4.201) to a scalar one as

$$v^T(H_B A - B)C(H_B A - B)^T v \leq v^T(HA - B)C(HA - B)^T v \quad (4.204)$$

for arbitrary vector v. It can be seen from the inequality (4.204) that the minimum value of $v^T[(HA-B)C(HA-B)^T]v$ is the left side of the inequality (4.204).

Now, all that we have to do is to find the minimum value of right side of the inequality (4.204) with respect to H and show that it is the same as the left side of (4.204). The right side of (4.204) can be written as

$$(w^T A - v^T B)C(w^T A - v^T B)^T \quad (4.205)$$

where $H^T v = w$. From $HE = F$ in (4.199), w should satisfy

$$E^T w = F^T v \quad (4.206)$$

Now, the following cost function to be optimized is introduced:

$$\phi = (w^T A - v^T B)C(w^T A - v^T B)^T + \lambda^T(E^T w - F^T v) \quad (4.207)$$

where λ is the vector of a Lagrange multiplier. In order to minimize ϕ, two necessary conditions

$$\frac{\partial \phi}{\partial w} = 0, \quad \frac{\partial \phi}{\partial \lambda} = 0 \quad (4.208)$$

are required, which give

$$w = (ACA^T)^{-1}(ACB^T v + E\lambda) \quad (4.209)$$

Substituting (4.209) into (4.206) yields

$$\lambda = (E^T \Pi^{-1} E)^{-1} [-E^T \Pi^{-1} A C B^T v + F^T v] \tag{4.210}$$

where $\Pi = ACA^T$. Thus, w is obtained as

$$w = \Pi^{-1} \{ A C B^T + E (E^T \Pi^{-1} E)^{-1} (-E^T \Pi^{-1} A C B^T + F^T) \} v \tag{4.211}$$

Since $H^T v = w$, H can be written as (4.200). This completes the proof.   ∎

By using the result of Lemma 4.3, the solution to the optimization problem (4.197) can be obtained according to the following correspondences:

$$
\begin{aligned}
A &\longleftarrow \bar{G}_N + \bar{D}_N \\
B &\longleftarrow O \\
C &\longleftarrow I \\
E &\longleftarrow \bar{C}_N \\
F &\longleftarrow I \\
\Pi &\longleftarrow (\bar{G}_N + \bar{D}_N)(\bar{G}_N + \bar{D}_N)^T
\end{aligned}
\tag{4.212}
$$

The filter gain L can be obtained from (4.193). In the following theorem, we summarize what we have done so far.

The following theorem provides the solution to the optimal criterion (4.196) with the unbiased condition (4.193).

Theorem 4.14. *Assume that $\{A, C\}$ of the given system is observable. The LEF filter is as follows:*

$$\hat{x}_{k|k-1} = (\bar{C}_N^T \Xi_N^{-1} \bar{C}_N)^{-1} \bar{C}_N^T \Xi_N^{-1} (Y_{k-1} - \bar{B}_N U_{k-1}) \tag{4.213}$$

where

$$
\begin{aligned}
\Xi_i &\triangleq \bar{G}_i \bar{G}_i^T + \bar{D}_i \bar{D}_i^T \\
&= \begin{bmatrix} \bar{G}_{i-1} \bar{G}_{i-1}^T + \bar{D}_{i-1} \bar{D}_{i-1}^T & 0 \\ 0 & I \end{bmatrix} + \begin{bmatrix} \bar{C}_{i-1} \\ C \end{bmatrix} A^{-1} G G^T A^{-T} \begin{bmatrix} \bar{C}_{i-1} \\ C \end{bmatrix}^T \\
&= \begin{bmatrix} \Xi_{i-1} & 0 \\ 0 & I \end{bmatrix} + \begin{bmatrix} \bar{C}_{i-1} \\ C \end{bmatrix} A^{-1} G G^T A^{-T} \begin{bmatrix} \bar{C}_{i-1} \\ C \end{bmatrix}^T
\end{aligned}
\tag{4.214}
$$

Note that the optimum gain matrix H_B (4.196) requires computation only on the interval $[0, N]$ once and is time-invariant for all horizons. It is noted that the estimation (4.213) holds for $k \geq k_0 + N$.

The LEF filter $\hat{x}_{k|k-1}$ has the unbiased property when applied to the nominal systems without disturbances (4.190) and (4.191), whereas the LEF filter $\hat{x}_{k|k-1}$ is obtained from systems (4.158) and (4.159) with an additive system and measurement disturbances w_k. The deadbeat property of the LEF filter is given in the following corollary.

Corollary 4.15.   *Assume that $\{A, C\}$ is observable and $N \geq n$. Then, the LEF filter $\hat{x}_{k|k-1}$ given by (4.213) provides the exact state when there are no disturbances on the horizon $[k - N, k]$.*

Proof. When there are no disturbances on the horizon $[k - N, k]$, $Y_{k-1} - \bar{B}_N U_{k-1}$ is determined by the current state x_k as $Y_{k-1} - \bar{B}_N U_{k-1} = \bar{C}_N x_k$. Therefore, the following is true:

$$
\begin{aligned}
\hat{x}_{k|k-1} &= H_B(Y_{k-1} - \bar{B}_N U_{k-1}) \\
&= (\bar{C}_N^T \Xi_N^{-1} \bar{C}_N)^{-1} \bar{C}_N^T \Xi_N^{-1} \bar{C}_N x_k \\
&= x_k
\end{aligned}
$$

This completes the proof.   ∎

Note that the above deadbeat property indicates finite convergence time and fast tracking ability of the LEF filter. Thus, it can be expected that the approach would be appropriate for fast estimation and detection of signals with unknown times of occurrence, which arise in many problems, such as fault detection and diagnosis of various systems, manoeuvre detection, target tracking of flying objects, etc.

In this section, the LEF filter $\hat{x}_{k|k-1}$ in batch form is represented in an iterative form, for computational advantage, and then shown to be similar in form to the MVF filter with unknown horizon initial state presented in [HKK01][KKH02].

Theorem 4.16.   *Assume that $\{A, C\}$ is observable and $N \geq n$. Then, the LEF filter $\hat{x}_{k|k-1}$ is given on the horizon $[k - N, k]$ as follows:*

$$
\hat{x}_{k|k-1} = \Omega_N^{-1} \check{x}_k \tag{4.215}
$$

where $k \geq k_0 + N$ and Ω_N and $\check{x}_k$ are obtained from following recursive equations:

$$
\Omega_{i+1} = [I + A^{-T}(\Omega_i + C^T C)A^{-1}GG^T]^{-1}A^{-T}(\Omega_i + C^T C)A^{-1} \tag{4.216}
$$

$$
\begin{aligned}
\check{x}_{k-N+i+1} = &[I + A^{-T}(\Omega_i + C^T C)A^{-1}GG^T]^{-1}A^{-T} \\
&[\check{x}_{k-N+i} + C^T y_{k-N+i} + (\Omega_i + C^T C)A^{-1}Bu_{k-N+i}] \tag{4.217}
\end{aligned}
$$

with $\check{x}_{k-N} = 0$ and $\Omega_0 = 0$.

Proof. If Q_w and R_v are replaced with unit matrices in the MVF filter, then the proof is the same as that of Theorem 4.5. ∎

It is surprising to observe that the LEF filter $\hat{x}_{k|k-1}$ (4.215) with an iterative form is the same as the MVF filter with unknown horizon initial state in [KKP99], where the covariances of the system noise and the measurement noise are taken as unit matrices.

To summarize, the LEF filter is linear with the most recent finite measurements and inputs, does not require *a priori* information about the horizon initial state, and has the unbiased property. Thus, it is surprising, in that a closed-form solution exists even with the unbiased condition. The LEF filter can be represented in both a standard batch form and an iterative form that has computational advantages. The LEF filter with the FIR structure for deterministic systems is similar in form to the existing RHUFF for stochastic systems with the unit covariance matrices of both the system noise and the measurement noise. Furthermore, owing to the FIR structure, the LEF filter is believed to be robust against temporary modelling uncertainties or numerical errors, whereas other minimax filters and H_∞ filters with an IIR structure may show poor robustness in these cases.

The LEF filter will be very useful for many signal processing problems where signals are represented by state-space models. In addition, the proposed LEF filter is useful for control problems and can substitute the commonly used H_∞ filters where initial state information is unknown, or can be ignored.

For the following structure and performance criterion:

$$\hat{x}_{k|k-1} = \sum_{i=k_0}^{k-1} H_{k-i} y_i + \sum_{i=k_0}^{k-1} L_{k-i} u_i \tag{4.218}$$

$$\min_{H,L} \max_{w_i} \frac{[x_k - \hat{x}_{k|k-1}]^T [x_k - \hat{x}_{k|k-1}]}{\sum_{i=k_0}^{k-1} w_i^T w_i} \tag{4.219}$$

the L_2-E IIR filter still holds in the form of an IIR structure.

$$\hat{x}_{k|k-1} = (\bar{C}_{k-k_0}^T \, \Xi_{k-k_0}^{-1} \, \bar{C}_{k-k_0})^{-1} \bar{C}_{k-k_0}^T \, \Xi_{k-k_0}^{-1} (Y_{[k_0,k-1]}$$
$$- \bar{B}_{k-k_0} U_{[k_0,k-1]}) \tag{4.220}$$

Note that the estimation (4.220) holds for $k \geq k_0 + n$, and $Y_{[k_0,k-1]}$ and $U_{[k_0,k-1]}$ are defined as follows:

$$Y_{[k_0,k-1]} \triangleq [y_{k_0}^T \ \ y_{k_0+1}^T \ \ \cdots \ \ y_{k-1}^T]^T$$
$$U_{[k_0,k-1]} \triangleq [u_{k_0}^T \ \ u_{k_0+1}^T \ \ \cdots \ \ u_{k-1}^T]^T$$

However, in this case, the computational time increases as time elapses. A recursive form of the batch form (4.220) is introduced as follows:

$$\hat{x}_{k|k-1} = \Omega_k^{-1} \tilde{x}_k \tag{4.221}$$

and also is given in the following form as

$$\hat{x}_{k+1|k} = A\hat{x}_{k|k-1} + Bu_k + AP_kC^T(I + CP_kC^T)^{-1}(y_k - C\hat{x}_{k|k-1}) \quad (4.222)$$
$$P_{k+1} = A(I + P_kC^TC)^{-1}P_kA^T + GG^T \quad (4.223)$$

Note that (4.221), (4.222), and (4.223) hold for $k \geq k_0 + n$.

The starting state and P_k of (4.222) and (4.223) can be obtained from (4.221) and Ω_k. It is surprising to see that the LEF filter with the IIR type for deterministic systems is similar in form to the Kalman filter for stochastic systems with unit system and measurement noise covariances.

4.5.3 H_∞ Finite Impulse Response Filter

For the design of an FIR filter, we can consider an H_∞ performance criterion as

$$\inf_{H,L} \sup_{w_i \neq 0} \frac{\sum_{i=k-N}^{k}\{x_i - \hat{x}_{i|i-1}\}^T\{x_i - \hat{x}_{i|i-1}\}}{\sum_{i=k-N}^{k} w_i^T w_i} \quad (4.224)$$

If disturbances are all zero on $[k-N\ ,\ k]$ and are not zero outside this interval, then the numerator in (4.224) is not zero while the denominator in (4.224) is zero. It means that the cost function (4.224) approaches ∞. Thus, the finite horizon H_∞ performance criterion (4.224) does not work in the FIR structure.

Thus, the following infinite horizon H_∞ performance criterion is employed to design an H_∞ FIR filter:

$$\inf_{H,L} \sup_{w_i \neq 0} \frac{\sum_{i=k_0}^{\infty}\{x_i - \hat{x}_{i|i-1}\}^T\{x_i - \hat{x}_{i|i-1}\}}{\sum_{i=k_0}^{\infty} w_i^T w_i} \quad (4.225)$$

Both the filter horizon and the performance horizon in (4.224) are both $[k - N,\ k]$ whereas in (4.225) they are $[k - N,\ k]$ and $[k_0\ ,\ \infty]$ respectively.

Similarly to LEF filters, we assume that the filter is of the FIR form (4.187). Additionally, the unbiased condition (4.189) is adopted.

The starting point in this section is to derive the transfer function $T_{ew}(z)$. Exogenous input w_k satisfies the following state model on W_{k-1}:

$$W_k = A_w W_{k-1} + B_w w_k \quad (4.226)$$

where

$$A_w = \begin{bmatrix} 0 & I & 0 & \cdots & 0 \\ 0 & 0 & I & \ddots & 0 \\ \vdots & \vdots & \ddots & \ddots & \vdots \\ 0 & 0 & \cdots & 0 & I \\ 0 & 0 & \cdots & 0 & 0 \end{bmatrix} \in R^{pN \times pN}, \quad B_w = \begin{bmatrix} 0 \\ 0 \\ 0 \\ \vdots \\ I \end{bmatrix} \in R^{pN \times p} \quad (4.227)$$

Thus, we have $W(z) = (zI - A_w)^{-1}B_w w(z)$.

It follows from (4.184) that

$$Y_{k-1} - \bar{B}_N U_{k-1} = \bar{C}_N x_k + (\bar{G}_N + \bar{D}_N) W_{k-1} \tag{4.228}$$

Pre-multiplying (4.228) by H and using the unbiasedness constraint $H\bar{C}_N = I$ gives

$$e_k = \hat{x}_{k|k-1} - x_k = H(\bar{G}_N + \bar{D}_N) W_{k-1} \tag{4.229}$$

From (4.226) and (4.229) we can obtain $T_{ew}(z)$ as follows:

$$T_{ew}(z) = H(\bar{G}_N + \bar{D}_N)(zI - A_w)^{-1} B_w \tag{4.230}$$

Using Lemma 2.13, we have the following theorem for the optimal H_∞ FIR filter:

Theorem 4.17. *Assume that the following LMI problem is feasible:*

$$\min_{F,X} \gamma_\infty \text{ subject to}$$

$$\begin{bmatrix} -X & XA_w & XB_w & 0 \\ A_w^T X & -X & 0 & S_0 \\ B_w^T X & 0 & -\gamma_\infty I & 0 \\ 0 & S_0^T & 0 & -\gamma_\infty I \end{bmatrix} < 0$$

where

$$S_0 = (\bar{G}_N + \bar{D}_N)^T M F^T + (\bar{G}_N + \bar{D}_N)^T H_0^T \tag{4.231}$$
$$H_0 = (\bar{C}_N^T \bar{C}_N)^{-1} \bar{C}_N^T \tag{4.232}$$

and columns of M consist of the basis of the null space of $\bar{C}_N^T$. Then, the optimal gain matrix of the H_∞ FIR filter (4.187) is given by

$$H = FM^T + H_0$$

Proof. According to the bounded real lemma, the condition $\|T_{ew}(z)\|_\infty < \gamma_\infty$ is equivalent to

$$\begin{bmatrix} -X & XA_w & XB_u & 0 \\ A_u^T X & -X & 0 & (\bar{G}_N + \bar{D}_N)^T H^T \\ B_u^T X & 0 & -\gamma_\infty I & 0 \\ 0 & H(\bar{G}_N + \bar{D}_N) & 0 & -\gamma_\infty I \end{bmatrix} < 0 \tag{4.233}$$

The equality constraint $H\bar{C}_N = I$ can be eliminated by computing the null space of $\bar{C}_N^T$. If we use following correspondences in Lemma A.3:

$$A \rightarrow \bar{C}_N^T$$
$$X \rightarrow H^T$$
$$Y \rightarrow I$$
$$A^\perp \rightarrow \bar{C}_N(\bar{C}_N^T \bar{C}_N)^{-1}$$
$$X = A^\perp Y + MV \rightarrow H^T = \bar{C}_N(\bar{C}_N^T \bar{C}_N)^{-1} + MF$$

then it can be seen that all solutions to the equality constraint $H\bar{C}_N = I$ are parameterized by

$$H = FM^T + H_0 \qquad (4.234)$$

where F is a matrix containing the independent variables. Replacing H by $FM^T + H_0$, the LMI condition in (4.233) is changed into the one in the Theorem 4.17. This completes the proof. ∎

4.5.4* H_2/H_∞ Finite Impulse Response Filters

The H_2 filter is to estimate the state using the measurements of y_i so that the H_2 norm from w_i to the estimation error is minimized.

Here, we consider the H_2 FIR filter. Using the result (2.268) in Section 2.7, we have the following theorem for an H_2 FIR filter:

Theorem 4.18. *Assume that the following LMI problem is feasible:*

$$\min_{F,W} \ \mathrm{tr}(W)$$

$$\text{subject to}$$

$$\begin{bmatrix} W & S_0^T \\ S_0 & I \end{bmatrix} > 0$$

where S_0 and H_0 are given in (4.231) and (4.232), and columns of M consist of the basis of the null space of $\bar{C}_N^T$.

Then the optimal gain matrix of the H_2 FIR filter (4.187) is given by

$$H = FM^T + H_0, \ \ L = -H\bar{B}_N$$

Proof. The H_2 norm of the transfer function $T_{ew}(z)$ in (4.230) is obtained by

$$\|T_{ew}(z)\|_2^2 = \mathrm{tr}\big(H(\bar{G}_N + \bar{D}_N)P(\bar{G}_N + \bar{D}_N)^T H^T\big)$$

where

$$P = \sum_{i=0}^{\infty} A_u^i B_u B_u^T (A_u^T)^i$$

Since $A_u^i = 0$ for $i \geq N$, we obtain

$$P = \sum_{i=0}^{\infty} A_u^i B_u B_u^T (A_u^T)^i = \sum_{i=0}^{N-1} A_u^i B_u B_u^T (A_u^T)^i = I$$

Thus we have

$$\|T_{ew}(z)\|_2^2 = \mathrm{tr}\big\{H(\bar{G}_N + \bar{D}_N)(\bar{G}_N + \bar{D}_N)^T H^T\big\} \qquad (4.235)$$

Introduce a matrix variable W such that

$$W > H(\bar{G}_N + \bar{D}_N)(\bar{G}_N + \bar{D}_N)^T H^T \qquad (4.236)$$

Then $\mathrm{tr}(W) > \|T_H(z)\|_2^2$. By the Schur complement, (4.236) is equivalently changed into

$$\begin{bmatrix} W & H(\bar{G}_N + \bar{D}_N) \\ (\bar{G}_N + \bar{D}_N)^T H^T & I \end{bmatrix} > 0 \qquad (4.237)$$

Hence, by minimizing $\mathrm{tr}(W)$ subject to $H\bar{C}_N = I$ and the above LMI, we can obtain the optimal gain matrix H for the H_2 FIR filter. The equality constraint $H\bar{C}_N = I$ can be eliminated in exactly the same way as in the H_∞ FIR filter. This completes the proof. ∎

From the viewpoint that the square of the H_2 norm is the error variance due to white noise with unit intensity, we can easily show that $\|T_{ew}(z)\|_2^2$ in (4.235) is the error variance as follows:

$$\begin{aligned} \|T_{ew}(z)\|_2^2 &= E\{e_k^T e_k\} = \mathrm{tr}\big(E\{e_k e_k^T\}\big) \\ &= \mathrm{tr}\big(H(\bar{G}_N + \bar{D}_N)E\{W_{k-1}W_{k-1}^T\}(\bar{G}_N + \bar{D}_N)H^T\big) \\ &= \mathrm{tr}\big(H(\bar{G}_N + \bar{D}_N)(\bar{G}_N + \bar{D}_N)H^T\big) \end{aligned}$$

Using the LMI representation for the H_2 FIR filter, we can obtain the mixed H_2/H_∞ FIR filters. Let us define γ_2^* to be the $\|T_{ew}(z)\|_2^2$ due to the optimal H_2 FIR filter. Under the constraint that the estimation error is guaranteed to be bounded above by $\alpha\gamma_2^*$ for $\alpha > 1$, we try to find out the optimal H_∞ filter. We have the following theorem for the mixed H_2/H_∞ FIR filter:

Theorem 4.19. *Assume that the following LMI problem is feasible:*

$$\min_{W,X,F} \gamma_\infty$$

subject to

$$\mathrm{tr}(W) < \alpha\gamma_2^*, \quad \text{where} \quad \alpha > 1$$

$$\begin{bmatrix} W & S_0^T \\ S_0 & I \end{bmatrix} > 0 \qquad (4.238)$$

$$\begin{bmatrix} -X & XA_u & XB_u & 0 \\ A_u^T X & -X & 0 & S_0 \\ B_u^T X & 0 & -\gamma_\infty I & 0 \\ 0 & S_0^T & 0 & -\gamma_\infty I \end{bmatrix} < 0 \qquad (4.239)$$

where S_0 and H_0 are given in (4.231) and (4.232) and M^T is the basis of the null space of $\bar{C}_N^T$. Then, the gain matrix of the mixed H_2/H_∞ FIR filter of the form (4.187) is given by

$$H = FM + H_0$$

Proof. So clear, hence omitted.

The above mixed H_2/H_∞ FIR filtering problem allows us to design the optimal FIR filter with respect to the H_∞ norm while assuring a prescribed performance level in the H_2 sense. By adjusting $\alpha > 0$, we can trade off the H_∞ performance against the H_2 performance.

The H_2 FIR filter can be obtained analytically from [KKH02]

$$H_B = (\bar{C}_N^T \Xi_N^{-1} \bar{C}_N)^{-1} \bar{C}_N^T \Xi_N^{-1}$$

Thus we have

$$\gamma_2^* = \text{tr}(H_B \Xi_N H_B^T)$$

where Ξ_N are obtained from (4.214).

It is noted that γ_∞ in Theorem 4.19 is minimized for a guaranteed H_2 norm. On the contrary, γ_2 can be minimized for a guaranteed H_∞ norm as follows:

$$\min_{W,X,F} \gamma_2$$

subject to

$$\text{tr}(W) < \gamma_2$$

(4.238)

$$\begin{bmatrix} -X & XA_u & XB_u & 0 \\ A_u^T X & -X & 0 & S_0 \\ B_u^T X & 0 & -\alpha\gamma_\infty I & 0 \\ 0 & S_0^T & 0 & -\alpha\gamma_\infty I \end{bmatrix} < 0 \quad (4.241)$$

(4.240)

for $\alpha > 1$.

Example 4.3

To illustrate the validity of the H_∞ and mixed H_2/H_∞ FIR filters, numerical examples are given for a linear discrete time-invariant state-space model from [KKP99]

$$x_{k+1} = \begin{bmatrix} 0.9950 & 0.0998 \\ -0.0998 & 0.9950 + \delta_k \end{bmatrix} x_k + \begin{bmatrix} 0.1 \\ 0.1 \end{bmatrix} u_k + \begin{bmatrix} 1 & 0 \\ 1 & 0 \end{bmatrix} w_k \quad (4.242)$$

$$y_k = [1 \quad 0]x_k + [0 \quad 1]w_k$$

where δ_k is a model uncertain parameter. In this example, H_2 and H_2 FIR filters are Kalman and MVF filters respectively. We have designed a mixed H_2/H_∞ filter with $N = 10$, $\alpha = 1.3$, and $\delta_k = 0$. Table 4.1 compares the H_2 and H_∞ norms of the conventional IIR filters and the FIR filters. It is shown that performances of the mixed H_2/H_∞ FIR filter lie between those of the H_2 FIR filter and the H_∞ FIR filter. In the case that there is no disturbance and the real system is exactly matched to the model, the performances of IIR filters seem better than those of FIR filters. However, it is not guaranteed to be true in real applications. As mentioned previously, it is general that the FIR filter is robust against temporary modelling uncertainties, since it utilizes only finite measurements on the most recent horizon. To illustrate this fact and the fast convergence, we applied the mixed FIR H_2/H_∞ filter to a system which actually has temporary uncertainty. The uncertain model parameter δ_k is considered as

$$\delta_k = \begin{cases} 1, \ 50 \le k \le 100 \\ 0, \ \text{otherwise} \end{cases}$$

Figure 4.9(a) compares the estimation errors of the mixed H_2/H_∞ FIR filter with those of the H_2 and the H_∞ filter of IIR type in the case that the exogenous input w_k is given by

$$w_k = 0.1 \begin{bmatrix} e^{-\frac{k}{30}} \\ e^{-\frac{k}{30}} \end{bmatrix}$$

and Figure 4.9(b) compares the estimation errors in the case that the exogenous input w_k is given by

$$w_k = \begin{bmatrix} w_{1,k} \\ w_{2,k} \end{bmatrix}, \ \text{where } w_{1,k} \sim (0,1), w_{2,k} \sim (0,1)$$

In both cases, it is clearly shown that the mixed H_2/H_∞ FIR filter is more robust against the uncertainty and faster in convergence. Therefore, it is expected that the FIR filter can be usefully used in real applications. ∎

Table 4.1. H_2 and H_∞ norm at $N = 10$ with $\alpha = 1.3$

	H_∞ norm	H_2 norm
H_∞ filter	2.0009	2.0223
MV filter	2.9043	1.8226
H_∞ FIR filter	4.2891	3.7295
MVF filter	5.4287	2.7624
Mixed H_2/H_∞ FIR filter	4.4827	3.1497

4.6 References

The dual filter to the receding horizon LQC appeared and its stability was first discussed in [KP77c]. The terminology "Kalman filter with the frozen gain" was used in [BLW91]. The guaranteed performance using the duality and the monotonicity condition for the filter in Theorem 4.1 appeared in [JHK04],

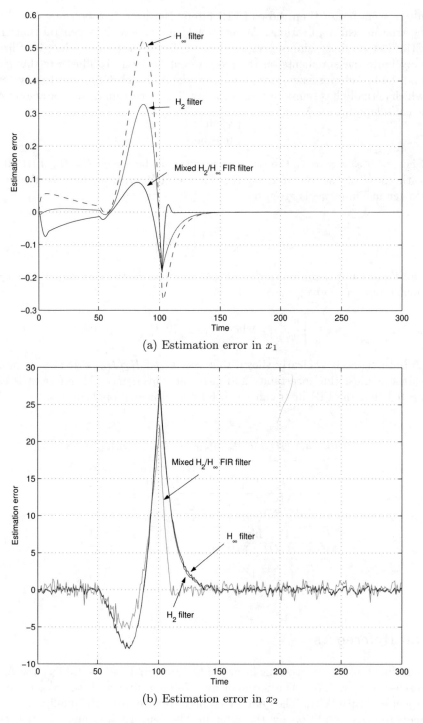

(a) Estimation error in x_1

(b) Estimation error in x_2

Fig. 4.9. Mixed H_2/H_∞ FIR filter for system (4.242)

where the upper bound of the estimation error in Theorem 4.2 is derived in parallel with Theorem 3.13. The special cases for the trace minimization problem in Lemma 4.3 appeared in many references. General results are proved in this book for further uses.

There are several known results on FIR filters for limited models. For special discrete stochastic systems, linear FIR filters were introduced in [Jaz68][Sch73][Bux74] [Bie75][BK85]. For deterministic discrete-time systems with zero noises, a least-squares filter with an FIR structure was given in [LL99]. Some filters without using the initial state information were introduced by a modification from the Kalman filter [DX94], [KKP99]. These filters handle the infinite covariance of the initial state without any consideration of optimality, which requires some conditions such as nonsingularity of the system matrix A. In this chapter, FIR filters are introduced for general models having no limited conditions, such as zero system noises and the nonsingularity of the system matrix A.

The MVF filter with nonsingular A in Theorem 4.4 and its recursive form within each horizon in Theorem 4.5 appeared in [KKH02]. The MVF filter with general A in Theorem 4.6, the batch form of the Kalman filter in Theorem 4.7, and a recursive form within each horizon in Theorem 4.10 are first introduced in [JHK04]. The recursive computation for the initial value of the Kalman filter in Theorems 4.8 and 4.9 appeared in [HKK99] and their proofs in this chapter are somewhat different from the originals.

The dual filter to the receding horizon H_∞ control and its stability in Theorem 4.11 and the H_∞ norm preserving property in Theorem 4.12 are first introduced in [JHK04]. The special cases for the eigenvalue minimization problem in Lemma 4.13 appeared in many references. General results are proved in this book for further uses. The deadbeat filter appears in [Kuc91].

The LEF filter for deterministic systems in Theorem 4.14 and its recursive form in Theorem 4.16 appeared in [HKK02]. The H_∞ FIR filter in Theorem 4.17 is first introduced in this book.

4.7 Problems

4.1. Suppose that, in dual IIR filters (4.10), the difference e_{i_0} between the real state x_{i_0} and the estimated one $\hat{x}_{i_0}$ at the initial time satisfies the following inequality:

$$e_{i_0} e_{i_0}^T \leq P_{N+1} + (A - LC)^{-(i-i_0)}(P_d - P_{N+1})(A - LC)^{-(i-i_0)T} \ (4.243)$$

for all $i = i_0, i_0 + 1, \cdots$ and a fixed matrix $P_d > P_{N+1}$, where $L = AP_N C^T (R_v + CP_N C^T)^{-1}$. Show that $(x_i - \hat{x}_{i|i-1})(x_i - \hat{x}_{i|i-1})^T < P_d$ in the case that disturbances do not show up.

4.2. Consider the dual IIR filter (4.10).

(1) If P_i is the value of the optimal error variance when

$$J_{i_0,i} \overset{\triangle}{=} E[(x_i - \hat{x}_{i|i-1})^T (x_i - \hat{x}_{i|i-1})|y_1, y_2, \cdots, y_{i-1}] \qquad (4.244)$$

is minimized, show that the inequality $J^*_{i_0,i+1} - J^*_{i_0,i} \leq 0$ holds under the following condition:

$$(A - L_iC)P_i(A - L_iC)^T + GQ_wG^T + L_iR_vL_i^T - P_i \leq 0 \quad (4.245)$$

where $L_i = AP_iC^T(R_v + CP_iC^T)^{-1}$ and $J^*_{i_0,i} = min(4.244)$.
(2) Show that $A - L_iC$ should be Hurwitz in the inequality (4.245).
(3) Show the following matrix inequality:

$$\left[I + C(e^{jw}I - A)^{-1}L_i\right](R_v + CP_iC^T)\left[I + L_i^T(e^{-jw}I - A^T)^{-1}C^T\right] \geq R_v$$

4.3. Consider the following model:

$$x_{k+1} = \begin{bmatrix} 0.9950 & 0.0998 \\ -0.0998 & 0.9950 + \delta_k \end{bmatrix} x_k + \begin{bmatrix} 0.1 \\ 0.1 \end{bmatrix} u_k + \begin{bmatrix} 1 & 0 \\ 1 & 0 \end{bmatrix} w_k \quad (4.246)$$

$$y_k = [1 \quad 0]x_k + v_k \qquad (4.247)$$

where uncertain model parameter δ_k is defined in (4.146), $Q_w = 0.1I$, and $R_v = 0.1$. The initial state is taken as $x_0^T = [2 \quad 2]$.

(1) Derive the steady-state Kalman filter and find its poles.
(2) The control input is taken as a constant value by 1. Via simulation, compare the transient response of the Kalman filter with that of the MVF filter with horizon size 5. Repeat the simulations while changing the horizon size and discuss how the transient response of FIR filters is changed.

4.4. Assume that w_k and v_k are not uncorrelated in the system (4.1) and (4.2) and their covariances are given by

$$E\begin{bmatrix} w_k \\ v_k \end{bmatrix}[w_k^T \quad v_k^T] = \begin{bmatrix} Q_w & S_{wv} \\ S_{wv}^T & R_v \end{bmatrix} \qquad (4.248)$$

Find out the MVF filter for the above case.

4.5. Let $P_{1,i}$ and $P_{2,i}$ be the solutions of two DREs

$$P_{i+1} = AP_iA^T - AP_iC^T[CP_iC^T + R]^{-1}CP_iA^T + Q \qquad (4.249)$$

with the same A, C and R matrices but different Q matrices Q_1, Q_2 and with initial conditions $P_1(0) = P_1 \geq 0$ and $P_2(0) = P_2 \geq 0$ respectively. Show that the difference between the two solutions $\tilde{P}_i = P_{2,i} - P_{1,i}$ satisfies the following equation:

$$\tilde{P}_{i+1} = A_c\tilde{P}_iA_c^T - \Lambda_i(R + CP_{2,i}C^T)\Lambda_i + \tilde{Q} \qquad (4.250)$$

where

$$A_c \triangleq A - AP_{1,i}C^T(R + CP_{1,i}C^T)^{-1}C$$
$$\Lambda_i \triangleq A_c\tilde{P}_iC^T(R + CP_{2,i}C^T)^{-1}$$
$$\tilde{Q} \triangleq Q_1 - Q_2$$

4.6. Consider an MVF filter and an augmented linear model (4.27) without control efforts

$$Y_{k-1} = \bar{C}_Nx_k + \bar{G}_NW_{k-1} + V_{k-1} \qquad (4.251)$$

(1) Show that the minimum variance unbiased FIR filter $\hat{x}_{k|k-1} = HY_{k-1}$ for the criterion

$$E[\|HY_{k-1} - x_k\|^2]$$

is determined so that each singular value $\sigma_i\left(H(\bar{G}_NQ_N\bar{G}_N^T + R_N)^{\frac{1}{2}}\right)$ is minimized. $(\sigma_1(\cdot) \geq \sigma_2(\cdot) \geq \cdots)$

(2) In order to handle the high risk more efficiently, we may employ the following criterion:

$$2\gamma^2 \log E \exp(\frac{1}{2\gamma^2}\|HY_{k-1} - x_k\|^2) \qquad (4.252)$$

Show that, in order to minimize the performance criterion (4.252), we have only to solve the following LMI:

$$\min_H \; \gamma^2 \log \det \begin{bmatrix} I & \gamma^{-1}M^T \\ \gamma^{-1}M & I \end{bmatrix}^{-1}$$
$$\text{subject to} \begin{bmatrix} I & \gamma^{-1}M^T \\ \gamma^{-1}M & I \end{bmatrix} > 0$$
$$H\bar{C}_N = I$$

where M is defined by

$$M \triangleq H\begin{bmatrix} \bar{G}_N & I \end{bmatrix}\begin{bmatrix} Q_w & 0 \\ 0 & R_v \end{bmatrix} \qquad (4.253)$$

(3) Show the following inequality:

$$E[\|HY_{k-1} - x_k\|^2] \leq 2\gamma^2 \log E \exp(\frac{1}{2\gamma^2}\|HY_{k-1} - x_k\|^2) \quad (4.254)$$

Note that it can be seen from the inequality (4.254) that the optimal solution to (4.252) is not an optimal one to a quadratic error variance.

4.7. Show that an unbiased condition (4.86) for general A is reduced to one (4.37), i.e. $H\bar{C}_N = I$, under the assumption that A is nonsingular.

4.8. Consider an MVF smoother in the following form:

$$\hat{x}_{k-h|k} = \sum_{i=k-N}^{k-1} H_{k-i}y_i + \sum_{i=k-N}^{k-1} L_{k-i}u_i \qquad (4.255)$$

As in an FIR filter, the FIR smoother is required to be unbiased and minimize the error variance $E[(\hat{x}_{k-h|k} - x_{k-h})^T(\hat{x}_{k-h|k} - x_{k-h})]$. In this problem, we assume that the system matrix A is invertible.

(1) Show that the following linear equation is obtained from (4.1) and (4.2):

$$Y_{k-1} = \hat{C}_N x_{k-h} + \hat{B}_N U_{k-1} + \hat{G}_N W_{k-1} + V_{k-1}$$

where

$$\hat{C}_N \triangleq \begin{bmatrix} \bar{C}_{N-h} \\ \tilde{C}_h \end{bmatrix}, \quad \hat{G}_N \triangleq \begin{bmatrix} \bar{G}_{N-h} & 0 \\ 0 & \tilde{G}_h \end{bmatrix}, \quad \hat{B}_N \triangleq \begin{bmatrix} \bar{B}_{N-h} & 0 \\ 0 & \tilde{B}_h \end{bmatrix}$$

(2) Obtain the optimal FIR smoother with the above notation.
(3) Assume $h = N$. Show that the above FIR smoother can be calculated from the following recursive equation:

$$\hat{x}_{k-N|k} = (C^T R_v^{-1} C + \hat{P}_N)^{-1}(C^T R_v^{-1} y_{k-N} + \hat{w}_N) \qquad (4.256)$$

where $\hat{P}_i$ and $\hat{w}_i$ are obtained from

$$\hat{P}_{i+1} = A^T C^T R_v^{-1} C A + A^T \hat{P}_i A$$
$$- A^T(C^T R_v^{-1} C + \hat{P}_i)G\left\{ Q_w^{-1} + G^T A^T(C^T R_v^{-1} C + \hat{P}_i)AG \right\}^{-1}$$
$$\times G^T A^T(C^T R_v^{-1} C + \hat{P}_i)A \qquad (4.257)$$
$$\hat{w}_{i+1} = A^T C^T R_v^{-1} y_{k-N+i} + A\hat{w}_i$$
$$- A^T(C^T R_v^{-1} C + \hat{P}_i)G\left\{ Q_w^{-1} + G^T A^T(C^T R_v^{-1} C + \hat{P}_i)AG \right\}^{-1}$$
$$\times G^T A^T(C^T R_v^{-1} y_{k-N+i} + \hat{w}_i) \qquad (4.258)$$

with $\hat{P}_0 = 0$ and $\hat{w}_0 = 0$.

4.9. In MVF filters, we assume that the horizon size is greater than or equal to the system order. This assumption can be relaxed if the pseudo inverse is used as $\hat{x}_{k|k-1} = \Omega_N^+ \check{x}_k$, where Ω_N and $\check{x}_k$ are obtained from (4.68) and (4.72) respectively. Show that $\hat{x}_{k|k-1} = \Omega_N^+ \check{x}_k$ is the minimum norm solution to the following performance criterion:

$$\min_{x_k} \|Y_{k-1} - \bar{B}_N U_{k-1} - \bar{C}_N x_k\|_{\Xi_N^{-1}}^2 \tag{4.259}$$

where Y_{k-1}, U_{k-1}, $\bar{B}_N$, $\bar{C}_N$ and Ξ_k are defined as in (4.27) and (4.34).

4.10. Consider an autoregressive (AR) model signal given by

$$x_{i+1} = \sum_{m=0}^{l} a_m x_{i-m} + v_{i+1} \tag{4.260}$$

where v_i is an uncorrelated process noise with zero mean and variance σ and a_m is a finite sequence with $a_{p-1} \neq 0$.

(1) Represent the AR process (4.260) as a state-space model with $X_i = [x_{i-l+1} \ \cdots \ x_1]^T$.
(2) The above signal (4.260) is transmitted through l different channels $h_1, ..., h_l$ and measured with some noises such as

$$Y_i = HX_i + W_i$$

where

$$Y_i = \begin{bmatrix} y_{1,i} \\ \vdots \\ y_{l,i} \end{bmatrix}, \quad H = \begin{bmatrix} h_1 \\ \vdots \\ h_l \end{bmatrix}, \quad W_i = \begin{bmatrix} w_{1,i} \\ \vdots \\ w_{l,i} \end{bmatrix}$$

as seen in Figure 4.10. Find out the MVF filter $\hat{x}_{k|k-1} = FY_{k-1}$ for the noises W_i with zero means and $E[W_i W_i^T] = Q$.

4.11. In Section 4.3, two MVF filters (4.90) and (4.62) and (4.63) are derived for a general matrix A and a nonsingular matrix A respectively. Show that (4.62) and (4.63) can be obtained from (4.90) under the assumption of a nonsingular matrix A.

4.12. Consider the following signal:

$$z_i = \frac{a_{o,i}}{\sqrt{2}} + \frac{b_{o,i}(-1)^i}{\sqrt{2}} + \sum_{k=1}^{N-1} \left[a_{k,i} \cos\left(\frac{2\pi ki}{T}\right) + b_{k,i} \sin\left(\frac{2\pi ki}{T}\right) \right]$$
$$+ v_i \tag{4.261}$$

where N is the number of harmonic components presented in z_i, v_i is the white Gaussian measurement noise with mean 0 and variance σ^2, and $\{a_k, b_k; k =$

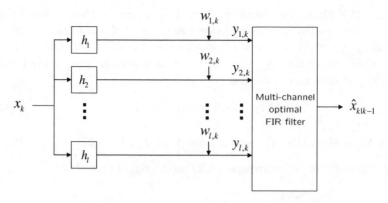

Fig. 4.10. Multi-channel optimal FIR filter

$0, 1, 2, \cdots, N-1\}$ are the Fourier coefficients of the kth harmonic components. Then, the state model for the signal z_i can be represented as

$$x_{i+1} = Ax_i \tag{4.262}$$
$$z_i = Cx_i + v_i \tag{4.263}$$

where

$$A = \begin{bmatrix} \begin{array}{cc|ccc} 1 & 0 & & & \\ 0 & -1 & & & \\ \hline & & \Theta_1 & & \\ & & & \ddots & \\ & & & & \Theta_{N-1} \end{array} \end{bmatrix}$$

$$\Theta_k = \begin{bmatrix} \cos k\theta & \sin k\theta \\ -\sin k\theta & \cos k\theta \end{bmatrix}$$

$$C = \begin{bmatrix} \frac{1}{\sqrt{2}} & \frac{1}{\sqrt{2}} & 1 & 0 & 1 & \cdots & 1 & 0 \end{bmatrix}$$

where $\theta = \frac{2\pi}{T}$. Here, the state x_i is given by

$$x_i = \begin{bmatrix} \tilde{x}_{1,i} \\ \tilde{x}_{2,i} \\ \tilde{x}_{3,i} \\ \tilde{x}_{4,i} \\ \vdots \\ \tilde{x}_{2N-1,i} \\ \tilde{x}_{2N,i} \end{bmatrix} = \begin{bmatrix} a_0 \\ (-1)^i b_0 \\ a_1 \cos\theta + b_1 \sin\theta \\ -a_1 \sin\theta + b_1 \cos\theta \\ \vdots \\ a_{N-1} \cos[(N-1)\theta] + b_{N-1} \sin[(N-1)\theta] \\ -a_{N-1} \sin[(N-1)\theta] + b_{N-1} \cos[(N-1)\theta] \end{bmatrix}$$

(1) Obtain the MVF filter.

(2) If the measured signal z_i is given by

$$z_i = \cos\left(\frac{\pi i}{6}\right) + v_i \tag{4.264}$$

where v_i is zero mean with covariance 0.001, using the result of (1), obtain the Fourier coefficient $\{a_k, b_k; k = 0, 1, 2, \cdots, N - 1\}$ with the following variables:

$$M = 6 \quad \text{(filter length)}$$
$$N = 12 \quad \text{(the number of harmonics)}$$
$$T = 24$$

Note:

$$a_k(i) = x_{2k+1}(i)\cos(k\theta i) - x_{2k+2}(i)\sin(k\theta i)$$
$$b_k(i) = x_{2k+1}(i)\sin(k\theta i) + x_{2k+2}(i)\cos(k\theta i), \quad (k = 1, 2, ..., N - 1)$$

(3) Plot the frequency response of the above MVF filter.

4.13. Consider the following optimal FIR smoother:

$$\hat{x}_{k-h|k} = \sum_{i=k-N}^{k-1} H_{k-i}y_i + \sum_{i=k-N}^{k-1} L_{k-i}u_i \tag{4.265}$$

(1) Determine H_i and L_i of the FIR smoother for the following performance criterion:

$$\min_{H,L} \max_{w_i} \frac{[x_{k-h} - \hat{x}_{k-h|k}]^T [x_{k-h} - \hat{x}_{k-h|k}]}{\sum_{i=1}^{N} w_{k-i}^T w_{k-i}} \tag{4.266}$$

(2) Determine H_i and L_i of the FIR smoother for the following performance criterion:

$$\inf_{H,L} \sup_{w_i \neq 0} \frac{\sum_{i=k_0}^{\infty} \{x_{i-h} - \hat{x}_{i-h|i}\}^T \{x_{i-h} - \hat{x}_{i-h|i}\}}{\sum_{i=k_0}^{\infty} w_i^T w_i} \tag{4.267}$$

Use the notation of Problem 4.8.

4.14. Show that an LEF filter has a guaranteed H_∞ norm as follows:

$$\sup_{w_k \in \ell_2} \frac{\|\hat{x}_{k|k} - x_k\|_2}{\|w_k\|_2} < \sqrt{N\lambda_{\max}\{(\bar{G}_N + \bar{D}_N)^T H^T H(\bar{G}_N + \bar{D}_N)\}} \tag{4.268}$$

where $\| \cdot \|$ is a two norm on $[0, \infty]$, and $\bar{G}_N$ and $\bar{D}_N$ are defined in (4.185) and (4.33) respectively.

4.15. Derive an L_2-E FIR filter again without a requirement of the nonsingularity of A.

4.16. Consider the linear discrete time-invariant state-space model as

$$x_{k+1} = \begin{bmatrix} 0.9950 & 0.0998 \\ -0.0998 & 0.9950 \end{bmatrix} x_k + \begin{bmatrix} 1 \\ 1 \end{bmatrix} u_k + \begin{bmatrix} 0.1 \\ 0.1 \end{bmatrix} w_k$$

$$y_k = \begin{bmatrix} 1 & 1 \end{bmatrix} x_k + v_k \tag{4.269}$$

where the initial state is taken as $x_o{}^T = [0\ 0]$ and the system noise covariance Q is 0.01 and the measurement noise covariance R is 0.01.

(1) Compute the error covariance of the FIR filter and Kalman filter on round-off digits of fourth, third and second with horizon size 5.
(2) From your simulation result in (1), explain why the error covariance of an FIR filter is smaller than the error covariance of a Kalman filter.

5

Output Feedback Receding Horizon Controls

5.1 Introduction

In this chapter, it is assumed that the output can be measured and the input is known. Controls that use only the measured output and the known control are called output feedback controls, which are an important class of controls. There are different approaches for output feedback controls.

The first approach is a state-observer-based output feedback control. It can be called a blind combination approach, where output feedback controls can be obtained by merely combining state feedback controls with state estimators. There exist many well-known state feedback controls, such as LQC in Chapter 2 or LQ RHC in Chapter 3, and many state estimators, such as Luenburger observers for deterministic systems and Kalman filters for stochastic systems in Chapter 2 or FIR filters in Chapter 4. This approach is easy to implement. However, it has no built-in optimality. It is noted that the stability of overall output feedback systems is determined by those of the controller and the observer.

The second approach is a predictor-based output feedback control. It can be called a predictor-based optimization approach. One can obtain predictors for the future states from measured outputs and known inputs and these predictors are considered in the performance criterion. The optimal control for linear systems turns out to be a function of the current estimated state. GPC can be approached this way and provided as output feedback controls, as seen in exercises.

The third approach is a global optimization approach. Usually, the optimal control may not provide the output feedback structure. Thus, we introduce controls with a built-in output feedback structure as a requirement. The performance criterion is optimized subject to controls with a finite memory feedback structure. Fortunately, the optimal control for linear systems is given in the form of an output feedback control.

The performance criterion can be defined on the infinite horizon or the receding horizon. In all the above approaches, the infinite horizon and the receding horizon can be utilized.

In addition to static and dynamic controls, output feedback controls can be given as a linear combination of finite measurements of inputs and outputs, which is called the finite memory control (FMC). An unbiased condition is introduced in various filters in Chapter 4. Even this unbiased condition can be extended to output feedback controls, particularly for FMCs. In this chapter, we focus on output feedback controls with a finite memory structure and an unbiased condition.

The organization of this chapter is as follows. In Section 5.2, a blind combination approach is covered. The predictor-based approach is given in Section 5.3. A special case of I/O systems is introduced for GPC. They are transformed to the state-space model. The direct approaches appeared in the exercises. The global optimization approach over receding horizons with the quadratic cost for linear stochastic systems results in LQFMC, which is given in Section 5.5. The global optimization approach over receding horizons with the minimax cost for linear deterministic systems results in the L_2-E and H_∞ FMCs, which are given in Section 5.6.

In this chapter, *the finite horizons for the filter and the control are denoted by N_f and N_c in order to distinguish them.*

5.2 State-observer-based Output Feedback Controls

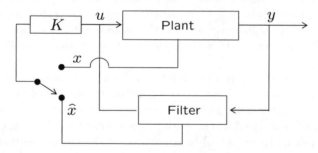

Fig. 5.1. State-observer-based output feedback controls

In the blind combination approach, output feedback receding horizon controls can be obtained by merely combining the state feedback receding horizon controls with some filters or observers.

Stochastic Systems

Consider the following system:

$$x_{i+1} = Ax_i + Bu_i + Gw_i \tag{5.1}$$
$$y_i = Cx_i + v_i \tag{5.2}$$

where $x_i \in \Re^n$, $u_i \in \Re^l$ and $y_i \in \Re^q$ are the state, the input, and the measurement respectively. At the initial time k_0, the state x_{k_0} is a random variable with a mean $\bar{x}_{k_0}$ and a covariance P_{k_0}. The system noise $w_i \in \Re^p$ and the measurement noise $v_i \in \Re^q$ are zero mean white Gaussian and mutually uncorrelated. The covariances of w_i and v_i are denoted by Q_w and R_v respectively, which are assumed to be positive definite matrices. These noises are uncorrelated with the initial state x_{k_0}.

The Kalmam filters are the most popular filters or observers for linear stochastic systems in Chapter 2. We can also use Kalman filters with frozen gains or MVF filters in Chapter 4. For a typical example, combining a receding horizon LQ control in Chapter 3 with an MVF filter in Chapter 4 can be taken. Thus, we have

$$u_k = -R^{-1}B^T[I + K_1BR^{-1}B^T]^{-1}K_1A\hat{x}_{k|k-1} \tag{5.3}$$

where $\hat{x}_{k|k-1}$ is obtained from

$$\hat{x}_{k|k-1} = (\bar{C}_{N_f}^T \Xi_{N_f}^{-1} \bar{C}_{N_f})^{-1} \bar{C}_{N_f}^T \Xi_{N_f}^{-1}(Y_{k-1} - \bar{B}_{N_f}U_{k-1}) \tag{5.4}$$

Here, Y_{k-1}, U_{k-1}, $\bar{C}_{N_f}$, $\bar{B}_{N_f}$ and Ξ_{N_f} are defined in (4.76), (4.77), (4.31), (4.32) and (4.214) respectively and K_1 is obtained from

$$K_j = A^T K_{j+1} A - A^T K_{i+1} B[R + B^T K_{j+1} B]^{-1} B^T K_{j+1} A + Q$$
$$= A^T K_{j+1}[I + BR^{-1}B^T K_{j+1}]^{-1} A + Q$$
$$K_{N_c} = Q_f$$

Note that it is assumed in (5.4) that A is invertible.

This method is easy to implement. However, it has no optimality. The stability of the observer-based output feedback control for the stochastic system can be investigated such as in Section 5.5.4.

Deterministic Systems

Consider the following systems:

$$x_{i+1} = Ax_i + Bu_i + B_w w_i \tag{5.5}$$
$$y_i = Cx_i + C_w w_i \tag{5.6}$$

where $x_i \in \Re^n$, $u_i \in \Re^l$, and $y_i \in \Re^q$ are the state, the input, and the measurement respectively. Note that $C_w B_w^T = 0$.

The most popular filters or observers are H_∞ filters for the linear deterministic systems in Chapter 2. We can also use H_∞ filters with frozen gains or the H_∞ FIR filters in Chapter 4. For a typical example, combining a receding horizon H_∞ control in Chapter 3 with an H_∞ FIR filter in Chapter 4 can be taken. Thus, we have

$$u_k = -R^{-1}B^T \Lambda_{k+1,k+N_c}^{-1} M_{k+1,k+N_c} A\hat{x}_{k|k-1} \qquad (5.7)$$

where

$$M_j = A^T \Lambda_{j+1}^{-1} M_{j+1} A + Q, \quad j = 1, \cdots, N_c - 1$$
$$M_{N_c} = Q_f$$
$$\Lambda_{j+1} = I + M_{j+1}(BR^{-1}B^T - \gamma^{-2}B_w R_w^{-1} B_w^T)$$

$\hat{x}_{k|k-1}$ in (5.7) is calculated as

$$\hat{x}_{k|k-1} = [FM + (\bar{C}_{N_f}^T \bar{C}_{N_f})^{-1}\bar{C}_{N_f}^T](Y_{k-1} - \bar{C}_{N_f}^T U_{k-1})$$

where the columns of M is the basis of the null space of $\bar{C}_N^T$ and F is obtained from the following LMI:

$$\min_{F,X>0} \gamma_\infty$$

subject to

$$\begin{bmatrix} -X & XA_u & XB_u & 0 \\ A_u^T X & -X & 0 & S_0 \\ B_u^T X & 0 & -\gamma_\infty I & 0 \\ 0 & S_0^T & 0 & -\gamma_\infty I \end{bmatrix} < 0$$

Here, A_u and B_u are defined in (4.227) and $S_0 = (\bar{G}_{N_f} + \bar{D}_{N_f})^T M^T F^T + (\bar{G}_{N_f} + \bar{D}_{N_f})^T ((\bar{C}_{N_f}^T \bar{C}_{N_f})^{-1}\bar{C}_{N_f}^T)^T$.

The stability of the observer-based output feedback control for the deterministic system can be investigated such as in Section 5.6.4.

5.3 Predictor-based Output Feedback Controls

For the predictor-based optimization approach, first a predictor is obtained and then the optimization is carried out with a performance criterion based on the predictor, as shown in Figure 5.2.

Stochastic Systems

In the system (5.1) and (5.2), the predictor estimates the future state using the measurements and known inputs, and can be represented as

$$\check{x}_{k+i|k} = x(k+i|y_0, y_1, \cdots, y_k, u_0, u_1, \cdots, u_k, u_{k+1}, \cdots, u_{k+i-1}) \quad (5.8)$$

If stochastic noises are considered in plant models, then the predictor can be expressed using the expectation

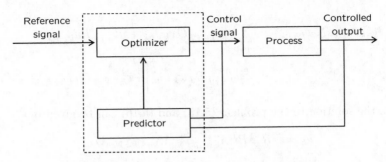

Fig. 5.2. Predictor-based output feedback control

$$\check{x}_{k+i|k} = E\left[x(k+i|y_0, y_1, \cdots, y_k, u_0, u_1, \cdots, u_k, u_{k+1}, \cdots, u_{k+i-1})\right] \quad (5.9)$$

For the linear stochastic system (5.1) and (5.2), assume that $\check{x}_{k|k}$ is a filter of x_k. Then the predictor $\check{x}_{k+i|k}$ can be obtained:

$$\check{x}_{k+i+1|k} = A\check{x}_{k+i|k} + Bu_{k+i|k} \quad (5.10)$$

where $\check{x}_{k|k} = \hat{x}_k$ and $\hat{x}_k$ is a certain given filter.

Thus, a performance criterion is as follows:

$$\sum_{j=0}^{N_c-1} \left[\check{x}_{k+j|k}^T Q\check{x}_{k+j|k} + u_{k+j|k}^T Ru_{k+j|k}\right] + \check{x}_{k+N_c|k}^T F\check{x}_{k+N_c|k} \quad (5.11)$$

Then, the solution to the problem (5.10) and (5.11) can be given by

$$u_k = -R^{-1}B^T[I + K_1BR^{-1}B^T]^{-1}K_1A\hat{x}_k \quad (5.12)$$

where K_i is given by

$$\begin{aligned}
K_i &= A^T K_{i+1} A - A^T K_{i+1} B[R + B^T K_{i+1} B]^{-1} B^T K_{i+1} A + Q \\
&= A^T K_{i+1}[I + BR^{-1}B^T K_{i+1}]^{-1}A + Q
\end{aligned} \quad (5.13)$$

with the boundary condition

$$K_{N_c} = F \quad (5.14)$$

Deterministic Systems

The predictor $\check{x}_{k+i|k}$ can be represented as

$$\check{x}_{k+i+1|k} = A\check{x}_{k+i|k} + Bu_{k+i|k} + B_w w_{k+i|k} \quad (5.15)$$

where $\check{x}_{k|k} = \hat{x}_k$. The performance criterion is given by

$$J(\hat{x}_k, u_{k+\cdot|k}, w_{k+\cdot|k}) = \sum_{j=0}^{N_c-1} \left[\check{x}_{k+j|k}^T Q \check{x}_{k+j|k} + u_{k+j|k}^T R u_{k+j|k} \right.$$
$$\left. - \gamma^2 w_{k+j|k}^T R_w w_{k+j|k} \right] + \check{x}_{k+N_c|k}^T Q_f \check{x}_{k+N_c|k} \qquad (5.16)$$

Then, the solution to the problem (5.15) and (5.16) can be given by

$$u_k = -R^{-1} B^T \Lambda_{k+1,k+N_c}^{-1} M_{k+1,k+N_c} A \hat{x}_k$$

where

$$M_j = A^T \Lambda_{j+1}^{-1} M_{j+1} A + Q$$
$$M_{N_c} = Q_f$$
$$\Lambda_{j+1} = I + M_{j+1}(BR^{-1}B^T - \gamma^{-2} B_w R_w^{-1} B_w^T)$$

for $j = 1, \cdots, N_c - 1$.

For stochastic and deterministic systems, stabilities of the predictor-based output feedback controls are determined according to what kinds of filters are used.

5.4 A Special Case of Input–Output Systems of General Predictive Control

In addition to the stochastic state-space system (5.1) and (5.2), GPC has used the following CARIMA model:

$$A(q^{-1})y_i = B(q^{-1})\triangle u_{i-1} + C(q^{-1})w_i \qquad (5.17)$$

where $A(q^{-1})$ and $B(q^{-1})$ are given in (3.195) and (3.196) and $C(q^{-1})$ is given as follows:

$$C(q^{-1}) = 1 + c_1 q^{-1} + \cdots + c_n q^{-n} \qquad (5.18)$$

It is noted that the $C(q^{-1})$ can often be

$$1 + c_1 q^{-1} + \cdots + c_p q^{-p}, \quad p \le n \qquad (5.19)$$

The CARIMA model (5.17) can be rewritten as

$$A(q^{-1})y_i = \widetilde{B}(q^{-1})\triangle u_i + C(q^{-1})w_i \qquad (5.20)$$

where

$$\widetilde{B}(q^{-1}) = b_1 q^{-1} + b_2 q^{-2} \cdots + b_n q^{-n+1}$$

From (5.20) we can obtain

$$A(q^{-1})(y_i - w_i) = \tilde{B}(q^{-1})\triangle u_i + [C(q^{-1}) - A(q^{-1})]w_i \qquad (5.21)$$

where

$$C(q^{-1}) - A(q^{-1}) = (c_1 - a_1)q^{-1} + \cdots + (c_n - a_n)q^{-n} \qquad (5.22)$$

Therefore, (5.21) has a similar structure to (3.200), such as

$$x_{i+1} = \bar{A}x_i + \bar{B}\triangle u_i + \bar{D}_w w_i$$
$$y_i = \bar{C}x_i + w_i \qquad (5.23)$$

where $\bar{A}, \bar{B}, \bar{C}$, and $\bar{D}_w$ are given by

$$\bar{A} = \begin{bmatrix} -a_1 & 1 & 0 & \cdots & 0 \\ -a_2 & 0 & 1 & \cdots & 0 \\ \vdots & \vdots & \vdots & \ddots & \vdots \\ -a_{n-1} & 0 & 0 & \cdots & 1 \\ -a_n & 0 & 0 & \cdots & 0 \end{bmatrix}, \ \bar{B} = \begin{bmatrix} b_1 \\ b_2 \\ \vdots \\ b_{n-1} \\ b_n \end{bmatrix} \qquad (5.24)$$

$$\bar{C} = \begin{bmatrix} 1 & 0 & 0 & \cdots & 0 \end{bmatrix}, \ \bar{D}_w = \begin{bmatrix} c_1 - a_1 \\ \vdots \\ c_n - a_n \end{bmatrix}$$

Stochastic Systems

In (5.23), w_i is a zero mean white noise with a covariance q_w. It is noted that the system noise and measurement noise are correlated as seen in (5.23). The performance criterion for (5.17) can be given as

$$\mathrm{E}\left\{ \sum_{j=1}^{N_c} \left[q(y_{k+j} - y_{k+j}^r)^2 + r(\triangle u_{k+j-1})^2|_{[k_s,k]} \right] \right\}$$

$$+ \mathrm{E}\left\{ \sum_{j=N_c+1}^{N_p} q_f(y_{k+j} - y_{k+j}^r)^2|_{[k_s,k]} \right\} \qquad (5.25)$$

where $\triangle u_{k+N_c} = \ldots = \triangle u_{k+N_p-1} = 0$. Expectation is taken under the condition that the measurement data of y_i and $\triangle u_i$ on $[k_s, \quad k]$ are given. The performance index (5.25) can be changed to

$$\sum_{j=1}^{N_c} \left[q(\breve{y}_{k+j|k} - y_{k+j}^r)^2 + r(\triangle u_{k+j-1})^2 \right]$$

$$+ \sum_{j=N_c+1}^{N_p} q_f(\breve{y}_{k+j|k} - y_{k+j}^r)^2 \qquad (5.26)$$

We can have predictors

$$\check{x}_{k+j+1|k} = \bar{A}\check{x}_{k+j|k} + \bar{B}\triangle u_{k+j|k} \quad j \geq 1 \tag{5.27}$$

$$\check{y}_{k+j|k} = \bar{C}\check{x}_{k+j|k} \tag{5.28}$$

But, if $j = 0$, the system noise and measurement noise are correlated. So, by using the result of Problem 2.15, we can change the correlated state-space model (5.23) into the uncorrelated state-space model:

$$x_{i+1} = \widetilde{A}x_i + \bar{B}\triangle u_i + \bar{D}_w y_i$$
$$y_i = \bar{C}x_i + w_i \tag{5.29}$$

where $\widetilde{A} = \bar{A} - \bar{D}_w\bar{C}$. Therefore, if $j = 0$, then we can have a predictor

$$\check{x}_{k+1|k} = \widetilde{A}\check{x}_{k|k} + \bar{B}\triangle u_k + \bar{D}_w y_k \tag{5.30}$$

Using (5.27), (5.28), and (5.30), we can obtain

$$\check{y}_{k+j|k} = \bar{C}\bar{A}^{j-1}\widetilde{A}\check{x}_{k|k} + \sum_{i=0}^{j-1} \bar{C}\bar{A}^{j-i-1}\bar{B}\triangle u_{k+i} + \bar{C}\bar{A}^{j-1}\bar{D}_w y_k \tag{5.31}$$

where $\check{x}_{k|k} = \hat{x}_{k|k}$ could be a starting point.

GPC can be obtained from the predictor (5.31) and the performance criterion (5.26). The GPC can be given similar to (3.329) as follows:

$$\triangle U_k = \left[\overline{W}^T\bar{Q}\overline{W} + \overline{W}_f^T\bar{Q}_f\overline{W}_f + \bar{R}\right]^{-1}\left\{\overline{W}^T\bar{Q}\left[Y_k^r - \overline{V}\hat{x}_{k|k} - \overline{N}y_k\right]\right.$$

$$\left. + \overline{W}_f^T\bar{Q}_f\left[Y_{k+N_c}^r - \overline{V}_f\hat{x}_{k|k} - \overline{N}_f y_k\right]\right\} \tag{5.32}$$

where

$$Y_k^r = \begin{bmatrix} y_{k+1}^r \\ \vdots \\ y_{k+N_c}^r \end{bmatrix}, \quad \overline{V} = \begin{bmatrix} \bar{C}\widetilde{A} \\ \bar{C}\bar{A}\widetilde{A} \\ \vdots \\ \bar{C}\bar{A}^{N_c-1}\widetilde{A} \end{bmatrix}, \quad \triangle U_k = \begin{bmatrix} \triangle u_k \\ \vdots \\ \triangle u_{k+N_c-1} \end{bmatrix}$$

$$Y_{k+N_c}^r = \begin{bmatrix} y_{k+N_c+1}^r \\ \vdots \\ y_{k+N_p}^r \end{bmatrix}, \quad \overline{V}_f = \begin{bmatrix} \bar{C}\bar{A}^{N_c}\widetilde{A} \\ \vdots \\ \bar{C}\bar{A}^{N_p-1}\widetilde{A} \end{bmatrix}$$

$$\overline{W} = \begin{bmatrix} \bar{C}\bar{B} & \cdots & 0 \\ \vdots & \ddots & \vdots \\ \bar{C}\bar{A}^{N_c-1}\bar{B} & \cdots & \bar{C}\bar{B} \end{bmatrix}, \quad \overline{W}_f = \begin{bmatrix} \bar{C}\bar{A}^{N_c}\bar{B} & \cdots & \bar{C}\bar{A}\bar{B} \\ \vdots & \ddots & \vdots \\ \bar{C}\bar{A}^{N_p-1}\bar{B} & \cdots & \bar{C}\bar{A}^{N_p-N_c}\bar{B} \end{bmatrix}$$

$$\bar{Q} = [\text{diag}(\overbrace{q\ q\ \cdots\ q}^{N_c})], \quad \bar{Q}_f = [\text{diag}(\overbrace{q_f\ q_f\ \cdots\ q_f}^{N_p-N_c})]$$

$$\bar{R} = [\text{diag}(\overbrace{r \; r \; \cdots \; r}^{N_c})], \quad \overline{N} = \begin{bmatrix} \bar{C}\bar{D}_w \\ \bar{C}\bar{A}\bar{D}_w \\ \vdots \\ \bar{C}\bar{A}^{N_c-1}\bar{D}_w \end{bmatrix}, \quad \overline{N}_f = \begin{bmatrix} \bar{C}\bar{A}^{N_c}\bar{D}_w \\ \vdots \\ \bar{C}\bar{A}^{N_p-1}\bar{D}_w \end{bmatrix}$$

Therefore, the GPC $\triangle u_k$ is given by

$$\triangle u_k = \begin{bmatrix} I & 0 & \cdots & 0 \end{bmatrix} \left[\overline{W}^T \overline{Q} \overline{W} + \overline{W}_f^T \overline{Q}_f \overline{W}_f + \bar{R} \right]^{-1} \left\{ \overline{W}^T \overline{Q} \right.$$

$$\left. \times \left[Y_k^r - \overline{V} \hat{x}_{k|k} - \overline{N} y_k \right] + \overline{W}_f^T \overline{Q}_f \left[Y_{k+N_c}^r - \overline{V}_f \hat{x}_{k|k} - \overline{N}_f y_k \right] \right\} \quad (5.33)$$

A starting point $\hat{x}_{k|k}$ can be obtained from an FIR filter:

$$\hat{x}_{k|k} = \sum_{i=k-N}^{k} H_{k-i} y_i + \sum_{i=k-N}^{k-1} L_{k-i} u_i \quad (5.34)$$

where H_{k-i} and L_{k-i} can be obtained similarly as in Chapter 4.

Deterministic Systems

We assume w_i in (5.17) is a deterministic disturbance. We can have predictors

$$\check{x}_{k+j+1|k} = \bar{A}\check{x}_{k+j|k} + \bar{B}\triangle u_{k+j|k} + \bar{D}_w w_{k+j|k}, \quad j \geq 1 \quad (5.35)$$

$$\check{y}_{k+j|k} = \bar{C}\check{x}_{k+j|k} + w_{k+j|k} \quad (5.36)$$

where a filter $\check{x}_{k+1|k}$ could be a starting point. From now on, $\triangle u_{k+j|k}$ and $w_{k+j|k}$ will be used as $\triangle u_{k+j}$ and w_{k+j} respectively. We consider the following performance criterion:

$$\sum_{j=1}^{N_c} \left[q(\check{y}_{k+j} - y_{k+j}^r)^2 + r(\triangle u_{k+j-1})^2 - \gamma(w_{k+j-1})^2 \right]$$

$$+ \sum_{j=N_c+1}^{N_p} q_f(\check{y}_{k+j} - y_{k+j}^r)^2 \quad (5.37)$$

where $\triangle u_{k+N_c} = \ldots = \triangle u_{k+N_p} = 0$ and $w_{k+N_c} = \ldots = w_{k+N_p} = 0$. From the state-space model (5.23), we have

$$\check{y}_{k+j} = \bar{C}\bar{A}^j \check{x}_{k|k} + \sum_{i=0}^{j-1} \bar{C}\bar{A}^{j-i-1}\bar{B}\triangle u_{k+i} + \sum_{i=0}^{j-1} \bar{C}\bar{A}^{j-i-1}\bar{D}w_{k+i}$$

$$+ w_{k+j} \quad (5.38)$$

where $\check{x}_{k|k} = \hat{x}_{k|k}$ could be a starting point. The performance index (5.37) can be represented by

$$J(x_k, \triangle U_k, W_k) = \left[Y_k^r - V\hat{x}_{k|k} - W\triangle U_k - HW_k\right]^T \bar{Q}[Y_k^r - V\hat{x}_{k|k} - W\triangle U_k$$
$$- HW_k]\triangle U_k^T \bar{R}\triangle U_k - W_k^T \bar{\Gamma} W_k + [Y_{k+N_c}^r - V_f\hat{x}_{k|k} - W_f\triangle U_k$$
$$- H_f W_k]^T \bar{Q}_f \left[Y_{k+N_c}^r - V_f\hat{x}_{k|k} - W_f\triangle U_k - H_f W_k\right] \qquad (5.39)$$

where

$$Y_k^r = \begin{bmatrix} y_{k+1}^r \\ \vdots \\ y_{k+N_c}^r \end{bmatrix}, \ V = \begin{bmatrix} \bar{C}\bar{A} \\ \vdots \\ \bar{C}\bar{A}^{N_c} \end{bmatrix}, \ \triangle U_k = \begin{bmatrix} \triangle u_k \\ \vdots \\ \triangle u_{k+N_c-1} \end{bmatrix}$$

$$Y_{k+N_c}^r = \begin{bmatrix} y_{k+N_c+1}^r \\ \vdots \\ y_{k+N_p}^r \end{bmatrix}, \ V_f = \begin{bmatrix} \bar{C}\bar{A}^{N_c+1} \\ \vdots \\ \bar{C}\bar{A}^{N_p} \end{bmatrix}, \ W_k = \begin{bmatrix} w_k \\ \vdots \\ w_{k+N_c-1} \end{bmatrix}$$

$$W = \begin{bmatrix} \bar{C}\bar{B} & \cdots & 0 \\ \vdots & \ddots & \vdots \\ \bar{C}\bar{A}^{N_c-1}\bar{B} & \cdots & \bar{C}\bar{B} \end{bmatrix}, \ W_f = \begin{bmatrix} \bar{C}\bar{A}^{N_c}\bar{B} & \cdots & \bar{C}\bar{A}\bar{B} \\ \vdots & \ddots & \vdots \\ \bar{C}\bar{A}^{N_p-1}\bar{B} & \cdots & \bar{C}\bar{A}^{N_p-N_c}\bar{B} \end{bmatrix}$$

$$\bar{R} = \left[\mathrm{diag}(\overbrace{r\ r\ \cdots\ r}^{N_c})\right], \ \bar{\Gamma} = \left[\mathrm{diag}(\overbrace{\gamma\ \gamma\ \cdots\ \gamma}^{N_c})\right]$$

$$\bar{Q}_f = \left[\mathrm{diag}(\overbrace{q_f\ q_f\ \cdots\ q_f}^{N_p-N_c})\right], \ \bar{Q} = \left[\mathrm{diag}(\overbrace{q\ q\ \cdots\ q}^{N_c})\right]$$

$$H = \begin{bmatrix} \bar{C}\bar{D}_w & I & 0 & \cdots & 0 \\ \bar{C}\bar{A}\bar{D}_w & \bar{C}\bar{D}_w & I & \cdots & 0 \\ \vdots & & \ddots & \ddots & \vdots \\ \bar{C}\bar{A}^{N_c-2}\bar{D}_w & \cdots & & \cdots & I \\ \bar{C}\bar{A}^{N_c-1}\bar{D}_w & \cdots & & \cdots & \bar{C}\bar{D}_w \end{bmatrix}$$

$$H_f = \begin{bmatrix} \bar{C}\bar{A}^{N_c}\bar{D}_w & \cdots & \bar{C}\bar{A}\bar{B} \\ \vdots & \ddots & \vdots \\ \bar{C}\bar{A}^{N_p-1}\bar{B} & \cdots & \bar{C}\bar{A}^{N_p-N_c}\bar{B} \end{bmatrix}$$

If we compare (3.332) and (5.39), the condition that the solution to the saddle point exists is

$$H^T \bar{Q} H - \bar{\Gamma} + H_f^T \bar{Q}_f H_f < 0 \qquad (5.40)$$

the maximizing disturbance W_k^* is given by

$$W_k^* = \mathcal{V}_1^{-1}\mathcal{V}_2 \qquad (5.41)$$

where

$$\mathcal{V}_1 = H^T \bar{Q} H - \bar{\Gamma} + H_f^T \bar{Q}_f H_f$$
$$\mathcal{V}_2 = H^T \bar{Q} \left[Y_k^r - V\hat{x}_{k|k} - W\triangle U_k\right] + H_f^T \bar{Q}_f \left[Y_{k+N_c}^r - V_f\hat{x}_{k|k} - W_f\triangle U_k\right]$$

and the optimal control vector $\triangle U_k^*$ is given by

$$\triangle U_k^* = -\mathcal{P}_1^{-1}\mathcal{P}_2 \tag{5.42}$$

where

$$\mathcal{P}_1 = -(H^T\bar{Q}W + H_f^T\bar{Q}_fW_f)^T\mathcal{V}_1^{-1}(H^T\bar{Q}W + H_f^T\bar{Q}_fW_f) + W^T\bar{Q}W$$
$$+ \bar{R}_N + W_f^T\bar{Q}_fW_f$$

$$\mathcal{P}_2 = (H^T\bar{Q}W + H_f^T\bar{Q}_fW_f)^T\mathcal{V}_1^{-1}[H^T\bar{Q}(Y_k^r - V\hat{x}_{k|k})$$
$$+ H_f^T\bar{Q}_f(Y_{k+N_c}^r - V_f\hat{x}_{k|k})] - W^T\bar{Q}(Y_k^r - V\hat{x}_{k|k}) - W_f^T\bar{Q}_f(Y_{k+N_c}^r - V_f\hat{x}_{k|k})$$

Therefore, the GPC $\triangle u_k$ is given by

$$\triangle u_k = -\begin{bmatrix} I & 0 & \cdots & 0 \end{bmatrix}\mathcal{P}_1^{-1}\mathcal{P}_2 \tag{5.43}$$

A starting point $\hat{x}_{k|k}$ can be also obtained similarly as in an FIR filter (5.34).

5.5 Finite Memory Control Based on Minimum Criterion

5.5.1 Finite Memory Control and Unbiased Condition

Finite Memory Control

Output feedback controls with a finite memory structure can be represented using measurements and inputs during a filter horizon $[k - N_f, \ k]$ as

$$u_k = \sum_{i=k-N_f}^{k-1} H_{k-i}y_i + \sum_{i=k-N_f}^{k-1} L_{k-i}u_i \tag{5.44}$$

where H_{k-j} and L_{k-j} are gain matrices with respect to y_i and u_i respectively. The control (5.44) can be represented in a simpler matrix form:

$$u_k = HY_{k-1} + LU_{k-1} \tag{5.45}$$

where H, L, Y_{k-1}, and U_{k-1} are defined as

$$H \triangleq [H_{N_f} \ \ H_{N_f-1} \ \ \cdots \ \ H_1] \tag{5.46}$$

$$L \triangleq [L_{N_f} \ \ L_{N_f-1} \ \ \cdots \ \ L_1] \tag{5.47}$$

$$Y_{k-1} \triangleq [y_{k-N_f}^T \ \ y_{k-N_f+1}^T \ \ \cdots \ \ y_{k-1}^T]^T \tag{5.48}$$

$$U_{k-1} \triangleq [u_{k-N_f}^T \ \ u_{k-N_f+1}^T \ \ \cdots \ \ u_{k-1}^T]^T \tag{5.49}$$

Controls of the type (5.44) are visualized in Figure 5.3 and will be obtained

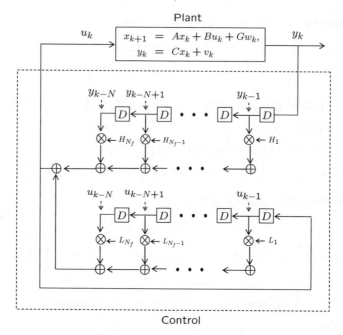

Fig. 5.3. Block diagram of FMC

from a performance criterion on the finite horizon. To this end, we introduce intermediate controls on the finite horizon $[k, \ k + N_c]$. Intermediate output feedback controls with the finite memory structure can be represented using measurements and inputs during a filter horizon $[k - N_f + j, \ k + j]$ as

$$u_{k+j} = \sum_{i=k+j-N_f}^{k+j-1} H_{k+j-i}^{(j)} y_i + \sum_{i=k+j-N_f}^{k+j-1} L_{k+j-i}^{(j)} u_i \qquad (5.50)$$

where the gains $H_{\cdot}^{(j)}$ and $L_{\cdot}^{(j)}$ may be dependent on j, and x_{k+j} is the trajectory generated from u_{k+j}. Note that even though the control (5.50) uses the finite measured outputs and inputs on the recent time interval as FIR filters, this is not of the FIR form, since controls appear on both sides of (5.50). So this kind of the control will be called an output feedback finite memory control, or simply an FMC rather than an FIR control.

The FMC can be expressed as a linear function of the finite inputs Y_{k-1} and inputs U_{k-1} on the horizon $[k - N_f \ k]$ as follows:

$$u_{k+j} = H^{(j)} Y_{k+j-1} + L^{(j)} U_{k+j-1} \qquad (5.51)$$

where $H^{(j)}$, $L^{(j)}$, Y_{k+j-1}, U_{k+j-1} are defined as

$$H^{(j)} \triangleq [H_{N_f}^{(j)} \; H_{N_f-1}^{(j)} \; \cdots \; H_1^{(j)}] \tag{5.52}$$

$$L^{(j)} \triangleq [L_{N_f}^{(j)} \; L_{N_f-1}^{(j)} \; \cdots \; L_1^{(j)}] \tag{5.53}$$

$$Y_{k+j-1} \triangleq [y_{k+j-N_f}^T \; y_{k+j-N_f+1}^T \; \cdots \; y_{k+j-1}^T]^T \tag{5.54}$$

$$U_{k+j-1} \triangleq [u_{k+j-N_f}^T \; u_{k+j-N_f+1}^T \; \cdots \; u_{k+j-1}^T]^T \tag{5.55}$$

It is a general rule of thumb that the FIR systems are robust against temporary modelling uncertainties or round-off errors. Therefore, the FMC may have such properties due to the finite memory structure.

Unbiased Condition

We start off from the case that the system matrix A is nonsingular and then proceed to the case of the general matrix A in the next section.

As in FIR filters in Chapter 4, we are interested in the measured output y_k and the input u_k on the finite recent horizon, from which the system (5.1) and (5.2) is represented as

$$Y_{k+j-1} = \bar{C}_{N_f} x_{k+j} + \bar{B}_{N_f} U_{k+j-1} + \bar{G}_{N_f} W_{k+j-1} + V_{k+j-1} \tag{5.56}$$

where W_{k+j} and V_{k+j} are given by

$$W_{k+j} \triangleq [w_{k+j-N_f}^T \; w_{k+j-N_f+1}^T \; \cdots \; w_{k+j-1}^T]^T \tag{5.57}$$

$$V_{k+j} \triangleq [v_{k+j-N_f}^T \; v_{k+j-N_f+1}^T \; \cdots \; v_{k+j-1}^T]^T \tag{5.58}$$

and $\bar{C}_{N_f}$, $\bar{B}_{N_f}$, and $\bar{G}_{N_f}$ are obtained from (4.31), (4.32), and (4.33) respectively.

Combining (5.51) and (5.56) yields

$$u_{k+j} = H(\bar{C}_{N_f} x_{k+j} + \bar{B}_{N_f} U_{k+j-1} + \bar{G}_{N_f} W_{k+j-1} + V_{k+j-1})$$
$$+ LU_{k+j-1} \tag{5.59}$$

If we assume that the full information of the state is available, then the desirable control is represented in the form of a state feedback control

$$u_{k+j}^* = -\mathcal{K}_{j+1} x_{k+j} \tag{5.60}$$

where $0 \leq j \leq N_f - 1$ and $\mathcal{K}_{j+1}$ can be chosen to be optimized for a certain performance criterion. It is desirable that the intermediate FMC (5.59) can track the state feedback control (5.60) on the average. Since any output feedback control cannot be better in view of performance than the optimal state feedback control, it is desirable that the control (5.59) should be unbiased from the optimal state feedback control (5.60). Thus, we require a constraint that the control (5.59) must be unbiased from the desired state feedback control (5.60) as

$$E[u_{k+j}] = E[u^*_{k+j}] \tag{5.61}$$

for all states and all controls up to $k + j - 1$. This is called the unbiased condition. If there exist no noises on the horizon $[k + j - N_f, k + j - 1]$, then $\hat{x}_{k|k-1}$ and x_k become deterministic values and $\hat{x}_{k|k-1} = x_k$. This is a *deadbeat* property, as shown in Figure 5.4.

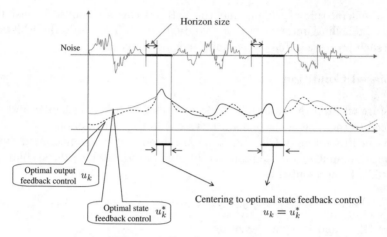

Fig. 5.4. Deadbeat property

Taking the expectation on both sides of (5.59) and (5.60), the following relations are obtained:

$$E[u_{k+j}] = H^{(j)}\bar{C}_{N_f}E[x_{k+j}] + (H^{(j)}\bar{B}_{N_f} + L^{(j)})U_{k+j-1} \tag{5.62}$$

$$E[u^*_{k+j}] = -\mathcal{K}_{j+1}E[x_{k+j}] \tag{5.63}$$

To be unbiased from the desirable state feedback control $u_{k+j} = -\mathcal{K}_{j+1}x_{k+j}$ irrespective of the input and the state, the following constraints must be met

$$H^{(j)}\bar{C}_{N_f} = -\mathcal{K}_{j+1} \tag{5.64}$$

$$H^{(j)}B_{N_f} = -L^{(j)} \tag{5.65}$$

for all j. It is noted that, once $H^{(j)}$ is found out independently from (5.65), $L^{(j)}$ is obtained from (5.65) automatically.

5.5.2 Linear Quadratic Finite Memory Control

Consider the following performance criterion:

$$E\left[\sum_{j=0}^{N_c-1} \left[x^T_{k+j}Qx_{k+j} + u^T_{k+j}Ru_{k+j}\right] + x^T_{k+N_c}Fx_{k+N_c}\right] \tag{5.66}$$

If we assume that the full information of the state is available, then it is well known that the optimal state feedback control can be written

$$u_{k+j}^* = -R^{-1}B^T[I + K_{j+1}BR^{-1}B^T]^{-1}K_{j+1}Ax_j$$
$$= -[R + B^T K_{j+1}B]^{-1}B^T K_{j+1}Ax_j \qquad (5.67)$$

where K_i satisfies (5.13).

For the performance criterion (5.66), $\mathcal{K}_{j+1}$ in (5.64) will be taken as

$$\mathcal{K}_{j+1} = [R + B^T K_{j+1}B]^{-1}B^T K_{j+1}A \qquad (5.68)$$

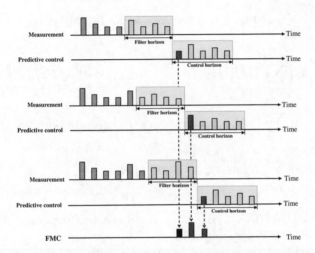

Fig. 5.5. Finite memory controls

In this section, the output feedback control with a finite memory structure and an unbiased condition will be obtained from the usual receding horizon LQG criterion (5.66) subject to the control (5.50) for the stochastic systems (5.1) and (5.2). These output feedback controls with finite memory structure for the cost criterion (5.66) can be called LQFMCs, which are depicted in Figure 5.5. The time indices related with (5.50) and (5.66) can be seen in Figure 5.6.

From Lemma 2.9, a quadratic performance criterion can be represented in a perfect square expression for any control:

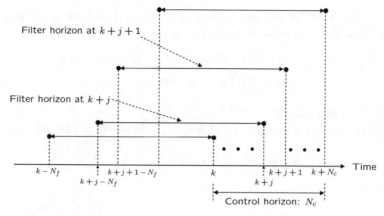

Fig. 5.6. Time variables for LQFMC

$$E\left[\sum_{j=0}^{N_c-1}\left[x_{k+j}^T Q x_{k+j} + u_{k+j}^T R u_{k+j}\right] + x_{k+N_c}^T F x_{k+N_c}\right]$$

$$= E\left[\sum_{j=0}^{N_c-1}\left\{-\mathcal{K}_{j+1}x_{k+j} + u_{k+j}\right\}^T \hat{R}_j\left\{-\mathcal{K}_{j+1}x_{k+j} + u_{k+j}\right\}\right]$$

$$+ \text{tr}\left[\sum_{j=0}^{N_c-1} K_{j+1}GQ_wG^T\right] + E\left[x_k^T K_0 x_k\right] \tag{5.69}$$

where $\hat{R}_j \triangleq [R + B^T K_{j+1}B]$, $\mathcal{K}_{j+1}$ is defined in (5.68), and K_{j+1} is the solution to Riccati Equation (5.13).

It is noted that the result of Lemma 2.9 holds for any inputs. By using Lemma 2.9, the performance criterion can be changed to the following form:

$$H_B = \arg\min_{H^{(0)},\cdots,H^{(N_c-1)}} E\left[\sum_{j=0}^{N_c-1}\left[x_{k+j}^T Q x_{k+j} + u_{k+j}^T R u_{k+j}\right] + x_{k+N_c}^T F x_{k+N_c}\right]$$

$$= \arg\min_{H^{(0)},\cdots,H^{(N_c-1)}} E\left[\sum_{j=0}^{N_c-1}\left\{\mathcal{K}_{j+1}x_{k+j} + u_{k+j}\right\}^T \hat{R}_j\left\{\mathcal{K}_{j+1}x_{k+j} + u_{k+j}\right\}\right]$$

$$+ \text{tr}\left[\sum_{j=0}^{N_c-1} K_{j+1}GQ_wG^T\right] + E\left[x_k^T K_0 x_k\right] \tag{5.70}$$

It can be seen that the last two terms in (5.70) do not depend on a control gain $H^{(j)}$. We have only to minimize the first term in (5.70) in order to obtain the solution.

Since, under the unbiasedness constraint, the following equation is obtained:

$$u_{k+j} + \mathcal{K}_{j+1}x_{k+j} = H^{(j)}\bar{G}_{N_f}W_{k+j-1} + H^{(j)}V_{k+j-1} \qquad (5.71)$$

the first term in (5.70) is obtained as

$$E\left[\{\mathcal{K}_{j+1}x_{k+j} + u_{k+j}\}^T \hat{R}_j \{\mathcal{K}_{j+1}x_{k+j} + u_{k+j}\}\right]$$

$$= E\left[\{H^{(j)}\bar{G}_{N_f}W_{k+j-1} + H^{(j)}V_{k+j-1}\}^T \hat{R}_j \{H^{(j)}\bar{G}_{N_f}W_{k+j-1} + H^{(j)}V_{k+j-1}\}\right]$$

$$= \text{tr}\left[\hat{R}_j^{\frac{1}{2}} H^{(j)}\bar{G}_{N_f}Q_{N_f}\bar{G}_{N_f}^T H^{(j)T}\hat{R}_j^{\frac{1}{2}} + \hat{R}_j^{\frac{1}{2}} H^{(j)}R_{N_f}H^{(j)T}\hat{R}_j^{\frac{1}{2}}\right] \qquad (5.72)$$

The objective now is to obtain the optimal gain matrix $H_B^{(j)}$, subject to the unbiasedness constraint (5.65), in such a way that the cost function (5.72) has a minimum variance as follows:

$$H_B^{(j)} = \arg\min_{H^{(j)}} \text{tr}\left[\hat{R}_j^{\frac{1}{2}} H^{(j)}\bar{G}_{N_f}Q_{N_f}\bar{G}_{N_f}^T H^{(j)T}\hat{R}_j^{\frac{1}{2}}\right.$$

$$\left. + \hat{R}_j^{\frac{1}{2}} H^{(j)}R_{N_f}H^{(j)T}\hat{R}_j^{\frac{1}{2}}\right] \qquad (5.73)$$

subject to

$$H^{(j)}\bar{C}_{N_f} = -\mathcal{K}_{j+1} \qquad (5.74)$$

where

$$Q_{N_f} = [\text{diag}(\overbrace{Q_w \ Q_w \ \cdots \ Q_w}^{N_f})] \quad \text{and} \quad R_{N_f} = [\text{diag}(\overbrace{R_v \ R_v \ \cdots \ R_v}^{N_f})] (5.75)$$

By using the result of Lemma 4.3, the solution to the optimization problem (5.73) and (5.74) can be obtained according to the following correspondences:

$$
\begin{aligned}
A &\longleftarrow \bar{G}_{N_f} \\
B &\longleftarrow O \\
C &\longleftarrow Q_{N_f} \\
D &\longleftarrow R_{N_f} \\
E &\longleftarrow \bar{C}_{N_f} \\
F &\longleftarrow -\hat{R}_j^{\frac{1}{2}}\mathcal{K}_{j+1} \\
H &\longleftarrow \hat{R}_j^{\frac{1}{2}} H^{(j)}
\end{aligned} \qquad (5.76)
$$

$L^{(j)}$ is determined from (5.65). It follows finally that

$$u_{k+j} = -H^{(j)}Y_{k+j-1} + L^{(j)}U_{k+j-1}$$

$$= -\mathcal{K}_{j+1}(\bar{C}_{N_f}^T \Xi_{N_f}^{-1}\bar{C}_{N_f})^{-1}\bar{C}_{N_f}^T \Xi_{N_f}^{-1}(Y_{k+j-1} - \bar{B}_{N_f}U_{k+j-1}) \quad (5.77)$$

If j is fixed to zero in (5.77), then we obtain the optimal solution for LQFMC in the following theorem.

Theorem 5.1.   *When* $\{A, C\}$ *is observable and* $N_f \geq n_o$, *the LQFMC* u_k *is given as follows:*

$$u_k = -\mathcal{K}_1(\bar{C}_{N_f}^T \Xi_{N_f}^{-1} \bar{C}_{N_f})^{-1} \bar{C}_{N_f}^T \Xi_{N_f}^{-1}(Y_{k-1} - \bar{B}_{N_f} U_{k-1}) \quad (5.78)$$

in the case that A *is nonsingular.* Y_{k-1}, U_{k-1}, $\bar{C}_{N_f}$, $\bar{B}_{N_f}$, Ξ_{N_f}, *and* $\mathcal{K}_1$ *are given by* (5.54), (5.55), (4.31), (4.32), (4.214), *and* (5.68) *respectively.*   ∎

The dimension of Ξ_{N_f} in (5.78) may be large. So a numerical error during inverting the matrix can happen. To avoid the handling of the large matrix, the optimal gain matrix for a nonsingular A can be obtained from the following recursive equations:

$$H_B = -[R + B^T K_1 B]^{-1} B^T K_1 A \Omega_{N_f}^{-1} \Psi_{N_f} \quad (5.79)$$

where Ω_{N_f} is defined as $\bar{C}_{N_f}^T \Xi_{N_f}^{-1} \bar{C}_{N_f}$ as in Chapter 4 and is written recursively as

$$\Omega_{i+1} = [I + A^{-T}(\Omega_i + C^T R_v^{-1} C)A^{-1} G Q_w G^T]^{-1} A^{-T}(\Omega_i + C^T R_v^{-1} C)A^{-1}$$

with $\Omega_0 = 0$. We define

$$\Psi_{i+1} \triangleq \bar{C}_{i+1}^T \bar{\Xi}_{i+1}^{-1} = [I + A^{-T}(\Omega_i + C^T R_v^{-1} C)A^{-1} G Q_w G^T]^{-1} A^{-T}$$
$$\times \begin{bmatrix} \bar{C}_i \\ C \end{bmatrix}^T \begin{bmatrix} \bar{\Xi}_i & 0 \\ 0 & R_v \end{bmatrix}^{-1} \quad (5.80)$$

with $\Psi_0 = 0$ and then recursive equations for Ψ_i can be easily derived from the structure of the matrix Ξ_{N_f} in (5.78) as follows:

$$\Psi_{i+1} = [I + A^{-T}(\Omega_i + C^T R_v^{-1} C)A^{-1} G Q_w G^T]^{-1} A^{-T} [\Psi_i \mid C^T R_v^{-1}] \quad (5.81)$$

which is left as an exercise. From Theorem 5.1, it can be known that the LQFMC u_k (5.78) processes the finite measurements and inputs on the horizon $[k - N_f, k]$ linearly and has the properties of unbiasedness from the optimal state feedback control *by design*. Note that the optimal gain matrix H_B (5.79) requires a computation only on the interval $[0, N_f]$ once and is time-invariant for all horizons. This means that the LQFMC is time-invariant. The LQFMC may be robust against temporary modelling uncertainties or round-off errors due to the finite memory structure.

So far, we have derived the LQFMC in a closed form. Besides the analytical form, the LQFMC can be represented in LMI form to get a numerical solution.

First, from the unbiased condition $H\bar{C}_{N_f} = -\mathcal{K}_1$, the control gain H of the LQFMC can be parameterized as

$$H = FM^T + H_0 \quad (5.82)$$

where $H_0 = -\mathcal{K}_1(\bar{C}_{N_f}^T \bar{C}_{N_f})^{-1} \bar{C}_{N_f}^T$ and the columns of M consist of the basis of the null space of $\bar{C}_{N_f}^T$. Let us introduce a matrix variable W such that

$$W > \hat{R}^{\frac{1}{2}} H \bar{G}_{N_f} Q_{N_f} \bar{G}_{N_f}^T H \hat{R}^{\frac{1}{2}} + \hat{R}^{\frac{1}{2}} H R_{N_f} H^T \hat{R}^{\frac{1}{2}} \qquad (5.83)$$

where $\hat{R}$ is given by

$$\hat{R} = R + B^T K_1 B$$

Thus, from (5.83) we have

$$\min_{W,F} \gamma \qquad (5.84)$$

subject to

$$\text{tr}(W) < \gamma$$

$$\begin{bmatrix} W & \hat{R}^{\frac{1}{2}}(FM^T + H_0) \\ (FM^T + H_0)^T \hat{R}^{\frac{1}{2}} & (\bar{G}_{N_f} Q_{N_f} \bar{G}_{N_f}^T + R_{N_f})^{-1} \end{bmatrix} > 0 \qquad (5.85)$$

From the optimal H we obtain $L = -H\bar{B}_{N_f}$.

5.5.3* Linear Quadratic Finite Memory Control with General A

Now we turn to the case where the system matrix A is general.

The system (5.1) and (5.2) will be represented in a batch form on the time interval $[k+j-N_f, k+j]$ called the horizon. On the horizon $[k+j-N_f, k+j]$, measurements are expressed in terms of the state x_{k+j} at the time $k+j$ and inputs on $[k+j-N_f, k+j]$ as follows:

$$Y_{k+j-1} = \tilde{C}_{N_f} x_{k+j-N_f} + \tilde{B}_{N_f} U_{k+j-1} + \tilde{G}_{N_f} W_{k+j-1} + V_{k+j-1} \qquad (5.86)$$

where

$$Y_{k+j-1} \triangleq [y_{k+j-N_f}^T \ y_{k+j-N_f+1}^T \ \cdots \ y_{k+j-1}^T]^T \qquad (5.87)$$

$$U_{k+j-1} \triangleq [u_{k+j-N_f}^T \ u_{k+j-N_f+1}^T \ \cdots \ u_{k+j-1}^T]^T \qquad (5.88)$$

$$W_{k+j-1} \triangleq [w_{k+j-N_f}^T \ w_{k+j-N_f+1}^T \ \cdots \ w_{k+j-1}^T]^T \qquad (5.89)$$

$$V_{k+j-1} \triangleq [v_{k+j-N_f}^T \ v_{k+j-N_f+1}^T \ \cdots \ v_{k+j-1}^T]^T$$

and $\tilde{C}_{N_f}$, $\tilde{B}_{N_f}$, and $\tilde{G}_{N_f}$ are obtained from (4.78), (4.79), and (4.80) respectively. The noise term $\tilde{G}_{N_f} W_{j-1} + V_{j-1}$ in (4.75) can be shown to be zero mean with a covariance Π_N given by

$$\Pi_{N_f} = \tilde{G}_{N_f} Q_{N_f} \tilde{G}_{N_f}^T + R_{N_f} \qquad (5.90)$$

where Q_{N_f} and R_{N_f} are given in (5.75).

The state x_{k+j} can be represented using the initial state x_{k+j-N} on the horizon as

$$
\begin{aligned}
x_{k+j} &= A^{N_f} x_{k+j-N_f} + \left[A^{N_f-1}G \ A^{N_f-2}G \cdots G \right] W_{k+j-1} \\
&+ \left[A^{N_f-1}B \ A^{N_f-2}B \cdots B \right] U_{k+j-1}
\end{aligned} \tag{5.91}
$$

Augmenting (5.86) and (5.91) yields the following linear model:

$$
\begin{aligned}
\begin{bmatrix} Y_{k+j-1} \\ 0 \end{bmatrix} &= \begin{bmatrix} \tilde{C}_{N_f} & 0 \\ A^{N_f} & -I \end{bmatrix} \begin{bmatrix} x_{k+j-N_f} \\ x_{k+j} \end{bmatrix} + \begin{bmatrix} & \tilde{B}_{N_f} & \\ A^{N_f-1}B & \cdots & B \end{bmatrix} U_{k+j-1} \\
&+ \begin{bmatrix} & \tilde{G}_{N_f} & \\ A^{N_f-1}G & \cdots & G \end{bmatrix} W_{k+j-1} + \begin{bmatrix} V_{k+j-1} \\ 0 \end{bmatrix}
\end{aligned} \tag{5.92}
$$

By using Equation (5.92), the FMC (5.51) can be rewritten as

$$
\begin{aligned}
u_{k+j} &= \left[H^{(j)} \ -\mathcal{M}_{j+1} \right] \begin{bmatrix} Y_{k+j-1} \\ 0 \end{bmatrix} + L^{(j)} U_{k+j-1} \\
&= \left[H^{(j)} \ -\mathcal{M}_{j+1} \right] \begin{bmatrix} \tilde{C}_{N_f} & 0 \\ A^{N_f} & -I \end{bmatrix} \begin{bmatrix} x_{k+j-N_f} \\ x_{k+j} \end{bmatrix} \\
&+ \left[H^{(j)} \ -\mathcal{M}_{j+1} \right] \begin{bmatrix} & \tilde{B}_{N_f} & \\ A^{N_f-1}B & \cdots & B \end{bmatrix} U_{k+j-1} \\
&+ \left[H^{(j)} \ -\mathcal{M}_{j+1} \right] \begin{bmatrix} & \tilde{G}_{N_f} & \\ A^{N_f-1}G & \cdots & G \end{bmatrix} W_{k+j-1} \\
&+ \left[H^{(j)} \ -\mathcal{M}_{j+1} \right] \begin{bmatrix} V_{k+j-1} \\ 0 \end{bmatrix} + L^{(j)} U_{k+j-1}
\end{aligned} \tag{5.93}
$$

where $\mathcal{M}_{j+1}$ is chosen later. Taking the expectation on both sides of (5.93) yields the following equation:

$$
\begin{aligned}
E[u_{k+j}] &= (H^{(j)}\tilde{C}_{N_f} - \mathcal{M}_{j+1}A^{N_f})E[x_{k+j-N_f}] + \mathcal{M}_{j+1}E[x_{k+j}] \\
&+ \left[H^{(j)} \ -\mathcal{M}_{j+1} \right] \begin{bmatrix} & \tilde{B}_{N_f} & \\ A^{N_f-1}B & \cdots & B \end{bmatrix} U_{k+j-1} + L^{(j)} U_{k+j-1}
\end{aligned}
$$

To satisfy the unbiased condition, i.e. $E[u_{k+j}] = E[u^*_{k+j}]$, irrespective of states and the input on the horizon $[k + j - N_f, k + j]$, $\mathcal{M}_{j+1}$ should be equal to $-\mathcal{K}_{j+1}$ and the following constraints are required:

$$
H^{(j)}\tilde{C}_{N_f} = -\mathcal{K}_{j+1}A^{N_f}, \tag{5.94}
$$

$$
L_i^{(j)} = -\left[H^{(j)} \ \mathcal{K}_{j+1} \right] \begin{bmatrix} & \tilde{B}_{N_f} & \\ A^{N_f-1}B & \cdots & B \end{bmatrix} \tag{5.95}
$$

This will be called the unbiasedness constraint. It is noted that a variable j ranges from 0 to $N_c - 1$. Under the unbiasedness constraint, we have

$$u_{k+j} + \mathcal{K}_{j+1}x_{k+j} = H^{(j)}\tilde{G}_{N_f}W_{k+j-1} - \mathcal{K}_{j+1}\left[A^{N_f-1}G \cdots G\right]W_{k+j-1}$$
$$+ H^{(j)}V_{k+j-1} \tag{5.96}$$

The objective now is to obtain the optimal gain matrix $H_B^{(j)}$, subject to the unbiasedness constraint (5.94) and (5.95), in such a way that the cost function (5.66) is minimized.

We can see that the last two terms in (5.69) do not depend on a control gain $H^{(j)}$. We have only to minimize the first term in (5.69) in order to obtain the solution.

The following relation for the first term in (5.69) is obtained:

$$E\left[\left\{\mathcal{K}_{j+1}x_{k+j} + u_{k+j}\right\}^T \hat{R}_j \left\{\mathcal{K}_{j+1}x_{k+j} + u_{k+j}\right\}\right]$$

$$= E\left[\left\{H^{(j)}\tilde{G}_{N_f}W_{k+j-1} - \mathcal{K}_{j+1}\left[A^{N_f-1}G \cdots G\right]W_{k+j-1} + H^{(j)}V_{k+j-1}\right\}^T \hat{R}_j\right.$$

$$\times \left\{H^{(j)}\tilde{G}_{N_f}W_{k+j-1} - \mathcal{K}_{j+1}\left[A^{N_f-1}G \cdots G\right]W_{k+j-1} + H^{(j)}V_{k+j-1}\right\}\right]$$

$$= \text{tr}\left[\hat{R}_j^{\frac{1}{2}}\left(H^{(j)}\tilde{G}_{N_f} - \mathcal{K}_{j+1}\left[A^{N_f-1}G \cdots G\right]\right)\right.$$

$$\times Q_{N_f}\left(H^{(j)}\tilde{G}_{N_f} - \mathcal{K}_{j+1}\left[A^{N_f-1}G \cdots G\right]\right)^T \hat{R}_j^{\frac{1}{2}}$$

$$\left.+ \hat{R}_j^{\frac{1}{2}}H^{(j)}R_{N_f}H^{(j)T}\hat{R}_j^{\frac{1}{2}}\right] \tag{5.97}$$

Then, from (5.69) and (5.94), the optimization problem to solve is summarized as

$$H_B^{(j)} = \arg\min_{H^{(j)}} \text{tr}\left[\hat{R}_j^{\frac{1}{2}}\left(H^{(j)}\tilde{G}_{N_f} - \mathcal{K}_{j+1}\left[A^{N_f-1}G \cdots G\right]\right)\right.$$

$$\times Q_{N_f}\left(H^{(j)}\tilde{G}_{N_f} - \mathcal{K}_{j+1}\left[A^{N_f-1}G \cdots G\right]\right)^T \hat{R}_j^{\frac{1}{2}}$$

$$\left.+ \hat{R}_j^{\frac{1}{2}}H^{(j)}R_{N_f}H^{(j)T}\hat{R}_j^{\frac{1}{2}}\right] \tag{5.98}$$

subject to
$$H^{(j)}\tilde{C}_{N_f} = -\mathcal{K}_{j+1}A^{N_f} \tag{5.99}$$

By using the result of Lemma 4.3, the solution $H_B^{(j)}$ to the optimization problem (5.98) and (5.99) can be obtained according to the following correspondences:

$$
\begin{aligned}
A &\longleftarrow \tilde{G}_{N_f} \\
B &\longleftarrow \hat{R}_j^{\frac{1}{2}} \mathcal{K}_{j+1}\left[A^{N_f-1}G \cdots G\right] \\
C &\longleftarrow Q_{N_f} \\
D &\longleftarrow R_{N_f} \\
E &\longleftarrow \tilde{C}_{N_f} \\
F &\longleftarrow -\hat{R}_j^{\frac{1}{2}} \mathcal{K}_{j+1} \\
H &\longleftarrow \hat{R}_j^{\frac{1}{2}} H^{(j)} \\
W_{1,1} &\longleftarrow \tilde{C}_{N_f}^T R_{N_f}^{-1} \tilde{C}_{N_f} \\
W_{1,2} &\longleftarrow \tilde{C}_{N_f}^T R_{N_f}^{-1} \tilde{G}_{N_f} \\
W_{2,2} &\longleftarrow \tilde{G}_{N_f}^T R_{N_f}^{-1} \tilde{G}_{N_f} + Q_{N_f}^{-1}
\end{aligned}
\tag{5.100}
$$

Then, $L^{(j)}$ is determined from (5.95).

If j is replaced by 0, the LQFMC is obtained at the current time k. Therefore, the LQFMC u_k with the optimal gain matrices H and L is summarized in the following theorem.

Theorem 5.2.   *When $\{A, C\}$ is observable and $N_f \geq n_o$, the LQFMC u_k is given as follows:*

$$
\begin{aligned}
u_k = &-\mathcal{K}_1\left[A^{N_f}\ A^{N_f-1}G\ A^{N_f-2}G \cdots AG\ G\right]\begin{bmatrix} W_{1,1} & W_{1,2} \\ W_{1,2}^T & W_{2,2} \end{bmatrix}^{-1} \\
&\times \begin{bmatrix} \tilde{C}_{N_f}^T \\ \tilde{G}_{N_f}^T \end{bmatrix} R_{N_f}^{-1}\left(Y_{k-1} - \tilde{B}_{N_f} U_{k-1}\right) \\
&-\mathcal{K}_1\left[A^{N_f-1}B\ A^{N_f-2}B\ A^{N_f-3}B \cdots AB\ B\right]U_{k-1}
\end{aligned}
\tag{5.101}
$$

W_{11}, W_{12}, and W_{22} are defined in (5.100). Y_{k-1}, U_{k-1}, $\tilde{C}_{N_f}$, $\tilde{B}_{N_f}$ and $\mathcal{K}_1$ are given by (5.48), (5.49), (4.78), (4.79) and (5.68) respectively. ∎

By using the recursive form of the MVF filter in Theorem 4.10, the LQFMC (5.7) in Theorem 5.1 can be obtained as

$$
u_k = -\mathcal{K}_1 \beta_k
\tag{5.102}
$$

where

$$
\begin{aligned}
\beta_{k-N_f+i+1} &= A\beta_{k-N_f+i} + AP_i C^T(R_v + CP_i C^T)^{-1}(y_{k-N_f+i} - C\beta_{k-N_f+i}) \\
&= (A - AP_i C^T(R_v + CP_i C^T)^{-1}C)\beta_{k-N_f+i} \\
&\quad + AP_i C^T(R_v + CP_i C^T)^{-1}y_{k-N_f+i}
\end{aligned}
\tag{5.103}
$$

$$
P_{i+1} = AP_i A^T + GQ_w G^T - AP_i C^T(R_v + CP_i C^T)^{-1}CP_i A^T
\tag{5.104}
$$

Here, $P_0 = (C^T R_v^{-1} C + \hat{P}_{N_f})^{-1}$ and $\beta_{k-N_f} = P_0(C^T R_v^{-1} y_{k-N} + \hat{\omega}_{N_f})$. $\hat{P}_{N_f}$ and $\hat{\omega}_{N_f}$ are obtained from the following recursive equations:

$$\hat{P}_{i+1} = A^T C^T R_v^{-1} C A + A^T \hat{P}_i A$$
$$- A^T (C^T R_v^{-1} C + \hat{P}_i) A G \Big\{ Q_w^{-1} + G^T A^T (C^T R_v^{-1} C + \hat{P}_i) A G \Big\}^{-1}$$
$$G^T A^T (C^T R_v^{-1} C + \hat{P}_i) A \tag{5.105}$$

and

$$\hat{\omega}_{i+1} = A^T C^T R_v y_{k-i} + A \hat{\omega}_i$$
$$- A^T (C^T R_v^{-1} C + \hat{P}_i) A G \Big\{ Q_w^{-1} + G^T A^T (C^T R_v^{-1} C + \hat{P}_i) A G \Big\}^{-1}$$
$$G^T A^T (C^T R_v^{-1} y_{k-i} + \hat{\omega}_i) \tag{5.106}$$

5.5.4 Properties of Linear Quadratic Finite Memory Control

Closed-Loop System

When the LQFMC is adopted, the system order for the internal state increases. Now we investigate poles of closed-loop systems

Substituting (5.45) into the model (5.5) yields the following equation:

$$x_{k+1} = A x_k + B u_k + G w_k = A x_k + B(H Y_{k-1} + L U_{k-1}) + G w_k \tag{5.107}$$

It is noted that (5.107) is a dynamic equation with respect to x_k and will be combined with the dynamics of Y_{k-1} and U_{k-1} into the overall closed-loop system.

U_k and Y_k are represented in a form of dynamic equations as follows:

$$U_k = \bar{A}_u U_{k-1} + \bar{B}_u u_k = \bar{A}_u U_{k-1} + \bar{B}_u (H Y_{k-1} + L U_{k-1}) \tag{5.108}$$
$$Y_k = \bar{A}_y Y_{k-1} + \bar{B}_y y_k = \bar{A}_y Y_{k-1} + \bar{B}_y (C x_k + v_k) \tag{5.109}$$

where

$$\bar{A}_u = \begin{bmatrix} 0 & I & 0 & \cdots & 0 \\ 0 & 0 & I & \ddots & 0 \\ \vdots & \vdots & \ddots & \ddots & \vdots \\ 0 & 0 & \cdots & 0 & I \\ 0 & 0 & \cdots & 0 & 0 \end{bmatrix} \in R^{pN_f \times pN_f}, \qquad \bar{B}_u = \begin{bmatrix} 0 \\ 0 \\ 0 \\ \vdots \\ I \end{bmatrix} \in R^{pN_f \times p} \tag{5.110}$$

and $\bar{A}_y$ and $\bar{B}_y$ are of the same form as $\bar{A}_u$ and $\bar{B}_u$ respectively except for dimension. From (5.107), (5.108), and (5.109), it can be easily seen that x_{k+1}, U_k, and W_k are augmented into one state and then the following augmented state-space can be obtained:

$$\begin{bmatrix} x_{k+1} \\ U_k \\ Y_k \end{bmatrix} = \begin{bmatrix} A & BL & BH \\ 0 & \bar{A}_u + \bar{B}_u L & \bar{B}_u H \\ \bar{B}_y C & O & \bar{A}_y \end{bmatrix} \begin{bmatrix} x_k \\ U_{k-1} \\ Y_{k-1} \end{bmatrix} + \begin{bmatrix} G & 0 \\ 0 & 0 \\ 0 & \bar{B}_y \end{bmatrix} \begin{bmatrix} w_k \\ v_k \end{bmatrix}$$

$$\stackrel{\triangle}{=} A_a \begin{bmatrix} x_k \\ U_{k-1} \\ Y_{k-1} \end{bmatrix} + B_a \begin{bmatrix} w_k \\ v_k \end{bmatrix} \tag{5.111}$$

Now, instead of the matrix A_a in (5.111), we consider the following matrix:

$$A_b \stackrel{\triangle}{=} \begin{bmatrix} A & BL & BH & B\mathcal{K}_1 \\ 0 & \bar{A}_u + \bar{B}_u L & \bar{B}_u H & \bar{B}_u \mathcal{K}_1 \\ \bar{B}_y C & 0 & \bar{A}_y & \bar{B}_y C \\ 0 & 0 & 0 & A \end{bmatrix}$$

of which eigenvalues consist of ones of the matrix A_a in (5.111) and a matrix A. Suppose that H is related to L according to (5.94) and (5.95). Premultiplying and postmultiplying A_b by some transformation matrix, we have

$$\begin{bmatrix} I & 0 & 0 & 0 \\ 0 & I & 0 & 0 \\ 0 & -M_1 & I & 0 \\ I & -M_2 & 0 & I \end{bmatrix} \begin{bmatrix} A & BL & BH & B\mathcal{K}_1 \\ 0 & \bar{A}_u + \bar{B}_u L & \bar{B}_u H & \bar{B}_u \mathcal{K}_1 \\ \bar{B}_y C & 0 & \bar{A}_y & \bar{B}_y C \\ 0 & 0 & 0 & A \end{bmatrix} \begin{bmatrix} I & 0 & 0 & 0 \\ 0 & I & 0 & 0 \\ 0 & M_1 & I & 0 \\ -I & M_2 & 0 & I \end{bmatrix} =$$

$$\begin{bmatrix} A - B\mathcal{K}_1 & O & BH & B\mathcal{K}_1 \\ -\bar{B}_u \mathcal{K}_1 & \bar{A}_u & \bar{B}_u H & \bar{B}_u \mathcal{K}_1 \\ 0 & -M_1 \bar{A}_u + \bar{A}_y M_1 + \bar{B}_y C M_2 & \bar{A}_y & \bar{B}_y C \\ 0 & -M_2 \bar{A}_u + A M_2 & 0 & A \end{bmatrix} \tag{5.112}$$

where $M_1 = \tilde{B}_N$ and $M_2 = [A^{N_f-1}B \quad A^{N_f-2}B \cdots\cdots B]$. Partitioning the matrix in (5.112), we define the following matrices:

$$A_{11} = \begin{bmatrix} A - B\mathcal{K}_1 & O \\ -\bar{B}_u K_1 & \bar{A}_u \end{bmatrix} \qquad A_{12} = \begin{bmatrix} BH & B\mathcal{K}_1 \\ \bar{B}_u H & \bar{B}_u \mathcal{K}_1 \end{bmatrix}$$

$$A_{21} = \begin{bmatrix} 0 & -M_1 \bar{A}_u + \bar{A}_y M_1 + \bar{B}_y C M_2 \\ 0 & -M_2 \bar{A}_u + A M_2 \end{bmatrix} \qquad A_{22} = \begin{bmatrix} \bar{A}_y & \bar{B}_y C \\ 0 & A \end{bmatrix}$$

In order to obtain eigenvalues of the matrix A_b, we should calculate

$$\det\left(\begin{bmatrix} \lambda I - A_{11} & -A_{12} \\ -A_{21} & \lambda I - A_{22} \end{bmatrix} \right) \tag{5.113}$$

Using the inverse of the block matrix in Appendix A, we have

$$\det(\lambda I - A_{11}) \det((\lambda I - A_{11}) - A_{12}(\lambda I - A_{22})^{-1} A_{21}) \tag{5.114}$$

Fortunately, it is observed that $A_{12}(\lambda I - A_{22})^{-1} A_{21} = 0$, which is proved as follows:

$$\begin{bmatrix} H & \mathcal{K}_1 \end{bmatrix} \begin{bmatrix} \lambda I - \bar{A}_y & -\bar{B}_y C \\ 0 & \lambda I - A \end{bmatrix}^{-1} \begin{bmatrix} \tilde{C}_{N_f} \\ A^{N_f} \end{bmatrix}$$

$$= \begin{bmatrix} H & \mathcal{K}_1 \end{bmatrix} \begin{bmatrix} (\lambda I - \bar{A}_y)^{-1} & (\lambda I - \bar{A}_y)^{-1} \bar{B}_y C (\lambda I - A)^{-1} \\ 0 & (\lambda I - A)^{-1} \end{bmatrix} \begin{bmatrix} \tilde{C}_{N_f} \\ A^{N_f} \end{bmatrix}$$

$$= H(\lambda I - \bar{A}_y)^{-1} \tilde{C}_{N_f} + H(\lambda I - \bar{A}_y)^{-1} \bar{B}_y C (\lambda I - A)^{-1} A^{N_f}$$

$$+ \mathcal{K}_1 (\lambda I - A)^{-1} A^{N_f} = 0$$

where the last equality comes from the fact $H\tilde{C}_{N_f} + \mathcal{K}_1 A^{N_f} = 0$. Finally, we see that the eigenvalues of the matrix A_b consist of ones of A_a and A. What we have done so far is summarized in the following theorem.

Theorem 5.3. *Suppose that H and L are related to each other according to (5.94) and (5.95). The overall system with the finite memory control (5.45) is represented as (5.111), whose eigenvalues consist of eigenvalues of $A - B\mathcal{K}_1$ and zeros.* ∎

The separation principle and the stability for the proposed LQFMC will be investigated from now on. It is very interesing to see that the globally optimal LQFMC (5.78) can be separated as a receding control and an FIR filter.

Theorem 5.4. *Assume that A is nonsingular. The LQFMC (5.78) can be represented as a receding horizon LQ control with the state replaced by an MVF filter as*

$$u_k = -[R + B^T K_1 B]^{-1} B^T K_1 A \hat{x}_k \qquad (5.115)$$

$$\hat{x}_k = (\bar{C}_{N_f}^T \Xi_{N_f}^{-1} \bar{C}_{N_f})^{-1} \bar{C}_{N_f}^T \Xi_{N_f}^{-1} [Y_{k-1} - \bar{B}_{N_f} U_{k-1}] \qquad (5.116)$$

where $\hat{x}_k$ is an actual state estimator.

Proof. This is obvious from (5.78). This completes the proof. ∎

Note that the result in Theorem 5.4 holds for a general system matrix A.

In Chapter 4, it is shown that $\hat{x}_k$ in (5.116) is an optimal minimum variance state estimator with an FIR structure. It is known that the FIR filter (5.116) is an unbiased filter that has the deadbeat property for systems with zero noise. From this property, we can show the following theorem.

Theorem 5.5. *If the final weighting matrix F in the cost function satisfies the following inequality:*

$$F \geq Q + D^T RD + (A - BD)^T F(A - BD) \quad \text{for some } D \qquad (5.117)$$

the closed-loop system driven by the proposed LQFMC is asymptotically stable.

Proof. According to Theorem 5.3, poles of the closed-loop system consist of eigenvalues of $A - B[R + B^T K_1 B]^{-1} B^T K_1 A$ and zeros.

From Theorem 3.8 in Chapter 3, $A - B[R + B^T K_1 B]^{-1} B^T K_1 A$ is Hurwitz for the terminal weighting matrix F satisfying the inequality (5.117). This completes the proof. ∎

If the power of noises is finite, the following bound is guaranteed for stochastic systems (5.1) and (5.2):

$$E[x_k x_k^T] < \infty \tag{5.118}$$

for all $k \geq k_0$ under the condition (5.117).

Example 5.1

To demonstrate the validity of the proposed LQFMC, a numerical example on the model of an F-404 engine is presented via simulation. The dynamic model is written as

$$x_{k+1} = \begin{bmatrix} 0.9305 + \delta_k & 0 & 0.1107 \\ 0.0077 & 0.9802 + \delta_k & -0.0173 \\ 0.0142 & 0 & 0.8953 + 0.1\delta_k \end{bmatrix} x_k \tag{5.119}$$

$$+ \begin{bmatrix} 0.4182 & 5.203 \\ 0.3901 & -0.1245 \\ 0.5186 & 0.0236 \end{bmatrix} u_k + \begin{bmatrix} 1 \\ 1 \\ 1 \end{bmatrix} w_k \tag{5.120}$$

$$y_k = \begin{bmatrix} 1 & 0 & 0 \\ 0 & 1 & 0 \end{bmatrix} x_k + v_k \tag{5.121}$$

where δ_k is an uncertain model parameter. The system noise covariance Q_w is 0.01^2 and the measurement noise covariance R_v is 0.01^2.

It will be shown by simulation that the proposed LQFMC is robust against temporary modelling uncertainties since it utilizes only finite outputs and inputs on the most recent horizon. To check the robustness, the proposed LQFMC and the LQG control are compared when a system has a temporary modelling uncertainty given by $\delta_k = 0.05$ on the interval $15 \leq k \leq 25$. The filter horizon length N_f and the control horizon length N_c of the LQFMC are both taken as 10. Figure 5.7 compares the robustness of two controls given temporary modelling uncertainties. As can be seen the robustness of the LQFMC is significant in Figure 5.7(a) for the first state.

Figure 5.7 shows that the deviation from zero of the LQFMC is significantly smaller than that of the LQG control on the interval where modeling uncertainty exists. In addition, it is shown that the recovery to the steady state is much faster than that of the LQG control after the temporary modelling uncertainty disappears. Therefore, it can be seen that the proposed LQFMC is more robust than the LQG control when applied to systems with a model parameter uncertainty. ∎

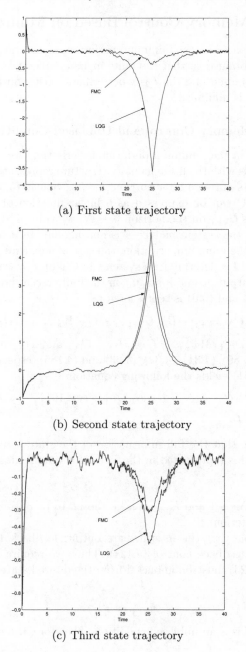

(a) First state trajectory

(b) Second state trajectory

(c) Third state trajectory

Fig. 5.7. State trajectory of LQG and LQFMC

5.6 Finite Memory Control Based on Minimax Criterion

As in the FMC based on the minimum performance criterion, a receding horizon control is obtained so that the finite memory structure and the unbiased
condition are required and the L_2-E performance criterion for a linear deterministic system is minimized.

5.6.1 Finite Memory Control and Unbiased Condition

As in the FMC in the minimum performance criterion, we start from output
feedback controls with the finite memory structure represented using measurements and inputs during a filter horizon $[k - N_f, k]$ as in (5.50). As mentioned
before, the FMC can be expressed as a linear function of the finite inputs
Y_{k-1} and inputs U_{k-1} on the horizon $[k - N_f \ k]$ as in (5.51). In this section,
we assume that the system matrix A is nonsingular. In a similar way to the
FMC based on the minimum criterion, the case of a general A can be treated.
As in FIR filters for deterministic systems in Chapter 4, we are interested in
the measured output y_k and input u_k on the finite recent horizon, from which
the system (5.5) and (5.6) is represented as

$$Y_{k+j-1} = \bar{C}_{N_f} x_{k+j} + \bar{B}_{N_f} U_{k+j-1} + \bar{G}_{N_f} W_{k+j-1} + \bar{D}_{N_f} W_{k+j-1} \quad (5.122)$$

and Y_{k+j-1}, U_{k+j-1}, W_{k+j-1}, $\bar{C}_{N_f}$, $\bar{B}_{N_f}$, $\bar{G}_{N_f}$ and $\bar{D}_{N_f}$ are obtained from
(5.87), (5.88), (5.89), (4.31), (4.32), (4.33) and (4.185) respectively. Combining
(5.122) and (5.51) yields the following equation:

$$u_{k+j} = H^{(j)}(\bar{C}_{N_f} x_{k+j} + \bar{B}_{N_f} U_{k+j-1} + \bar{G}_{N_f} W_{k+j-1} + \bar{D}_{N_f} W_{k+j-1})$$
$$+ L^{(j)} U_{k+j-1} \quad (5.123)$$

If we assume that the full information of the state is available, then the
desirable control is represented in the form of a state feedback control

$$u^*_{k+j} = -\mathcal{K}_{j+1} x_{k+j} \quad (5.124)$$

where $0 \le j \le N_f - 1$ and $\mathcal{K}_{j+1}$ can be chosen to be optimized for a certain
performance criterion.

It is desirable that the intermediate output feedback FMC (5.123) can
track the state feedback control (5.124). Thus, we require a constraint that
the control (5.123) must be unbiased from the desired state feedback control
(5.124) as

$$u_{k+j} = u^*_{k+j} \quad (5.125)$$

for all states. This is an unbiased condition. Therefore, the relation should be
satisfied

$$H^{(j)} \bar{C}_{N_f} = -\mathcal{K}_{j+1}, \ H^{(j)} \bar{B}_{N_f} = -L^{(j)} \quad (5.126)$$

in order to achieve (5.125).

5.6.2 L_2-E Finite Memory Controls

For linear deterministic systems (5.5) and (5.6), the predictor (5.10) is given by

$$\check{x}_{k+j+1|k} = A\check{x}_{k+j|k} + B\check{u}_{k+j|k} \tag{5.127}$$

where it is assumed that disturbances behave around the origin and thus have no net effects. Hereafter, for simplicity, x_{k+j} and u_{k+j} will be used instead of $\check{x}_{k+j|k}$ and $\check{u}_{k+j|k}$ respectively.

The performance criterion is given as

$$\min_{u_i} \max_{w_i} \frac{J - J^*}{\sum_{j=-N_f}^{-1} w_{k+j}^T w_{k+j}} \tag{5.128}$$

where

$$J\bigg|_{u=FMC} = \sum_{j=0}^{N_c-1} \left[x_{k+j}^T Q x_{k+j} + u_{k+j}^T R u_{k+j} \right] + x_{k+N_c}^T F x_{k+N_c} \tag{5.129}$$

$$J\bigg|_{u=SFC}^* = \min \sum_{j=0}^{N_c-1} \left[x_{k+j}^T Q x_{k+j} + u_{k+j}^T R u_{k+j} \right] + x_{k+N_c}^T F x_{k+N_c} \tag{5.130}$$

Note that SFC in (5.130) stands for state feedback controls. It is also noted that J (5.129) is the cost when the FMC as an output feedback control is applied under disturbances on $[k-N_f, k-1]$, and J^* (5.130) is the optimal cost when the state feedback control is used with no disturbances on $[k-N_f, k-1]$. As can be seen in (5.128), it is observed that the maximization is taken for disturbances in the past time.

The system (5.5) and (5.6) will be represented in a batch form on the time interval $[k+j-N_f, k+j]$ called the horizon. Disturbances are assumed to be zero at a future time, i.e. $w_k = w_{k+1} = \cdots = 0$. We introduce a simple linear operation that reflects this assumption. On the horizon $[k+j-N_f, k+j]$, measurements are expressed in terms of the state x_{k+j} at the time $k+j$ and inputs as follows:

$$Y_{k+j-1} = \bar{C}_{N_f} x_{k+j} + \bar{B}_{N_f} U_{k+j-1} + (\bar{G}_{N_f} + \bar{D}_{N_f}) T_{N_f-1-j} W_{k+j-1}$$

where

$$T_{N_f-1-j} = \left[\begin{array}{c|c} I_{N_f-j} & 0 \\ \hline 0 & 0_j. \end{array} \right] \tag{5.131}$$

and $T_{N_f-1-j} W_{k+j-1}$ is made up as follows:

$$T_{N_f-1-j} W_{k+j-1} = [\overbrace{w_{k+j-N_f}^T \ w_{k+j-N_f+1}^T \cdots \ w_{k-1}^T}^{N_f-j} \ \overbrace{w_k^T = 0 \cdots \ w_{k+j-1}^T = 0}^{j}]^T$$

Matrices $H^{(j)}$ and $L^{(j)}$ in the FMC (5.51) will be chosen to minimize the performance criterion (5.128).

Under the unbiased condition (5.65), we have

$$u_{k+j} + \mathcal{K}_{j+1}x_{k+j} = H^j(\bar{G}_{N_f} + \bar{D}_{N_f})T_{N_f-1-j}W_{k+j-1} \qquad (5.132)$$

The objective now is to obtain the optimal gain matrix $H_B^{(j)}$, subject to the unbiasedness constraints (5.94) and (5.95), in such a way that the cost function (5.128) is optimized.

$$H_B = \arg \min_{H^1,\cdots,H^n} \max_{w_i} \frac{\sum_{j=0}^{N_c-1} \left[x_{k+j}^T Q x_{k+j} + u_{k+j}^T R u_{k+j} \right] + x_{k+N_c}^T F x_{k+N_c}}{\sum_{j=-N_f}^{-1} w_{k+j}^T w_{k+j}}$$

$$= \arg \min_{H^1,\cdots,H^n} \max_{w_i} \frac{\sum_{j=0}^{N_c-1} \left\{ \mathcal{K}_{j+1}x_{k+j} + u_{k+j} \right\}^T \hat{R}_j \left\{ \mathcal{K}_{j+1}x_{k+j} + u_{k+j} \right\}}{\sum_{j=-N_f}^{-1} w_{k+j}^T w_{k+j}}$$

From (5.132), we have the following relation:

$$\left\{ \mathcal{K}_{j+1}x_{k+j} + u_{k+j} \right\}^T \hat{R}_j \left\{ \mathcal{K}_{j+1}x_{k+j} + u_{k+j} \right\}$$

$$= \left\{ H^{(j)}(\bar{G}_{N_f} + \bar{D}_{N_f})T_{N_f-1-j}W_{k+j-1} \right\}^T$$

$$\times \hat{R}_j \left\{ H^{(j)}(\bar{G}_{N_f} + \bar{D}_{N_f})T_{N_f-1-j}W_{k+j-1} \right\}$$

$$\leq \lambda_{\max} \left[\hat{R}_j^{\frac{1}{2}} H^{(j)}(\bar{G}_{N_f} + \bar{D}_{N_f})T_{N_f-1-j}T_{N_f-1-j}(\bar{G}_{N_f} + \bar{D}_{N_f})^T H^{(j)T} \hat{R}_j^{\frac{1}{2}} \right]$$

$$\times W_{k+j-1}^T W_{k+j-1} \qquad (5.133)$$

which comes from the following fact:

$$(Ax)^T(Ax) \leq \lambda_{\max}(A^TA)x^Tx = \lambda_{\max}(AA^T)x^Tx \qquad (5.134)$$

where the equality holds if x is the eigenvector corresponding to $\lambda_{\max}(A^TA)$. It is noted that there always exists a W_{k+j-1} such that the equality holds in (5.133).

Then, from (5.133) and the constraints (5.126), the optimization problem to solve is summarized as

$$H_B^{(j)} = \arg\min_{H^{(j)}} \lambda_{\max} \left[\hat{R}_j^{\frac{1}{2}} H^{(j)}(\bar{G}_{N_f} + \bar{D}_{N_f})T_{N_f-1-j}T_{N_f-1-j}^T \right.$$

$$\left. \times (\bar{G}_{N_f} + \bar{D}_{N_f})^T H^{(j)T} \hat{R}_j^{\frac{1}{2}} \right] \qquad (5.135)$$

$$\text{subject to}$$

$$H^{(j)}\bar{C}_{N_f} = -\mathcal{K}_{j+1} \qquad (5.136)$$

By using the result of Lemma 4.13, the solution to the optimization problem (5.135) and (5.136) can be obtained according to the following correspondences:

$$
\begin{aligned}
A &\longleftarrow \bar{G}_{N_f} + \bar{D}_{N_f} \\
B &\longleftarrow O \\
C &\longleftarrow T_{N_f-1-j} T^T_{N_f-1-j} \\
E &\longleftarrow \bar{C}_{N_f} \\
F &\longleftarrow -\hat{R}_j^{\frac{1}{2}} \mathcal{K}_{j+1} \\
H &\longleftarrow \hat{R}_j^{\frac{1}{2}} H^{(j)} \\
\Pi &\longleftarrow (\bar{G}_{N_f} + \bar{D}_{N_f}) T_{N_f-1-j} T^T_{N_f-1-j} (\bar{G}_{N_f} + \bar{D}_{N_f})^T
\end{aligned}
\tag{5.137}
$$

The filter gain $L^{(j)}$ can be obtained from (5.126). If j is replaced by 0, the optimal solution for the nonsingular A is obtained at the current time, which is summarized in the following theorem.

Theorem 5.6.   *When $\{A, C\}$ is observable and $N_f \geq n_o$, the L_2-E FMC u_k is given as follows:*

$$
u_k = -\mathcal{K}_1 (\bar{C}^T_{N_f} \Xi^{-1}_{N_f} \bar{C}_{N_f})^{-1} \bar{C}^T_{N_f} \Xi^{-1}_{N_f} (Y_{k-1} - \bar{B}_{N_f} U_{k-1}) \tag{5.138}
$$

in case that A is nonsingular. Y_{k-1}, U_{k-1}, $\bar{C}_{N_f}$, $\bar{B}_{N_f}$, Ξ_{N_f}, and $\mathcal{K}_1$ are given by (5.48), (5.49), (4.31), (4.32), (4.214), and (5.68) respectively. ∎

The Case of General A

On the horizon $[k + j - N_f, k + j]$, measurements are expressed in terms of the state x_{k+j} at the time $k + j$ and inputs as follows:

$$
\begin{aligned}
Y_{k+j-1} = {}& \tilde{C}_{N_f} x_{k+j-N_f} + \tilde{B}_{N_f} U_{k+j-1} + \tilde{G}_{N_f} T_{N_f-1-j} W_{k+j-1} \\
& + \bar{D}_{N_f} T_{N_f-1-j} W_{k+j-1}
\end{aligned}
\tag{5.139}
$$

where T_{N_f-1-j} is defined in (5.131).

The state x_{k+j} can be represented using the initial state x_{k+j-N} on the horizon as in (5.91) for deterministic systems. Augmenting (5.139) and (5.91) yields the following linear model:

$$
\begin{aligned}
\begin{bmatrix} Y_{k+j-1} \\ 0 \end{bmatrix} = {}& \begin{bmatrix} \tilde{C}_{N_f} & 0 \\ A^{N_f} & -I \end{bmatrix} \begin{bmatrix} x_{k+j-N_f} \\ x_{k+j} \end{bmatrix} + \begin{bmatrix} \tilde{B}_{N_f} \\ A^{N_f-1}B & \cdots & B \end{bmatrix} U_{k+j-1} \\
& + \begin{bmatrix} \tilde{G}_{N_f} + \bar{D}_{N_f} \\ A^{N_f-1}G & \cdots & G \end{bmatrix} T_{N_f-1-j} W_{k+j-1}
\end{aligned}
\tag{5.140}
$$

For the minimax performance criterion, the FMC is of the form (5.50) as in the minimum performance criterion. By using Equation (5.140), the control (5.50) can be rewritten as

$$u_{k+j} = \begin{bmatrix} H^{(j)} & -\mathcal{M}_{j+1} \end{bmatrix} \begin{bmatrix} Y_{k+j-1} \\ 0 \end{bmatrix} + L^{(j)}U_{k+j-1}$$

$$= \begin{bmatrix} H^{(j)} & -\mathcal{M}_{j+1} \end{bmatrix} \begin{bmatrix} \tilde{C}_{N_f} & 0 \\ A^{N_f} & -I \end{bmatrix} \begin{bmatrix} x_{k+j-N_f} \\ x_{k+j} \end{bmatrix}$$

$$+ \begin{bmatrix} H^{(j)} & -\mathcal{M}_{j+1} \end{bmatrix} \begin{bmatrix} & \tilde{B}_{N_f} \\ A^{N_f-1}B & \cdots & B \end{bmatrix} U_{k+j-1}$$

$$+ \begin{bmatrix} H^{(j)} & -\mathcal{M}_{j+1} \end{bmatrix} \begin{bmatrix} & \tilde{G}_{N_f} + \bar{D}_N \\ A^{N_f-1}G & \cdots & G \end{bmatrix} T_{N_f-1-j}W_{k+j-1}$$

$$+ L^{(j)}U_{k+j-1} \tag{5.141}$$

where $\mathcal{M}_{j+1}$ is chosen later. If it is assumed that disturbances do not show up in (5.141), then we have the following equation:

$$u_{k+j} = (H^{(j)}\tilde{C}_{N_f} - \mathcal{M}_{j+1}A^{N_f})x_{k+j-N_f} + \mathcal{M}_{j+1}x_{k+j}$$

$$+ \begin{bmatrix} H^{(j)} & -\mathcal{M}_{j+1} \end{bmatrix} \begin{bmatrix} & \tilde{B}_{N_f} \\ A^{N_f-1}B & \cdots & B \end{bmatrix} U_{k+j-1} + L^{(j)}U_{k+j-1}$$

To satisfy the unbiased condition, i.e. $u_{k+j} = u^*_{k+j}$, where $u^*_{k+j} = -\mathcal{K}_{j+1}x_{k+j}$, irrespective of the initial state and the input on the horizon $[k+j-N_f, k+j]$, $\mathcal{M}_{j+1}$ should be equal to $-\mathcal{K}_{j+1}$ and the following constraint is required:

$$H^{(j)}\tilde{C}_{N_f} = -\mathcal{K}_{j+1}A^{N_f} \tag{5.142}$$

$$L^{(j)}_i = -\begin{bmatrix} H^{(j)} & \mathcal{K}_{j+1} \end{bmatrix} \begin{bmatrix} & \tilde{B}_{N_f} \\ A^{N_f-1}B & \cdots & B \end{bmatrix} \tag{5.143}$$

These will be called the unbiasedness constraint. It is noted that a variable j ranges from 0 to $N_c - 1$.

The system (5.5) and (5.6) will be represented in a batch form on the time interval $[k+j-N_f, k+j]$ called the horizon. On the horizon $[k+j-N_f, k+j]$, measurements are expressed in terms of the state x_{k+j} at the time $k+j$ and inputs as follows:

$$Y_{k+j-1} = \tilde{C}_{N_f}x_{k+j-N_f} + \tilde{B}_{N_f}U_{k+j-1}$$

$$+ (\tilde{G}_{N_f} + \bar{D}_{N_f})T_{N_f-1-j}W_{k+j-1} \tag{5.144}$$

where T_{N_f-1-j} is defined in (5.131).

The state x_{k+j} can be represented using the initial state x_{k+j-N} on the horizon as

$$x_{k+j} = A^{N_f}x_{k+j-N_f} + \begin{bmatrix} A^{N_f-1}G & A^{N_f-2}G & \cdots & G \end{bmatrix} T_{N_f-1-j}W_{k+j-1}$$

$$+ \begin{bmatrix} A^{N_f-1}B & A^{N_f-2}B & \cdots & B \end{bmatrix} U_{k+j-1} \tag{5.145}$$

Augmenting (5.144) and (5.145) yields the following linear model:

$$\begin{bmatrix} Y_{k+j-1} \\ 0 \end{bmatrix} = \begin{bmatrix} \tilde{C}_{N_f} & 0 \\ A^{N_f} & -I \end{bmatrix} \begin{bmatrix} x_{k+j-N_f} \\ x_{k+j} \end{bmatrix} + \begin{bmatrix} & \tilde{B}_{N_f} \\ A^{N_f-1}B & \cdots & B \end{bmatrix} U_{k+j-1}$$

$$+ \begin{bmatrix} & (\tilde{G}_{N_f} + \bar{D}_{N_f})T_{N_f-1-j} & \\ A^{N_f-1}G & \cdots\cdots & G \end{bmatrix} W_{k+j-1} \qquad (5.146)$$

Matrices $H^{(j)}$ and $L^{(j)}$ in the FMC (5.51) will be chosen to minimize the performance criterion (5.128) without the assumption of nonsingularity of the system matrix A.

It can be observed that (5.146) is different from (5.140) only in that there is a term T_{N_f-1-j} attached to $(\tilde{G}_{N_f} + \bar{D}_{N_f})$.

As in the LQFMC, we have

$$u_{k+j} = u^*_{k+j} + H^{(j)}\tilde{G}_{N_f}W_{k+j-1} - K_{j+1}\left[A^{N_f-1}G \cdots G\right]W_{k+j-1}$$
$$+ H^{(j)}V_{k+j-1}$$
$$u_{k+j} - u^*_{k+j} = H^{(j)}\tilde{G}_{N_f}W_{k+j-1} - K_{j+1}\left[A^{N_f-1}G \cdots G\right]W_{k+j-1}$$
$$+ H^{(j)}V_{k+j-1}$$

Since

$$u_{k+j} + \mathcal{K}_{j+1}x_{k+j} = H^{(j)}(\tilde{G}_{N_f} + \bar{D}_{N_f})T_{N_f-1-j}W_{k+j-1}$$
$$- \mathcal{K}_{j+1}\left[A^{N_f-1}G \cdots G\right]W_{k+j-1} \qquad (5.147)$$

the following relation for the first term in (5.133) is obtained:

$$\left\{\mathcal{K}_{j+1}x_{k+j} + u_{k+j}\right\}^T \hat{R}_j \left\{\mathcal{K}_{j+1}x_{k+j} + u_{k+j}\right\}$$
$$= \left\{H^{(j)}(\tilde{G}_{N_f} + \bar{D}_{N_f})T_{N_f-1-j}W_{k+j-1} - \mathcal{K}_{j+1}\left[A^{N_f-1}G \cdots G\right]\right.$$
$$\times T_{N_f-1-j}W_{k+j-1}\Big\}^T \hat{R}_j \Big\{H^{(j)}(\tilde{G}_{N_f} + \bar{D}_{N_f})T_{N_f-1-j}W_{k+j-1}$$
$$-\mathcal{K}_{j+1}\left[A^{N_f-1}G \cdots G\right]T_{N_f-1-j}W_{k+j-1}\Big\}$$
$$\leq \text{tr}\left[\hat{R}_j^{\frac{1}{2}}\left(H^{(j)}(\tilde{G}_{N_f} + \bar{D}_{N_f}) - \mathcal{K}_{j+1}\left[A^{N_f-1}G \cdots G\right]\right)T_{N_f-1-j}\right.$$
$$\times Q_{N_f}T_{N_f-1-j}\left(H^{(j)}(\tilde{G}_{N_f} + \bar{D}_{N_f}) - \mathcal{K}_{j+1}\left[A^{N_f-1}G \cdots G\right]\right)^T \hat{R}_j^{\frac{1}{2}}\right]$$

where $\hat{R}_j \triangleq R + B^T K_{j+1}B$.

Then, from (5.133) and (5.126), the optimization problem that we have to solve is summarized as

$$H_B^{(j)} = \underset{H^{(j)}}{\arg\min} \ \lambda_{\max}\left[\hat{R}_j^{\frac{1}{2}}\left(H^{(j)}(\tilde{G}_{N_f} + \tilde{D}_{N_f}) - \mathcal{K}_{j+1}\left[A^{N_f-1}G \cdots G\right]\right)\right.$$

$$\times T_{N_f-1-j}Q_{N_f}T_{N_f-1-j}\left(H^{(j)}(\tilde{G}_{N_f} + \bar{D}_{N_f})\right.$$

$$\left.\left. - \mathcal{K}_{j+1}\left[A^{N_f-1}G \cdots G\right]\right)^T \hat{R}_j^{\frac{1}{2}}\right]$$

subject to

$$H^{(j)}\tilde{C}_{N_f} = -\mathcal{K}_{j+1}A^{N_f} \tag{5.148}$$

By using the result of Lemma 4.3, the solution to the optimization problem (4.197) can be obtained according to the following correspondences:

$$
\begin{aligned}
A &\longleftarrow & \tilde{G}_{N_f} + \bar{D}_{N_f} \\
B &\longleftarrow & \hat{R}_j^{\frac{1}{2}}\mathcal{K}_{j+1}\left[A^{N_f-1}G \cdots G\right] \\
C &\longleftarrow & T_{N_f-1-j}Q_{N_f}T_{N_f-1-j}^T \\
E &\longleftarrow & \bar{C}_{N_f} \\
F &\longleftarrow & I \\
H &\longleftarrow & \hat{R}_j^{\frac{1}{2}}H^{(j)} \\
W_{1,1} &\longleftarrow & \tilde{C}_{N_f}^T\tilde{C}_{N_f} \\
W_{1,2} &\longleftarrow & \tilde{C}_{N_f}^T\tilde{G}_{N_f} \\
W_{2,2} &\longleftarrow & \tilde{G}_{N_f}^T\tilde{G}_{N_f} + I
\end{aligned} \tag{5.149}
$$

If j is replaced by 0, i.e. the current time, then the optimal solution for the general A is obtained, which is summarized in the following theorem.

Theorem 5.7. *When $\{A, C\}$ is observable and $N_f \geq n_o$, the L_2-E FMC u_k is given as follows:*

$$u_k = -\mathcal{K}_1\left[\left[A^{N_f} \ A^{N_f-1}G \ A^{N_f-2}G \cdots AG \ G\right]\begin{bmatrix} W_{1,1} \ W_{1,2} \\ W_{1,2}^T \ W_{2,2} \end{bmatrix}^{-1}\right.$$

$$\left.\times \begin{bmatrix} \tilde{C}_{N_f}^T \\ \tilde{G}_{N_f}^T \end{bmatrix}R_{N_f}^{-1}\left(Y_{k-1} - \tilde{B}_{N_f}U_{k-1}\right)\right]$$

$$- \mathcal{K}_1\left[A^{N_f-1}B \ A^{N_f-2}B \ A^{N_f-3}B \cdots AB \ B\right]U_{k-1} \tag{5.150}$$

where W_{11}, W_{12}, and W_{22} are defined in (5.100) and Y_{k-1}, U_{k-1}, $\tilde{C}_{N_f}$, $\tilde{B}_{N_f}$, Ξ_{N_f}, and $\mathcal{K}_1$ are given by (5.48), (5.49), (4.78), (4.79), (4.214), and (5.68) respectively. ∎

The recursive form of the L_2-E FMC is the same as the LQFMC.

5.6.3 H_∞ Finite Memory Controls

In this section we introduce the H_∞ FMC for the following performance criterion:

$$\inf_{u_i} \sup_{w_i} \frac{\sum_{i=i_0}^{\infty} x_i^T Q x_i + u_i^T R u_i}{\sum_{i=i_0}^{\infty} w_i^T R_w w_i} \tag{5.151}$$

It is noted that the filter horizon and the performance horizon are $[k - N, \ k]$ and $[k_0, \ \infty]$ respectively.

Before deriving the FMC based on the minimax performance criterion, we introduce a useful result that is employed during obtaining the FMC for the system (5.5) and (5.6). As in the performance criterion for the LQC, the H_∞ performance criterion can be represented in a perfect square expression for arbitrary control from Lemma 2.11:

$$\sum_{i=i_0}^{\infty} \left[x_i^T Q x_i + u_i^T R u_i - \gamma^2 w_i^T R_w w_i \right]$$

$$= \sum_{i=i_0}^{\infty} \left[(u_i - u_i^*)^T \mathcal{V} (u_i - u_i^*) - \gamma^2 (w_i - w_i^*)^T \mathcal{W} (w_i - w_i^*) \right] \tag{5.152}$$

where w_i^* and u_i^* are given as

$$w_i^* = \gamma^{-2} R_w^{-1} B^T M [I + (BR^{-1}B^T - \gamma^{-2} B_w R_w^{-1} B_w^T)M]^{-1} A x_i \tag{5.153}$$

$$u_i^* = -R^{-1}B^T M [I + (BR^{-1}B^T - \gamma^2 B_w R_w^{-1} B_w^T)M]^{-1} A x_i \tag{5.154}$$

$$\mathcal{V} = \gamma^2 R_w - B_w^T M B_w \tag{5.155}$$

$$\mathcal{W} = I + R^{-1}B^T M (I - \gamma^{-2} B_w R_w^{-1} B_w^T M)^{-1} B \tag{5.156}$$

and M is given in (2.140) of Chapter 2.

From (5.152), the performance criterion (5.151) can be changed to

$$\inf_{u_i} \sup_{w_i} \frac{\sum_{i=i_0}^{\infty} (u_i - u_i^*) \mathcal{V} (u_i - u_i^*)^T}{\sum_{i=i_0}^{\infty} (w_i - w_i^*) \mathcal{W} (w_i - w_i^*)^T} \tag{5.157}$$

where w_i^* and u_i^* are defined in (5.153) and (5.154).

From w_k^* and the state-space system (5.5) and (5.6), the following new state-space equations are obtained:

$$\begin{aligned}
x_{i+1} &= A x_i + B u_i + B_w w_i \\
&= A x_i + B u_i + B_w (w_i - w_i^*) + B_w w_i^* \\
&= \left[I + \gamma^{-2} B_w R_w^{-1} B_w^T M [I - \gamma^{-2} B_w R_w^{-1} B_w^T M]^{-1} \right] (A x_i + B u_i) \\
&\quad + B_w (w_i - w_i^*) \\
&= [I - \gamma^{-2} B_w R_w^{-1} B_w^T M]^{-1} (A x_i + B u_i) + B_w (w_i - w_i^*) \\
&= A_a x_i + B_a u_i + B_w \triangle w_i
\end{aligned} \tag{5.158}$$

$$\begin{aligned}
y_i &= C x_i + C_w w_i \\
&= C x_i + C_w w_i - \gamma^{-2} C_w B_w^T M [I - \gamma^{-2} B_w R_w^{-1} B_w^T M]^{-1} (A x_i + B u_i) \\
&= C x_i + C_w \triangle w_i
\end{aligned} \tag{5.159}$$

where

$$A_a \triangleq [I - \gamma^{-2} B_w R_w^{-1} B_w^T M]^{-1} A \qquad (5.160)$$

$$B_a \triangleq [I - \gamma^{-2} B_w R_w^{-1} B_w^T M]^{-1} B \qquad (5.161)$$

$$\triangle w_i \triangleq w_i - w_i^* \qquad (5.162)$$

The control problem based on (5.151) is reduced to the H_∞ problem (5.157) from $\triangle w_i$ to $\triangle u_i$. Disturbances $\triangle w_i$ are deviated from the worst-case.

The system (5.5) and (5.6) will be represented in a batch form on the time interval $[k - N_f, \ k]$ called the horizon. On the horizon $[k - N_f, \ k]$, measurements are expressed in terms of the state x_k at the time k and inputs on $[k - N_f, \ k]$ as follows.

The new state-space equations (5.158)-(5.159) can be represented in a batch form on the time interval $[k - N_f, k]$ as

$$Y_{k-1} = \bar{C}_{N_f}^* x_k + \bar{B}_{N_f}^* U_{k-1} + (\bar{G}_{N_f}^* + \bar{C}_{d,N_f}^*) \triangle W_{k-1} \qquad (5.163)$$

where $\bar{C}_{N_f}^*$, $\bar{B}_{N_f}^*$, $\bar{G}_{N_f}^*$, $\bar{C}_{d,N_f}^*$, and $\triangle W_{k-1}$ are given by

$$\bar{C}_{N_f}^* \triangleq \begin{bmatrix} CA_a^{-N_f} \\ CA_a^{-N_f+1} \\ CA_a^{-N_f+2} \\ \vdots \\ CA_a^{-1} \end{bmatrix}, \bar{B}_{N_f} \triangleq - \begin{bmatrix} CA_a^{-1}B_a & CA_a^{-2}B_a & \cdots & CA_a^{-N_f}B_a \\ 0 & CA_a^{-1}B_a & \cdots & CA_a^{-N_f+1}B_a \\ 0 & 0 & \cdots & CA_a^{-N_f+2}B_a \\ \vdots & \vdots & \vdots & \vdots \\ 0 & 0 & \cdots & CA_a^{-1}B_a \end{bmatrix}$$

$$\bar{G}_{N_f} \triangleq - \begin{bmatrix} CA_a^{-1}B_w & CA_a^{-2}B_w & \cdots & CA_a^{-N_f}B_w \\ 0 & CA_a^{-1}B_w & \cdots & CA_a^{-N_f+1}B_w \\ 0 & 0 & \cdots & CA_a^{-N_f+2}B_w \\ \vdots & \vdots & \vdots & \vdots \\ 0 & 0 & \cdots & CA_a^{-1}B_w \end{bmatrix}$$

$$\triangle W_{k-1} \triangleq \begin{bmatrix} w_{k-N_f} - w_{k-N_f}^* \\ w_{k-N_f+1} - w_{k-N_f+1}^* \\ w_{k-N_f+2} - w_{k-N_f+2}^* \\ \vdots \\ w_{k-1} - w_{k-1}^* \end{bmatrix}, \quad \bar{C}_{d,N_f} \triangleq [\mathrm{diag}(\overbrace{C_z \ C_z \ \cdots \ C_z}^{N_f})]$$

u_k in (5.45) is represented as

$$u_k = HY_{k-1} + LU_{k-1} = H\bar{C}_{N_f}^* x_k + H\bar{B}_{N_f}^* U_{k-1} + LU_{k-1}$$
$$+ H(\bar{G}_{N_f}^* + \bar{C}_{d,N_f}^*)W_{k-1} \qquad (5.164)$$

With zero disturbance on $[k-N_f, k-1]$, we require the unbiased condition that u_k (5.164) be equal to the optimal state feedback control (2.146) in Chapter 2. Therefore, we have

$$H\bar{C}_{N_f}^* = -R^{-1}B^T M[I + (BR^{-1}B^T - \gamma^{-2}B_w R_w^{-1} B_w^T)M]^{-1}A \quad (5.165)$$

$$H\bar{B}_{N_f}^* = -L \quad (5.166)$$

Using the result of (A.3) in Appendix A, H satisfying (5.165) can be represented as

$$H = FM^T + H_0 \quad (5.167)$$

$$H_0 = -R^{-1}B^T M[I + (BR^{-1}B^T - \gamma^{-2}B_w R_w^{-1} B_w^T)M]^{-1}$$
$$\times A(\bar{C}_{N_f}^{*T}\bar{C}_{N_f}^*)^{-1}\bar{C}_{N_f}^{*T} \quad (5.168)$$

for an arbitrary matrix F and a matrix M whose columns span the bases of the null space of $\bar{C}_{N_f}^{*T}$.

Using the constraints (5.165) and (5.166) and considering disturbances gives

$$e_k \overset{\triangle}{=} u_k - u_k^* = H(\bar{G}_{N_f}^* + \bar{C}_{d,N_f}^*)\triangle W_{k-1} \quad (5.169)$$

The dynamics of $\triangle W_k$ are given in (4.226).

From the dynamics of $\triangle W_k$ and (5.169), we can obtain a transfer function $T_{ew}(z)$ from disturbances w_k to estimation error e_k as follows:

$$T_{ew}(z) = H(\bar{G}_{N_f}^* + \bar{C}_{d,N_f}^*)(zI - \bar{A}_w)^{-1}\bar{B}_w \quad (5.170)$$

In order to handle the weighted H_∞ performance criterion (5.157), we consider the following transfer function:

$$T_{ew}(z) = \mathcal{V}^{-\frac{1}{2}}H(\bar{G}_{N_f}^* + \bar{C}_{d,N_f}^*)(zI - \bar{A}_w)^{-1}\bar{B}_w\mathcal{W}^{\frac{1}{2}} \quad (5.171)$$

Using the result in Lemma 2.13, we can obtain the LMI for satisfying the H_∞ performance .

Theorem 5.8. *Assume that the following LMI is satisfied for $X > 0$ and F:*

$$\min_{X>0,F} \gamma_\infty$$

subject to

$$\begin{bmatrix} -X & X\bar{A}_w & X\bar{B}_w\mathcal{W}^{\frac{1}{2}} & 0 \\ \bar{A}_w^T X & -X & 0 & \Xi^T \\ \mathcal{W}^{\frac{1}{2}}\bar{B}_w^T X & 0 & -\gamma_\infty I & 0 \\ 0 & \Xi & 0 & -\gamma_\infty I \end{bmatrix} < 0$$

where $\Xi \overset{\triangle}{=} \mathcal{V}^{-\frac{1}{2}}H(\bar{G}_{N_f}^* + \bar{C}_{d,N_f}^*)$.
Then, the gain matrices of the H_∞ FMC are given by

$$H = FM^T + H_0, \quad L = -H\bar{B}_{N_f}^*$$

5.6.4* H_2/H_∞ Finite Memory Controls

We introduce the H_2 FMC in an LMI form. Since the optimal state feedback H_2 control is an infinite horizon LQ control, the H_2 FMC should be unbiased from the control $u_i = -\mathcal{K}_\infty x_i$, where

$$\mathcal{K}_\infty = R^{-1}B^T[I + K_\infty BR^{-1}B^T]^{-1}K_\infty A$$

and K_∞ is given in (2.109).

Theorem 5.9. *Assume that the following LMI problem is feasible:*

$$\min_{F,W} \ \mathrm{tr}(W)$$

subject to

$$\begin{bmatrix} W & (FM^T + H_0)(\bar{G}_{N_f} + \bar{D}_{N_f}) \\ (\bar{G}_{N_f} + \bar{D}_{N_f})^T(FM^T + H_0)^T & I \end{bmatrix} > 0 \quad (5.172)$$

where $H_0 = -\mathcal{K}_\infty(\bar{C}_{N_f}^T \bar{C}_{N_f})^{-1}\bar{C}_{N_f}^T$ and columns of M are the basis of the null space of $\bar{C}_{N_f}^T$. Then, the optimal gain matrices of the H_2 FMC are given by

$$H = FM^T + H_0, \ L = -H\bar{B}_{N_f}$$

Proof. In (5.132), j is set to zero and we have

$$u_k + \mathcal{K}_1 x_k = H(\bar{G}_{N_f} + \bar{D}_{N_f})W_{k-1} \quad (5.173)$$

Recalling that $W_k = \bar{A}_w W_{k-1} + \bar{B}_w w_k$, we can obtain the H_2 norm of the transfer function $T_{ew}(z)$ given by

$$\|T_{ew}(z)\|_2^2 = \mathrm{tr}\big(H(\bar{G}_{N_f} + \bar{D}_{N_f})P(\bar{G}_{N_f} + \bar{D}_{N_f})^T H^T\big)$$

where

$$P = \sum_{i=0}^{\infty} \bar{A}_w^i \bar{B}_w \bar{B}_w^T (\bar{A}_w^T)^i$$

Since $A_u^i = 0$ for $i \geq N_f$, we obtain

$$P = \sum_{i=0}^{\infty} \bar{A}_w^i \bar{B}_w \bar{B}_w^T (\bar{A}_w^T)^i = \sum_{i=0}^{N_f-1} \bar{A}_w^i \bar{B}_w \bar{B}_w^T (\bar{A}_w^T)^i = I$$

Thus we have

$$\|T_{ew}(z)\|_2^2 = \mathrm{tr}\big(H(\bar{G}_{N_f} + \bar{D}_{N_f})(\bar{G}_{N_f} + \bar{D}_{N_f})^T H^T\big)$$

Introduce a matrix variable W such that

$$W > H(\bar{G}_{N_f} + \bar{D}_{N_f})(\bar{G}_{N_f} + \bar{D}_{N_f})^T H^T \qquad (5.174)$$

Then, $\mathrm{tr}(W) > \|T_{ew}(z)\|_2^2$. By the Schur complement, (5.174) is equivalently changed into (5.172). Hence, by minimizing $\mathrm{tr}(W)$ subject to $H\bar{C}_{N_f} = -\mathcal{K}_\infty$ and the above LMI, we can obtain the optimal gain matrix H for the H_2 FMC. This completes the proof. ∎

Each performance criterion has its own advantages and disadvantages, so that there are trade-offs between them. In some cases we want to adopt two or more performance criteria simultaneously in order to satisfy specifications. In this section, we introduce two kinds of controls based on mixed criteria. The optimal H_2 and H_∞ norms are denoted by γ_2^* and γ_∞^*.

1. Minimize the H_∞ norm for a fixed guaranteed H_2 norm

In this case, the FMC should be unbiased from $u_i = -\mathcal{K}_a x_i$, where $\mathcal{K}_a$ is obtained from the optimization problem (2.284). Assume that the following LMI problem is feasible:

$$\min_{W,X,F} \gamma_\infty \qquad (5.175)$$

subject to

$$\mathrm{tr}(W) < \alpha\gamma_2^*$$

$$\begin{bmatrix} W & (FM^T + H_0)(\bar{G}_{N_f} + \bar{D}_{N_f}) \\ (\bar{G}_{N_f} + \bar{D}_{N_f})^T(FM^T + H_0)^T & I \end{bmatrix} > 0 \qquad (5.176)$$

$$\begin{bmatrix} -X & X\bar{A}_w & X\bar{B}_w\mathcal{W}^{-\frac{1}{2}} & 0 \\ \bar{A}_w^T X & -X & 0 & \Xi^T \\ \mathcal{W}^{-\frac{1}{2}}\bar{B}_w^T X & 0 & -\gamma_\infty I & 0 \\ 0 & \Xi & 0 & -\gamma_\infty I \end{bmatrix} < 0$$

where $\alpha > 1$, $H_0 = -\mathcal{K}_a(\bar{C}_{N_f}^{*T}\bar{C}_{N_f}^*)^{-1}\bar{C}_{N_f}^{*T}$, $\Xi = \mathcal{V}^{\frac{1}{2}}(FM^T + H_0)(\bar{G}_{N_f}^* + \bar{C}_{d,N_f}^*)$, and columns of M consist of the basis of the null space of $\bar{C}_{N_f}^{*T}$. Then, the gain matrices of the H_2/H_∞ FMC are given by

$$H = FM^T + H_0, \qquad L = -H\bar{B}_{N_f}^*$$

2. Minimize the H_2 norm for a fixed guaranteed H_∞ norm

In this case, the FMC should be unbiased from $u_i = -\mathcal{K}_b x_i$, where $\mathcal{K}_b$ is obtained from the optimization problem (2.280). H_0 in the above problem should be changed to $H_0 = -\mathcal{K}_b(\bar{C}_{N_f}^{*T}\bar{C}_{N_f}^*)^{-1}\bar{C}_{N_f}^{*T}$.

Instead of minimizing γ_∞ for a guaranteed H_2 norm, γ_2 is minimized for a guaranteed H_∞ norm as follows:

$$\min_{W,X,F} \gamma_2$$

subject to

$$\text{tr}(W) < \gamma_2$$

(5.176),

$$\begin{bmatrix} -X & X\bar{A}_w & X\bar{B}_w W^{-\frac{1}{2}} & 0 \\ \bar{A}_w^T X & -X & 0 & \Xi^T \\ W^{-\frac{1}{2}}\bar{B}_w^T X & 0 & -\alpha\gamma_\infty^* I & 0 \\ 0 & \Xi & 0 & -\alpha\gamma_\infty^* I \end{bmatrix} < 0$$

where $\alpha > 1$.

5.7 References

The predictor-based output feedback control has been known for many years. A linear finite memory structure and an unbiased condition were first introduced in [KH04].

The batch forms of the LQFMC for nonsingular A and general A in Theorems 5.1 and 5.2, and the separation principle in Theorem 5.4 are the discrete-time versions for the continuous-time results in [KH04]. The closed-loop system with the FMC in Theorem 5.3 appeared in [HK04]. The stability of the LQFMC in Theorem 5.5 appeared in [LKC98].

The L_2-E FMC in Theorems 5.6 and 5.7 appeared in [AHK04]. The H_∞ FMC with the unbiased condition in Theorem 5.8 appeared in [AHK04]. The H_2 FMC in Theorem 5.9 and the mixed H_2/H_∞ FMC in Section 5.6.4 also appeared in [AHK04].

5.8 Problems

5.1. Consider the CARIMA model (5.20) with $C(q^{-1}) = 1$. Assume that there exist $E_j(q^{-1})$ and $F_j(q^{-1})$ satisfying the following Diophantine equation:

$$1 = E_j(q^{-1})A(q^{-1}) + q^{-j}F_j(q^{-1}) \tag{5.177}$$

where

$$E_j(q^{-1}) = d_0^j + d_1^j q^{-1} + \dots + d_{j-1}^j q^{-j+1}$$
$$F_j(q^{-1}) = f_0^j + f_1^j q^{-1} + \dots + f_{n-1}^j q^{-n+1}$$

Also assume that there exist $G_j(q^{-1})$ and $H_j(q^{-1})$ satisfying the following equation:

$$E_j(q^{-1})\tilde{B}(q^{-1}) = G_j(q^{-1}) + q^{-j}H_j(q^{-1}) \tag{5.178}$$

where

$$G_j(q^{-1}) = g_0^j + g_1^j q^{-1} + \ldots + g_{j-1}^j q^{-j+1}$$
$$H_j(q^{-1}) = h_0^j + h_1^j q^{-1} + \ldots + h_n^j q^{-n}$$

(1) By using (5.177) and (5.178), show that the following output predictor form can be obtained:

$$y_{k+j} = G_j(q^{-1})\Delta u_{k+j-1} + H_j(q^{-1})\Delta u_{k-1} + F_j(q^{-1})y_k$$
$$+ E_j(q^{-1})w_{k+j} \tag{5.179}$$

It is noted that the input is divided into the past input and the future input.

(2) Show that $\hat{y}_{k+j|k}$ can be written as

$$\hat{y}_{k+j|k} = G_j(q^{-1})\Delta u_{k+j-1} + H_j(q^{-1})\Delta u_{k-1} + F_j(q^{-1})y_k \tag{5.180}$$

(3) (5.180) can be written as

$$Y_{k+1} = G\Delta U_{k-1} + F'(q^{-1})y_k + H'(q^{-1})\Delta u_{k-1} \tag{5.181}$$

where Y_k and ΔU_k are given in (??). Find the matrices G, $F'(q^{-1})$, and $H'(q^{-1})$.

(4) By using (5.181), represent the performance criterion (5.26) with $g_f = 0$ and $N_c - 1$ replaced by N_c. And find the optimal control.

5.2. Consider the CARIMA model (5.20). Assume that there exist $E_j(q^{-1})$ and $F_j(q^{-1})$ satisfying the following Diophantine equation:

$$C(q^{-1}) = E_j(q^{-1})A(q^{-1}) + q^{-j}F_j(q^{-1}) \tag{5.182}$$

where $E_j(q^{-1})$ and $F_j(q^{-1})$ are given in (5.178) and (5.178) respectively. Also assume that there exist $M_j(q^{-1})$ and $N_j(q^{-1})$ satisfying the following second Diophantine equation:

$$1 = C_j(q^{-1})M_j(q^{-1}) + q^{-j}N_j(q^{-1}) \tag{5.183}$$

(1) Show that

$$\hat{y}_{k+j} = M_j(q^{-1})E_j(q^{-1})\tilde{B}(q^{-1})\Delta u_{k+j-1} + M_j(q^{-1})F_j(q^{-1})y_k$$
$$+ N_j(q^{-1})y_k \tag{5.184}$$

where $M_j(q^{-1})E_j(q^{-1})\tilde{B}(q^{-1})\Delta u_{k+j-1}$ can be divided into $G(q^{-1})$ $\times \Delta u_{k+j-1} + G_p(q^{-1})\Delta u_{k-1}$ for some $G(q^{-1})$ and $G_p(q^{-1})$.

(2) By using (5.184), represent the performance criterion (5.26) with $g_f = 0$ and $N_c - 1$ replaced by N_c. And find the optimal control.

5.3. Consider the CARIMA model (5.20) with $C(q^{-1}) = 1$ where w_i is a disturbance.

(1) Represent (5.179) in a vector form.
(2) Find the minimax GPC that optimizes the following cost function

$$\min_u \max_w \left[\sum_{i=1}^{N_y} \left\{ [y_{k+i} - y_{k+i}^r]^T q_1 [y_{k+i} - y_{k+i}^r] + \lambda \triangle u_{k+i-1}^T \triangle u_{k+i-1} \right\} \right.$$
$$\left. - \gamma^2 w_{k+i-1}^T w_{k+i-1} \right]$$

(3) Obtain an asymptotic stability condition using a cost monotonicity condition.

5.4. Consider the LQFMC (5.78) and LQG (2.230).

(1) Calculate $E[x_k x_k^T]$ for the LQFMC and the finite-time LQG.
(2) Calculate $E[u_k u_k^T]$ for the LQFMC and the finite-time LQG.

5.5. Design an FMC in an LMI form that satisfies the specification $E[x_k x_k^T] \leq \gamma^2$.

5.6. If the system matrix A is nonsingular, show that the conditions (5.94) and (5.95) can be reduced to (5.64) and (5.65) by using matrix manipulations.

5.7. If the system matrix A is nonsingular, show by a similarity transformation that the closed-loop poles for the LQFMC consist of eigenvalues of of $A - BH\bar{C}_{N_f}$ and zeros. Check whether this property can be preserved in the case that the unbiased condition is not met. If so, what are the conditions on the FMC gains?

Hint: use the following similarity transformation T:

$$T = \begin{bmatrix} I & O & O \\ O & I & O \\ \bar{C}_{N_f} & -\bar{B}_{N_f} & I \end{bmatrix} \tag{5.185}$$

5.8. Consider a problem for finding closed poles of the LQFMC for a general matrix A.

(1) Suppose that P is given by

$$P = \begin{bmatrix} A & KM \\ C & M \end{bmatrix} \tag{5.186}$$

where all matrices are of appropriate sizes and $KC = 0$. Show that $\det(P) = \det(A)\det(M)$.

(2) By using the above result, show that the closed-loop poles for the LQFMC consist of eigenvalues of $A - BH\bar{C}_{N_f}$ and zeros.

5.9. Consider the following system:

$$x_{i+1} = Ax_i + Bu_i + Gw_i$$
$$y_i = Cx_i + v_i$$

where w_i and v_i are deterministic disturbances. Derive the L_2-E FMC for the following weighted L_2-E performance criterion:

$$\min_{u} \max_{w} \frac{J - J^*}{\sum_{j=-N_f+1}^{0} \begin{bmatrix} w_{k+j} \\ v_{k+j} \end{bmatrix}^T \begin{bmatrix} Q_w & S_{wv} \\ S_{wv}^T & R_v \end{bmatrix}^{-1} \begin{bmatrix} w_{k+j} \\ v_{k+j} \end{bmatrix}} \tag{5.187}$$

where J and J^* are defined in (5.129) and (5.130).

5.10. In the stochastic state-space model (5.1) and (5.2), parameters are given as

$$A = \begin{bmatrix} 0.1014 & 0.4056 & 0.0036 \\ 0.0362 & 0.6077 & 0.2135 \\ 0.8954 & 0.9927 & 0.0399 \end{bmatrix}, \quad G = \begin{bmatrix} 0.1669 \\ 0.2603 \\ 0.0211 \end{bmatrix} \tag{5.188}$$

$$C = \begin{bmatrix} 0.1342 & 0.0125 & 0.1295 \end{bmatrix}, \quad B = \begin{bmatrix} 0.8 & 0 & 0 \end{bmatrix}^T \tag{5.189}$$

Weighting matrices for a performance criterion are set to $Q = I_3$, $R = 1$, and $Q_f = 5I_3$.

(1) Consider the steady state LQG controls for $Q_w = 1$, 10, 50 with a fixed $R_v = 1$. Find poles of a closed-loop system. Check the maximum absolute value of the filter poles for each case.
(2) Design the LQFMC for $Q_w = 1$, $R_v = 1$, $N = 6$.
(3) Supposed that an LQG and an FMC are designed with $Q_w = 50$ and $Q_w = 1$ respectively. Compare the cost function (5.66) of the infinite horizon LQG with that of the FMC over the infinite horizon.

Note: LQG/LTR may degrade much performance in order to achieve a robustness, whereas the FMC guarantees, at least, best performance in a latest horizon.

5.11. Assume that w_k and v_k are not uncorrelated in the stochastic system (5.1) and (5.2) and their covariances are given by

$$E \begin{bmatrix} w_k \\ v_k \end{bmatrix} \begin{bmatrix} w_k^T & v_k^T \end{bmatrix} = \begin{bmatrix} Q_w & S_{wv} \\ S_{wv}^T & R_v \end{bmatrix}$$

Find the LQFMC for the above case.

5.12. Find an optimal FMC for the following performance criterion

$$\min_{u} \max_{w} \frac{\sum_{j=0}^{N_c}(u_{k+j} - u^*_{k+j})^T(u_{k+j} - u^*_{k+j})}{\sum_{j=-N_f+1}^{N_c} w^T_{k+j} w_{k+j}} \tag{5.190}$$

where u^* is an optimal state feedback control.

5.13. Derive the H_∞ FMC for the general A if possible.

5.14. Consider the system model (4.246)-(4.247). The system noise covariance is 0.02 and the measurement noise covariance is 0.02.
1) For $\delta_k = 0$, obtain an LQ FMC with the horizon length $N = 10$.
2) For $\delta_k = 0$, obtain an LQG controller.
3) For $\delta_k = 0$, perform simulations for the LQ FMC and LQG controller. And then compare them.
4) Perform simulations on the time interval $[0, 250]$ under the temporary uncertainty

$$\delta_k = \begin{cases} 0.1, \ 50 \le k \le 100 \\ 0, \quad \text{otherwise} \end{cases} \tag{5.191}$$

5.15. Derive the FMC to minimize the following cost function:

$$2\gamma^2 \log E \exp\left[\frac{1}{2\gamma^2} \sum_{i=0}^{N-1} [x^T_{k+i} Q x_{k+i} + u^T_{k+i} R u_{k+i}] + x^T_{k+N} Q_f x_{k+N}\right] \tag{5.192}$$

for a state-space model (5.1) and (5.2) and a fixed γ. Unbiasedness conditions (5.94) and (5.95) are required.

6

Constrained Receding Horizon Controls

6.1 Introduction

In many control systems, input variables cannot be arbitrarily large and may have some limitations, such as magnitude limits, since input devices are valves, motors and pumps, etc. Also, state values must be bounded in many cases. For example, in process plants, the temperature or the pressure as state variables must be within certain limits because of safety. As pointed before, many useful controls can be obtained from optimal controls on the infinite horizon. These controls for the infinite horizon are difficult to obtain, particularly for input and state constraints. However, optimal controls for the finite horizon are relatively easy to obtain for linear systems with input and state constraints by using SDP. Since receding horizon control is based on the finite horizon, receding horizon controls can handle input and state constraints. Although constrained systems are difficult to stabilize by using conventional controls, RHC provides a systematic way to stabilize the closed-loop system.

Similar to linear systems, the cost monotonicity holds for constrained systems under certain conditions, together with feasibility conditions. Stability can be guaranteed under a cost monotonicity condition. The cost monotonicity condition for the constrained RHC can be obtained by three approaches, i.e. the terminal equality constraint, the free terminal cost, and the terminal invariant set with or without a terminal cost.

It takes a relatively long time to calculate the constrained RHC. Thus, it is necessary to introduce fast algorithms that require less computation time. The feasibility is also an important issue for guaranteeing the existence of the optimal control satisfying constraints. Output feedback RHCs for constrained systems are necessary if the state is not available.

The organization of this chapter is as follows. In Section 6.2, input and output constraints, together with reachable and maximal output admissible sets, are discussed for the feasibility. In Section 6.3, constrained receding horizon LQ controls are represented in LMI forms. In Section 6.4, the RHC with a terminal equality and a terminal cost is introduced for constrained systems.

In particular, a cost monotonicity condition is presented in order to guarantee the stability. In Section 6.5, constrained RHCs with a terminal invariant set are introduced for constrained systems. Soft constrains are dealt with in Section 6.6. In Section 6.7, constrained output feedback receding horizon controls are introduced.

6.2 Reachable and Maximal Output Admissible Sets

In this section we will consider the following linear discrete time-invariant system:

$$x_{i+1} = Ax_i + Bu_i \tag{6.1}$$

with input and state constraints:

$$\begin{cases} -u_{lim} \le u_i \le u_{lim}, & i = 0, 1, \cdots, \infty \\ -g_{lim} \le Gx_i \le g_{lim}, & i = 0, 1, \cdots, \infty \end{cases} \tag{6.2}$$

where $u_{lim} \in \Re^m, G \in \Re^{n_g \times n}$, and $g \in \Re^{n_g}$. It is noted that the inequality in (6.2) holds component by component. Since Gx_i can be considered as an output, $-g_{lim} \le Gx_i \le g_{lim}$ in (6.2) is often called an output constraint. The system (6.1) with constraints (6.2) is said to be feasible if there exist sequences for an input and a state that satisfy (6.1) and (6.2). Constraints (6.2) can be written as

$$Eu_k \le e, \quad \bar{G}x_k \le g, \quad k = 0, 1, 2, \cdots \tag{6.3}$$

where

$$E = \begin{bmatrix} I \\ -I \end{bmatrix}, \quad e = \begin{bmatrix} u_{lim} \\ -u_{lim} \end{bmatrix}, \quad \bar{G} = \begin{bmatrix} G \\ -G \end{bmatrix}, \quad g = \begin{bmatrix} g_{lim} \\ -g_{lim} \end{bmatrix} \tag{6.4}$$

$\mathcal{U}$ and $\mathcal{X}$ are defined as

$$\mathcal{U} = \{u| - u_{lim} \le u \le u_{lim}\}$$
$$\mathcal{X} = \{x| - g_{lim} \le Gx \le g_{lim}\}$$

In order to observe how the input constraint has an effect on the system, it will be interesting to investigate all possible states that can be reached from the initial state by all available controls. For simplicity, the initial state is considered to be the origin. The reachable set is defined as

$$\mathcal{R}_N \overset{\triangle}{=} \left\{ x(N; x_0, u) \middle| x_0 = 0, \ u_i \in \mathcal{U}, \ 0 \le i \le N - 1 \right\} \tag{6.5}$$

In order to see whether the state constraint is violated or not, it will be interesting to investigate all possible initial states, from which all states belong to $\mathcal{X}$. For simplicity, the input is considered to be zero. The maximal output admissible set $\mathcal{O}_\infty$ is the largest output admissible set, namely the set of all initial states x_0, which makes the resultant state trajectory x_i satisfy $x_i \in \mathcal{X}$ for all $i > 0$.

$$\mathcal{O}_\infty \triangleq \{x_0 \mid x(i; x_0) \in \mathcal{X}, \quad \forall i = 0, 1, 2., \cdots\} \tag{6.6}$$

Reachable Sets

Consider the linear system

$$x_{k+1} = Ax_k + Bu_k \tag{6.7}$$

where $A \in \Re^n$, $B \in \Re^m$, $n \geq m$, and B is of full rank.

Let $\mathcal{R}_N$ denote the set of reachable states with unit total input energy

$$\mathcal{R}_N \triangleq \left\{ x_N \middle| x_0 = 0, \ \sum_{i=0}^{N-1} u_i^T u_i \leq 1 \right\} \tag{6.8}$$

Constraints on total input energy often appear in optimization problems.

Suppose that there exists a Lyapunov function $V_k = x_k^T P x_k$ such that

$$V_{k+1} - V_k \leq u_k^T u_k \tag{6.9}$$

for every x_k and u_k. Then, summing up from $k = 0$ to $k = N - 1$ yields

$$V_N - V_0 \leq \sum_{k=0}^{N-1} u_k^T u_k \tag{6.10}$$

Noting that $V_0 = x_0^T P x_0 = 0$ since $x_0 = 0$, we get

$$V_N \leq \sum_{k=0}^{N-1} u_k^T u_k \leq 1 \tag{6.11}$$

for every input u_k satisfying the constraint in (6.8). Thus, the set $\{x_N \mid x_N^T P x_N \leq 1\}$ contains the reachable set $\mathcal{R}_N$. It follows from (6.9) that we have

$$V_{k+1} - V_k = (Ax_k + Bu_k)^T P (Ax_k + Bu_k) - x_k^T P x_k \leq u_k^T u_k \tag{6.12}$$

leading to

$$\begin{bmatrix} x_k \\ u_k \end{bmatrix}^T \begin{bmatrix} A^T P A - P & A^T P B \\ B^T P A & B^T P B - I \end{bmatrix} \begin{bmatrix} x_k \\ u_k \end{bmatrix} \leq 0 \tag{6.13}$$

From the above equation, P can be obtained by

$$\begin{bmatrix} A^T P A - P & A^T P B \\ B^T P A & B^T P B - I \end{bmatrix} \leq 0 \tag{6.14}$$

It follows that

$$\mathcal{R}_N \subset \left\{ x_N \middle| x_N^T P x_N \leq 1 \right\} \tag{6.15}$$

Next, we consider another reachable set with unit input energy u_i. Thus, we are interested in the following set:

$$\mathcal{R}_N \triangleq \left\{ x_N \middle| x_0 = 0, \quad u_i^T u_i \leq 1 \right\} \tag{6.16}$$

Starting from the origin, $x_1 = B u_0$ belongs to the set $\{x_1 : x_1^T x_1 \leq \lambda_{\max}(B^T B)\}$ since $x_1^T x_1 = u_0^T B^T B u_0 \leq \lambda_{\max}(B^T B) u_0^T u_0 \leq \lambda_{\max}(B^T B)$. According to Appendix D, we have

$$x_1^T (B^T B)^{-1} x_1 \leq 1 \quad \Longleftrightarrow \quad x_1^T x_1 \leq \lambda_{\max}(B^T B) \tag{6.17}$$

It is noted that $B^T B$ is nonsingular. Assume that x_i belongs to the ellipsoid $x_i^T P_i^{-1} x_i \leq 1$. We want to find P_{i+1} such that $x_{i+1}^T P_{i+1}^{-1} x_{i+1} \leq 1$. First it is shown that $x_{i+1} x_{i+1}^T$ is bounded above as

$$x_{i+1} x_{i+1}^T = (A x_i + B u_i)(A x_i + B u_i)^T \tag{6.18}$$

$$= A x_i x_i^T A^T + B u_i u_i^T B^T + A x_i u_i^T B^T + B u_i x_i^T A^T \tag{6.19}$$

$$\leq 2[A x_i x_i^T A^T + B u_i u_i^T B^T] \tag{6.20}$$

$$\leq 2[A P_i A^T + B B^T] \tag{6.21}$$

where the first inequality comes from $PP^T + QQ^T \geq PQ^T + QP^T$, and $x_i x_i^T \leq P_i$ and $u_i u_i^T \leq I$ are used for the second inequality. Thus, we can choose $P_{i+1} = 2[A P_i A^T + B B^T]$, and thus $x_{i+1}^T P_{i+1}^{-1} x_{i+1} \leq 1$. It follows that

$$\mathcal{R}_N \subset \left\{ x_N \middle| x_N^T P_N^{-1} x_N \leq 1 \right\} \tag{6.22}$$

It is noted that P_1 is set to $B^T B$.

Maximal Output Admissible Sets

We are concerned with characterizing the initial states of the unforced linear system

$$x_{i+1} = A x_i \tag{6.23}$$

with the state constraint

$$\mathcal{X} = \{x | \Phi x \leq \psi\} \tag{6.24}$$

where $\Phi \in \Re^{q \times n}$ and $\psi \in \Re^{q \times 1}$ are a matrix and a vector with appropriate dimensions. We assume that the region is convex and has the origin inside.

A set $\mathcal{O}$ is output admissible if $x_0 \in \mathcal{O}$ implies that $x_i \in \mathcal{X}$ for all $k > 0$. The maximal output admissible set $\mathcal{O}_\infty$ is the largest output admissible set, namely the set of all initial states x_0, which makes the resultant state trajectory x_i remain inside forever, i.e. $x_i \in \mathcal{X}$ for all $i > 0$. This set is visualized in Figure 6.1. A set $\mathcal{X}$ is invariant if $x_0 \in \mathcal{X}$ implies that $x_i \in \mathcal{X}$ for all $i > 0$. It is noted that the maximal output admissible set is always invariant.

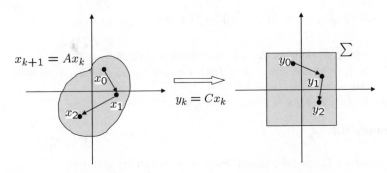

Fig. 6.1. Graphical representation of the maximal output admissible set

The maximal output admissible set is defined as

$$\mathcal{O}_\infty(A, \Phi, \psi) = \{x | \Phi A^k x \le \psi, \ \forall k = 0, 1, 2., \cdots \} \tag{6.25}$$

Define the set $\mathcal{O}_i$ as

$$\mathcal{O}_i(A, \Phi, \psi) = \{x | \Phi A^k x \le \psi, \ \forall k = 0, 1, 2., \cdots, i\} \tag{6.26}$$

Obviously, the set $\mathcal{O}_i(A, \Phi, \psi)$ satisfies the condition

$$\mathcal{O}_\infty \subset \mathcal{O}_{i_2} \subset \mathcal{O}_{i_1} \tag{6.27}$$

for all i_1 and i_2 such that $i_1 \le i_2$. Now, it may happen that the sequence of sets $\mathcal{O}_i$ ($i = 1, 2, \cdots$) stops getting smaller as i increases. That is, there may be the smallest value of i^* such that $\mathcal{O}_i = \mathcal{O}_{i^*}$ for all $i \ge i^*$. We call i^* the output admissibility index. In that case, it is clear that $\mathcal{O}_\infty = \mathcal{O}_{i^*}$.

We say $\mathcal{O}_\infty$ is finitely determined if, for some i, $\mathcal{O}_\infty = \mathcal{O}_i$. $\mathcal{O}_\infty$ is finitely determined if and only if $\mathcal{O}_i = \mathcal{O}_{i+1}$ for some i.

If (A, Φ) is observable, then it is known that $\mathcal{O}_\infty$ is convex and bounded. Additionally, $\mathcal{O}_\infty$ contains an origin in its interior if A is stable. It is also known that if A is a stable matrix, (A, Φ) is observable, $\mathcal{X}$ is bounded, and $\mathcal{X}$ contains zero, then $\mathcal{O}_\infty$ is finitely determined.

If $\mathcal{O}_\infty$ is finitely determined, then the algorithm will terminate for some value $i = i^*$ and

$$\mathcal{O}_\infty = \{x | \Phi A^i x \le \psi, \ 0 \le i \le i^*\} \tag{6.28}$$

The following algorithm describes how to obtain $\mathcal{O}_\infty$. Basically, this algorithm is based on linear programming, so it can be solved efficiently.

Algorithm for obtaining maximal output admissible set

1. Take $j = 0$.
2. For $i = 1$ to q, solve the following linear programming problem:

$$J_i = \max_z \; \varPhi^i A^{j+1} z \;\; \text{subject to} \;\; \varPhi A^k z \leq \psi, \forall k = 0, 1, \cdots, j$$

   where $\varPhi^i$ denotes the ith row of the matrix $\varPhi$.
3. If $J_i > \psi^i$ for $i = 1$ to q, take $j = j + 1$ and repeat the previous step.
4. If $J_i \leq \psi^i$ for $i = 1$ to q, then $\mathcal{O}_\infty = \mathcal{O}_j$.

Example 6.1

The unforced linear system and state constraints are given as

$$x_{i+1} = \begin{bmatrix} 0.9746 & 0.1958 \\ -0.1953 & 0.5839 \end{bmatrix} x_i$$

$$\varPhi = \begin{bmatrix} -0.8831 & -0.8811 \\ 0.8831 & 0.8811 \\ 1.0000 & 0 \\ -1.0000 & 0 \\ 0 & 1.0000 \\ 0 & -1.0000 \end{bmatrix} \qquad \psi = \begin{bmatrix} 0.5000 \\ 0.5000 \\ 1.5000 \\ 1.5000 \\ 0.3000 \\ 0.3000 \end{bmatrix}$$

Figure 6.2 shows the region corresponding to the $\mathcal{O}_\infty$ for the above system.

Maximal Output Admissible Set for Constrained Infinite Horizon Linear Quadratic Regulator

The cost function to be minimized is given by

$$J_\infty = \sum_{k=0}^{\infty} \left[x_k^T Q x_k + u_k^T R u_k \right] \tag{6.29}$$

subject to (6.3).

It is well known that the steady-state LQR for an unconstrained system is given by

$$u_k = K x_k \tag{6.30}$$

for some K. The input constraint is equivalently changed into the state constraint as follows:

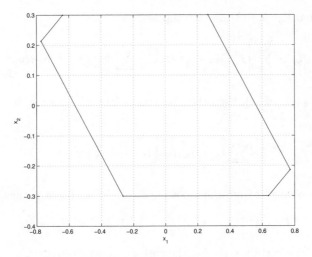

Fig. 6.2. Maximal output admissible sets

$$Eu_k \leq e \Longleftrightarrow EKx_k \leq e \tag{6.31}$$

Then the total constraint is given by

$$\Phi x \leq \psi \tag{6.32}$$

where

$$\Phi = \begin{bmatrix} EK \\ G \end{bmatrix} \qquad \psi = \begin{bmatrix} e \\ g \end{bmatrix} \tag{6.33}$$

The closed-loop system is represented by

$$x_{k+1} = A_{cl}x_k \tag{6.34}$$

$$y_k = Cx_k \tag{6.35}$$

where $A_{cl} = A + BK$. Because the control $u_k = Kx_k$ is optimal in the case that the system is unconstrained, it may not be optimal for the actual system, which has input and state constraints. However, if the initial state x_0 is in a certain region, then the input and the state constraints will never be active and thus the control $u_k = Kx_k$ for an unconstrained optimal problem is optimal for a constrained optimal problem. We will obtain such a region for x_0. We can conclude that if x_0 belongs to $\mathcal{O}_\infty(A_{cl}, \Phi, \psi)$, then (6.30) becomes optimal for the performance criterion (6.29) with constraints (6.32).

If x_0 does not belong to $\mathcal{O}_\infty(A_{cl}, \Phi, \psi)$, then we will use another control scheme. An optimal control is used for the finite horizon from the initial time in order to make the final state x_N on the horizon enter $\mathcal{O}_\infty(A_{cl}, \Phi, \psi)$, and the steady-state optimal control $u_k = Kx_k$ is used for the remaining infinite horizon.

Note that the J_∞ can be decomposed as

$$J_\infty = \sum_{k=0}^{N-1}\left[x_k^T Q x_k + u_k^T R u_k\right] + \sum_{k=N}^{\infty}\left[x_k^T Q x_k + u_k^T R u_k\right] \qquad (6.36)$$

If $x_N \in \mathcal{O}_\infty(A_{cl}, \Phi, \phi)$, then the cost function on $[N, \ \infty]$ can be represented as

$$\sum_{k=N}^{\infty}\left[x_k^T Q x_k + u_k^T R u_k\right] = x_N^T P x_N \qquad (6.37)$$

where P is obtained from the following steady-state Riccati equation:

$$P = A^T P A - A^T P B [R + B^T P B]^{-1} B^T P A + Q \qquad (6.38)$$

Therefore, in a certain the case, the cost J_∞ can be rewritten as

$$J_N = \sum_{k=0}^{N-1}\left[x_k^T Q x_k + u_k^T R u_k\right] + x_N^T P x_N \qquad (6.39)$$

Observe that the original infinite horizon cost J_∞ can be equivalently replaced by a finite horizon cost J_N.

Consider the following optimization problem with constraints:

$$\min_{u_i} J_N \qquad (6.40)$$

subject to

$$E u_k \le e, G x_k \le g, k = 0, 1, \cdots, N - 1 \qquad (6.41)$$

Assume that the problem is feasible. Let us denote the optimal control and the resultant state by u_k^* and x_k^* respectively. We will use the control strategy

$$\begin{cases} u_k = u_k^*, & k = 0, 1, \cdots, N - 1 \\ u_k = K x_k, & k = N, N + 1, \cdots \end{cases} \qquad (6.42)$$

It is clear that u_k and x_k, $k = 0, 1, \cdots, N - 1$, satisfy the given input and the state constraints. However, it is not guaranteed that u_k and x_k for $k = N, N + 1, \cdots$ satisfy the given constraints. We can conclude that if $x_N \in \mathcal{O}_\infty(A_{cl}, \Phi, \phi)$, u_k and x_k for $k = N, N + 1, \cdots$ satisfy the given constraints.

6.3 Constrained Receding Horizon Control with Terminal Equality Constraint

Input and State Constraints

The system (6.1) and the constraints (6.2) can be represented in the predictive form

$$x_{k+j+1} = Ax_{k+j} + Bu_{k+j} \tag{6.43}$$

with input and state constraints:

$$\left\{ \begin{array}{l} -u_{lim} \le u_{k+j} \le u_{lim}, \; j = 0, 1, \cdots, N-1 \\ -g_{lim} \le Gx_{k+j} \le g_{lim}, \; j = 0, 1, \cdots, N \end{array} \right\} \tag{6.44}$$

States on $[k, \; k+N]$ can be represented as

$$X_k = Fx_k + HU_k \tag{6.45}$$

where X_k, F, H and U_k are given by

$$U_k = \begin{bmatrix} u_k \\ u_{k+1} \\ \vdots \\ u_{k+N-1} \end{bmatrix}, \quad X_k = \begin{bmatrix} x_k \\ x_{k+1} \\ \vdots \\ x_{k+N-1} \end{bmatrix}, \quad F = \begin{bmatrix} I \\ A \\ \vdots \\ A^{N-1} \end{bmatrix} \tag{6.46}$$

$$H = \begin{bmatrix} 0 & 0 & 0 & \cdots & 0 \\ B & 0 & 0 & \cdots & 0 \\ AB & B & 0 & \cdots & 0 \\ \vdots & \vdots & \ddots & \vdots & \vdots \\ A^{N-2}B & A^{N-3}B & \cdots & B & 0 \end{bmatrix} \tag{6.47}$$

Each state on the horizon $[k, \; k+N]$ is given by

$$x_{k+j} = A^j x_k + \bar{B}_j U_k \tag{6.48}$$

where

$$\bar{B}_j \triangleq \begin{bmatrix} A^{j-1}B & A^{j-2}B & \cdots & B & O & \cdots & O \end{bmatrix} \tag{6.49}$$

Constraints (6.44) can be written as

$$\begin{bmatrix} -u_{lim} \\ -u_{lim} \\ \vdots \\ -u_{lim} \end{bmatrix} \le U_k \le \begin{bmatrix} u_{lim} \\ u_{lim} \\ \vdots \\ u_{lim} \end{bmatrix} \tag{6.50}$$

$$\begin{bmatrix} -g_{lim} \\ -g_{lim} \\ \vdots \\ -g_{lim} \end{bmatrix} \le \bar{G}_N(Fx_k + HU_k) \le \begin{bmatrix} g_{lim} \\ g_{lim} \\ \vdots \\ g_{lim} \end{bmatrix} \tag{6.51}$$

where

$$\bar{G}_N = \mathrm{diag}\{\underbrace{G, \cdots, G}_{N}\}$$

Input and state constraints in (6.50) and (6.51) can be converted to LMI form as follows:

$$\begin{matrix} e_{m \times j + l} U_k - u^l_{lim} \leq 0, \\ -e_{m \times j + l} U_k - u^l_{lim} \leq 0, \end{matrix} \quad l = 1, 2, \cdots, m, \ j = 0, 1, \cdots, N - 1 \ (6.52)$$

$$\begin{matrix} e^T_{n_g \times j + q} \bar{G}_N (F x_k + H U_k) - g^q_{lim} \leq 0 \\ -e^T_{n_g \times j + q} \bar{G}_N (F x_k + H U_k) - g^q_{lim} \leq 0 \end{matrix} j = 0, 1, \cdots, N, \ q = 1, 2, \cdots, n_g \ (6.53)$$

where e_s is a vector such that $e_s = [0, \cdots, 0, 1, 0, \cdots, 0]^T$ with the nonzero element in the sth position, and x^a implies the ath component of the vector x. The inequality (6.53) can be also written as

$$\begin{aligned} e_q (G(A^j x_k + \bar{B}_j U_k)) - g^q_{lim} \leq 0 \\ -e_q (G(A^j x_k + \bar{B}_j U_k)) - g^q_{lim} \leq 0 \end{aligned} \tag{6.54}$$

Solution in Linear Matrix Inequality Form

Here, we consider the following performance criterion and the terminal equality constraint:

$$J(x_k, k) = \sum_{j=0}^{N-1} (x^T_{k+j} Q x_{k+j} + u^T_{k+j} R u_{k+j}) \tag{6.55}$$

$$x_{k+N} = 0 \tag{6.56}$$

where $Q \geq 0$ and $R > 0$. If there is a feasible solution for the system (6.43) and the performance criterion (6.55) with constraints (6.56) at the initial time k_0, then the next solution is guaranteed to exist, since at least one feasible solution $u_{[k+1,k+N]} = [u_{[k+1,k+N-1]} \ 0]$ exists.

For (6.56), it is required to satisfy

$$A^N x_k + \bar{B} U_k = 0 \tag{6.57}$$

where $\bar{B} = \bar{B}_N$ is given by

$$\begin{bmatrix} A^{N-1} B \ A^{N-2} B \cdots B \end{bmatrix} \tag{6.58}$$

and U_k is defined in (6.47). We parameterize U_k in (6.57) in terms of known variables. If we use the following correspondences in Lemma A.3:

$$\begin{aligned} A &\rightarrow \bar{B} \\ X &\rightarrow U_k \\ Y &\rightarrow -A^N x_k \end{aligned}$$

then it can be seen that all solutions U_k to (6.57) are parameterized by

$$U_k = -\bar{B}^{-1}A^N x_k + M\hat{U}_k \tag{6.59}$$

where $\hat{U}_k$ is a matrix containing the independent variables, $\bar{B}^{-1}$ is the right inverse of $\bar{B}$, and columns of M are orthogonal to each other, spanning the null space of $\bar{B}$. It is noted that $\bar{B}^{-1} = \bar{B}^T(\bar{B}\bar{B}^T)^{-1}$. The control (6.59) can be combined with the receding horizon control with the terminal equality constraint in Section 3.5.2.

The optimization problem for the fixed terminal case is reduced to the following SDP:

$$\min \gamma_1 \tag{6.60}$$

$$\begin{bmatrix} \gamma_1 - \mathcal{V}_3 - \mathcal{V}_2\hat{U}_k - \hat{U}_k^T\mathcal{V}_1^{\frac{1}{2}} \\ -\mathcal{V}_1^{\frac{1}{2}}\hat{U}_k \qquad I \end{bmatrix} \geq 0$$

subject to

(6.52) and (6.53)

where

$$\mathcal{V}_1 = M^T W M$$
$$\mathcal{V}_2 = -2(A^N x_k - x_{k+N}^r)^T \bar{B}^{-T} W M + w^T M$$
$$\mathcal{V}_3 = (A^N x_k - x_{k+N}^r)^T \bar{B}^{-T} W B^{-1}(A^N x_k - x_{k+N}^r)$$
$$\qquad + [Fx_k - X_k^r]^T \bar{Q}_N[Fx_k - X_k^r] - w^T \bar{B}^{-1}(A^N x_k - x_{k+N}^r)$$
$$W = H^T \bar{Q}_N H + \bar{R}_N$$
$$w = 2x_k^T F^T \bar{Q}_N H$$

and $\bar{Q}_N$ and $\bar{R}_N$ are defined in (3.312).

It is noted that $\hat{U}_k$ is obtained from the SDP problem (6.60) and U_k is computed from $\hat{U}_k$ according to (6.59).

Theorem 6.1. *The optimization problem based on (6.43), (6.44), (6.55), and (6.56) can be formulated into an SDP (6.60).*

What remains to do is just to pick the first one up among U_k^* as

$$u_k^* = [1, 0, \cdots, 0] U_k^* \tag{6.61}$$

Cost Monotonicity

We already learned in Chapter 3 that the equality condition $x_{k+N} = 0$ provides the cost monotonicity condition, which also holds in constrained systems according to the following theorem.

Theorem 6.2. *Assume that the system is feasible on $[k, \sigma]$ with the terminal equality constraint. Then the optimal cost $J^*(x_\tau, \tau, \sigma)$ for the terminal equality constraint satisfies the following monotonicity relation:*

$$J^*(x_\tau, \tau, \sigma + 1) \leq J^*(x_\tau, \tau, \sigma), \quad \tau \leq \sigma \qquad (6.62)$$

Proof. The proof procedure is the same as Theorem 3.1, except that there exist the inputs and the corresponding states that satisfy the constraints. This completes the proof. ∎

Stability

We have the following result on the stability of the constrained RHCs.

Theorem 6.3. *Assume that the system is feasible at the initial time and some horizon N^* with the terminal equality constraint and Q is positive definite. The receding horizon LQ control with the terminal equality constraint stabilizes the closed-loop system (6.1) for all horizons $N \geq N^*$.*

Proof. The proof procedure is the same as Theorem 3.7, except that there exist inputs and corresponding states that satisfy the constraints. This completes the proof. ∎

6.4 Constrained Receding Horizon Control with Terminal Set Constraint

In a fixed terminal case, the condition $x_{k+N} = 0$ is too strong to have a solution satisfying all constraints. Thus, we relax this condition and require that the terminal state has only to enter some ellipsoid instead of going into the origin exactly, which is depicted in Figure 6.3.

Using the property of the invariant ellipsoid, we introduce a receding horizon dual-mode control that satisfies the constraints on the input and the state. Outside the ellipsoid, the receding horizon control is employed. Once the state enters the ellipsoid, the control is switched to a linear state feedback control, which makes the state stay in the ellipsoid forever and stabilizes the system. This is called a receding horizon dual-mode control.

Terminal Invariant Set for $u_i = Hx_i$

Now, we will try to find a stable linear feedback control that meets all input and state constraints within the terminal ellipsoid (6.78) and make the state stay inside forever.

We define an ellipsoid $\mathcal{E}_{Q_f, \alpha}$ centered at the origin:

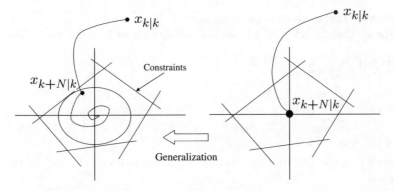

Fig. 6.3. Terminal ellipsoid constraint and terminal equality constraint

$$\mathcal{E}_{Q_f,\alpha} = \{x \in \mathcal{R}^n | x^T Q_f x < \alpha\} \qquad (6.63)$$

We will choose a stable linear feedback control $u_i = Hx_i$ so that it meets all given input and state constraints

$$-u_{lim} \le u_i \le u_{lim} \qquad (6.64)$$
$$-g_{lim} \le Gx_i \le g_{lim} \qquad (6.65)$$

within the ellipsoid (6.78) and make the state stay inside forever.

The condition

$$(A - BH)^T Q_f (A - BH) - Q_f < 0 \qquad (6.66)$$

leads to

$$\|x_{i+1}\|_{Q_f}^2 - \|x_i\|_{Q_f}^2 = x_i^T [(A - BH)^T Q_f (A - BH) - Q_f] x_i < 0 \qquad (6.67)$$

The matrix inequality (6.66) is converted to the LMI as

$$\begin{bmatrix} Q_f & A - BH \\ (A - BH)^T & Q_f^{-1} \end{bmatrix} < 0$$
$$\begin{bmatrix} X & AX - BY \\ XA^T - Y^T B^T & X \end{bmatrix} < 0 \qquad (6.68)$$

where $X = Q_f^{-1}$ and $Y = HX$.

Therefore, the state trajectory x_i, $i > 0$, with the state feedback control $u_i = Hx_i$ remains in the ellipsoid $\mathcal{E}_{Q_f,\alpha}$ and approaches zero asymptotically.

Now we will investigate if the input constraint (6.64) is satisfied in the region $\mathcal{E}_{Q_f,\alpha}$. First, we consider the jth element of u_i at time step i which satisfies the constraint

$$|u_i^j| \le u_{lim}^j, \qquad j \ge 0, j = 1, \cdots, m \qquad (6.69)$$

where u_{lim}^j is the jth element of u_{lim}.

Recall that $(x^T x) \geq (l^T x)^2$ for any vectors x and l with $l^T l = 1$. Then,

$$\max_{i \geq 0} |u_i^j|^2 = \max_{i \geq 0, x_i \in \mathcal{E}_{Q_f}} |(YX^{-1}x_i)^j|^2$$

$$\leq \max_{z \in \mathcal{E}_{Q_f,\alpha}} |(YX^{-1}z)^j|^2 = \max_{z \in \mathcal{E}_{Q_f,\alpha}} |(YX^{-\frac{1}{2}}X^{-\frac{1}{2}}z)^j|^2$$

$$= \max_{w^T w \leq \alpha} |(YX^{-\frac{1}{2}}w)^j|^2 = \alpha |E_j YX^{-\frac{1}{2}}|_2^2 = \alpha E_j YX^{-1}Y^T E_j^T \quad (6.70)$$

where E_j is the jth unit row vector with the nonzero element in the jth position

$$E_j \triangleq \begin{bmatrix} 0 \cdots 0 & 1 & 0 \cdots 0 \end{bmatrix}$$

Note that E_j plays a role in selecting the jth row of a matrix by pre-multiplication.

From (6.69) and (6.70), we have $\alpha E_j YX^{-1}Y^T E_j^T \leq u_{lim}^j$, which can be converted to the following LMI:

$$\begin{bmatrix} (u_{lim}^j)^2 & E_j Y \\ (E_j Y)^T & \frac{1}{\alpha}X \end{bmatrix} > 0 \quad (6.71)$$

for $j = 1, 2, \cdots, m$.

The state constraint (6.65) can be checked with a similar procedure.

$$\max_{i \geq 0} |(Gx_i)^j|^2 \leq \max_{z \in \mathcal{E}_{Q_f}} |(GX^{\frac{1}{2}}X^{-\frac{1}{2}}z)^j|^2 = \max_{|w|_2^2 = \alpha} |(GX^{\frac{1}{2}}w)^j|^2$$

$$= \alpha |E_j GX^{\frac{1}{2}}|_2^2 = \alpha E_j GXG^T E_j^T \quad (6.72)$$

From (6.72), the LMI for the state constraint is obtained:

$$\begin{bmatrix} (g_{lim}^j)^2 & E_j GX \\ X^T (E_j G)^T & \alpha^{-1}X \end{bmatrix} > 0 \quad (6.73)$$

For constrained systems, we should make the ellipsoid shrink so that the state feedback control $u_i = Hx_i$ asymptotically stabilizes the system while satisfying the input and the state constraints. Based on this fact, we introduce the following lemma.

Lemma 6.4. *Suppose that there exist $X > 0$ and Y satisfying the LMIs (6.68), (6.71), and (6.73). For the state feedback controller $u_i = YX^{-1}x_i = Hx_i$, the resultant state trajectory x_i always remains in the region $\mathcal{E}_{Q_f,\alpha}$ satisfying the constraints on the input and the state.*

According to Lemma 6.4, the input and state constraints are expressed by each component. This may be rewritten in another form to consider the constraints in terms of one LMI.

If we introduce another variable Z that satisfies

$$\alpha E_j Y X^{-1} Y^T E_j^T = \alpha (Y X^{-1} Y^T)_{j,j} \leq Z_{j,j} \leq u_{lim}^j$$

where $Z_{j,j}$ is (j,j) elements of the matrices Z, $\alpha Y X^{-1} Y^T \leq Z$ implies that the input constraint $-u_{lim} \leq u_i \leq u_{lim}$ is guaranteed in the region $\mathcal{E}_X$. Therefore, we obtain the following LMI:

$$\begin{bmatrix} Z & Y \\ Y^T & \frac{1}{\alpha} X \end{bmatrix} \geq 0, \text{ with } Z_{j,j} \leq (u_{lim}^j)^2 \tag{6.74}$$

If we introduce a variable V that satisfies

$$\alpha (GXG^T)_{j,j} \leq V_{j,j} \leq g_{lim}^j$$

then $\alpha GXG^T \leq V$ implies that the state constraint $-g_{lim} \leq Gx_i \leq g_{lim}$ is guaranteed in the region $\mathcal{E}_{Q_f,\alpha}$. Therefore, we obtain the following LMI:

$$\alpha GXG^T \leq V, \text{ with } V_{j,j} \leq (g_{lim}^j)^2 \tag{6.75}$$

In order for a dual-mode control to stabilize the system, the state must enter the ellipsoid within the given finite time

Constrained Receding Horizon Dual-mode Control with Fixed Invariant Set

The following performance is considered:

$$J_1(x_k, k) \triangleq \sum_{j=0}^{N_k-1} (x_{k+j}^T Q x_{k+j} + u_{k+j}^T R u_{k+j}) \tag{6.76}$$

with the input and state constraints

$$\begin{cases} -u_{lim} \leq u_{k+j} \leq u_{lim}, & j = 0, 1, \cdots, N_k - 1 \\ -g_{lim} \leq G x_{k+j} \leq g_{lim}, & j = 0, 1, \cdots, N_k \end{cases} \tag{6.77}$$

and the terminal set constraint

$$x_{k+N}^T P x_{k+N} \leq \alpha \tag{6.78}$$

which is represented as

$$\mathcal{E}_{P,\alpha} = \{x \in \mathcal{R}^n | x^T P x \leq \alpha\} \tag{6.79}$$

The terminal set (6.78) can be written as

$$(A^{N_k} x_k + \bar{B} U_k)^T P (A^{N_k} x_k + \bar{B} U_k) < \alpha$$

which is converted into the following LMI:

$$\begin{bmatrix} \alpha & (A^{N_k}x_k + \bar{B}U_k)^T \\ (A^{N_k}x_k + \bar{B}U_k) & P^{-1} \end{bmatrix} > 0 \tag{6.80}$$

According to (3.319) and (3.320) in Section 3.5.2, the problem minimizing (6.76) is reduced to the following LMI form:

$$\begin{bmatrix} \gamma_1 - w^T U_k - x_k^T F_k^T \bar{Q}_N F_k x_k & U_k^T \\ U_k & W^{-1} \end{bmatrix} \geq 0 \tag{6.81}$$

We try to find the minimum γ_1 subject to (6.81), (6.52), (6.53), and (6.80) to obtain an optimal control which steers the state into the ellipsoid (6.79).

Dual-mode control can be obtained as follows: at time $k = 0$, if $x_0 \in \mathcal{E}$, switch to a local linear control, i.e. employ a linear feedback law $u_k = \mathcal{K}x_k$ thereafter such that $x_k \in \mathcal{E}$. Otherwise, compute a control horizon pair $(u_{0+\cdot}, N_0)$ for the problem $J(x_0, 0, u_{0+\cdot}, N_0)$. Apply the control u_0 to the real system.

Assume that the control pair $(u_{k-1+\cdot}, N_{k-1})$ is found. At the time k, if $x_k \in \mathcal{E}$, switch to a local linear control, i.e. employ the linear feedback control law $\mathcal{K}x_k$ thereafter such that $x_k \in \mathcal{E}$. Otherwise, we compute another admissible control horizon pair $(u_{k+\cdot}, N_k)$, which is better than the preceding control horizon pair in the sense that

$$J(x_k, k, u_{k+\cdot}, N_k) \leq J(x_k, k, u'_{k+\cdot}, N_{k-1} - 1) \tag{6.82}$$

where $u'_{k+\cdot}$ is equal to $u_{k-1+\cdot}$ in the interval $[k, \ k - 1 + N_{k-1}]$ and is an admissible control for $J(x_k, k, u_{k+\cdot}, N_k)$. Finally, we apply the control u_k to the real system.

The control u_k always exists, since there is at least one feasible solution if N_k is set to $N_{k-1} - 1$. It is noted that N_k is time-varying, whereas N is fixed in previous sections.

Theorem 6.5. *Suppose that the system (6.1) is stabilizable. Then, for $N_k > 1$, there exists a control horizon pair (u_k, N_k) such that the following equation is satisfied:*

$$J(x_k, k, u_k, N_k) \leq J(x_k, k, u'_k, N_{k-1} - 1) \tag{6.83}$$

and there exists constant $0 < \eta < \infty$ satisfying

$$J(x_{k+1}, k + 1, u_{k+1}, N_{k+1}) \leq J(x_k, k, u_k, N_k) - \eta \tag{6.84}$$

for all k such that both x_k and x_{k+1} are in $\mathcal{E}_{P,\alpha}$.

Proof. The first statement is already explained in (6.82). Since the trajectory by (u_k, N_k) is the same as one by $(u_{k+1}, N_k - 1)$ on $[k + 1, k + N_k]$, we have

$$J(x_k, k, u_k, N_k) - J(x_{k+1}, k+1, u'_{k+1}, N_k - 1) \tag{6.85}$$

$$= x_k^T Q x_k + u_k^T R u_k \tag{6.86}$$

$$\geq \ \inf\{x_k^T Q x_k \mid x_k \in \mathcal{E}_{P,\alpha}\} \tag{6.87}$$

$$= \ \inf\{\frac{x^T Q x}{x^T P x} x^T P x \mid x \in \mathcal{E}_{P,\alpha}\} \tag{6.88}$$

$$\geq \frac{\lambda_{\min}(Q)}{\lambda_{\max}(P)}\alpha \tag{6.89}$$

for all k such that both x_k and x_{k+1} lie in $\mathcal{E}_{P,\alpha}$. Thus, η can be taken as

$$\eta = \frac{\lambda_{min}(Q)}{\lambda_{max}(P)}\alpha \tag{6.90}$$

This completes the proof. ∎

Theorem 6.6. *The dual-mode receding horizon controller is asymptotically stabilizing with a region of attraction X. For all $x_0 \in X$, there exists a finite time N^* such that $x_{N^*} \in \mathcal{E}_{P,\alpha}$.*

Proof. Suppose that there is no such k for which $x_k \in \mathcal{E}_{P,\alpha}$. From Theorem 6.5, it follows that there exists a $0 < \eta < \infty$ such that

$$J(x_{k+1}, k+1, u_{k+1+\cdot}, N_{k+1}) \leq J(x_k, k, u_{k+\cdot}, N_k) - \eta \tag{6.91}$$

for all k, which immediately implies that

$$J(x_k, k, u_{k+\cdot}, N_k) < 0 \tag{6.92}$$

for some $k > 0$. However, this contradicts the fact that

$$J(x_k, k, u_{k+\cdot}, N_k) \geq 0 \tag{6.93}$$

for all k. Therefore, there exists a finite k such that $x_k \in \mathcal{E}_{P,\alpha}$. x_k enters $\mathcal{E}_{P,\alpha}$ within a finite time and then approaches zero by a local linear stabilizing state feedback control. This completes the proof. ∎

The horizon N is varying with time in the above cases. However, we can use the fixed horizon N with a stability property. We introduce a dual-mode control with a stability property, which will be explained later.

6.5 Constrained Receding Horizon Control with Free Terminal Cost

Solution in Linear Matrix Inequality Form

The performance criterion can be written into two parts as follows:

$$J(x_k, k) = J_1(x_k, k) + J_2(x_{k+N}, k) \tag{6.94}$$

where

$$J_1(x_k, k) \triangleq \sum_{j=0}^{N-1} (x_{k+j}^T Q x_{k+j} + u_{k+j}^T R u_{k+j}) \tag{6.95}$$

$$J_2(x_{k+N}, k) \triangleq x_{k+N}^T Q_f x_{k+N} \tag{6.96}$$

and $Q \geq 0$, $R > 0$, and $Q_f > 0$. The constrained RHC is obtained by computing the above optimization problem at time k and repeating it at the next time.

According to (3.319) and (3.320) in Section 3.5.2, the problem minimizing (6.94) with constraints (6.2) can be reduced to the following problem:

$$U_k^* = \arg \min \quad \gamma_1 + \gamma_2 \tag{6.97}$$
$$\text{subject to}$$
$$(6.81)$$
$$\begin{bmatrix} \gamma_2 & [A^N x_k + \bar{B} U_k]^T \\ [A^N x_k + \bar{B} U_k] & Q_f^{-1} \end{bmatrix} \geq 0 \tag{6.98}$$

$$(6.52) \text{ and } (6.53)$$

where

$$\bar{Q}_N = \text{diag}\{\underbrace{Q, \cdots, Q}_{N}\} \quad \bar{R}_N = \text{diag}\{\underbrace{R, \cdots, R}_{N}\} \tag{6.99}$$

Now we can summarize what we have shown so far as follows:

Theorem 6.7. *The optimization problem based on (6.43), (6.44), and (6.94) can be formulated into an SDP as follows:*

$$\min_{\gamma_1, \gamma_2, U_k} \quad \gamma_1 + \gamma_2 \tag{6.100}$$

subject to (6.81), (6.98), (6.52), and (6.53).

If U_k is found from Theorem 6.7, then the optimal control at the current time k is given by (6.61).

For the free terminal cost, we have a similar result to the terminal equality constraint.

Cost Monotonicity

Theorem 6.8. *Assume that the system is feasible. If Q_f in (6.96) satisfies the inequality*

$$Q_f \geq Q + H^T R H + (A - BH)^T Q_f (A - BH) \tag{6.101}$$

for some $H \in \Re^{m \times n}$, then the optimal cost $J^*(x_\tau, \tau, \sigma)$ then satisfies the following monotonicity relation:

$$J^*(x_\tau, \tau, \sigma + 1) \leq J^*(x_\tau, \tau, \sigma) \tag{6.102}$$

Proof. The proof procedure is the same as Theorem 3.2, except that there exist the inputs and the corresponding states that satisfy the constraints. This completes the proof. ∎

It is noted that feasibility is assumed in Theorem 3.1. We introduce LMI forms to handle feasibility and the constraints.

We show that this inequality condition (6.101) can be converted into an LMI form.

From Chapter 3 we have the LMI form of the cost monotonicity condition

$$\begin{bmatrix} X & (AX - BY)^T & (Q^{\frac{1}{2}}X)^T & (R^{\frac{1}{2}}Y)^T \\ AX - BY & X & 0 & 0 \\ Q^{\frac{1}{2}}X & 0 & I & 0 \\ R^{\frac{1}{2}}Y & 0 & 0 & I \end{bmatrix} \geq 0 \tag{6.103}$$

where $X = Q_f^{-1}$ and $Y = HX$.

Then, we have to solve the following problem:

$$\min_{\gamma_1, \gamma_2, X, Y, U_k} \gamma_1 + \gamma_2$$

subject to
(6.81), (6.98), (6.52), (6.53), and (6.103).

It is noted that even though the the above optimization problem has a solution at current time k , we cannot guarantee that the optimization problem also has a solution at the next time $k + 1$, as seen in Figure 6.4.

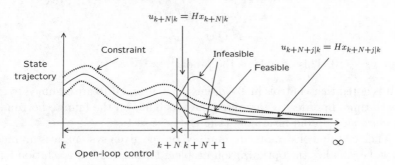

Fig. 6.4. How to choose H at the terminal time

Stability

We have the following results on the stability of the constrained RHCs.

Theorem 6.9. *Assume that the system is feasible for all k and some N^*, and Q is positive definite. The receding horizon LQ control with the free terminal cost under a cost monotonicity condition (6.102) stabilizes the closed-loop system (6.1) for $N \geq N^*$.*

Proof. The proof procedure is the same as Theorem 3.8, except that there exist the inputs and the corresponding states that satisfy the constraints. Since $Q > 0$, $x_i^T Q x_i \to 0$ implies $x_i \to 0$. This completes the proof. ∎

A feedback control $u_\sigma = H x_\sigma$ was obtained from Theorem 6.8. We can utilize the state feedback control every time.

Let $\sigma = k$; we must have

$$(A + BH)x_k \in \mathcal{X}, \quad H x_k \in \mathcal{U} \tag{6.104}$$

and

$$-g_{lim} \leq G(A - BH)x_k \leq g_{lim} \tag{6.105}$$
$$-u_{lim} \leq H x_k \leq u_{lim} \tag{6.106}$$

It is noted that H used in $u_k = H x_k$ is obtained at the terminal time k.

Theorem 6.10. *If the system is feasible and Q_f is positive definite, then the state feedback control $u_i = H x_i$ obtained from (6.103), (6.105) and (6.106) stabilizes the closed-loop system while satisfying the constraints.*

Proof. Let $V(x_i) = x_i^T Q_f x_i$. Since $Q_f > 0$, $V(x_i)$ can be a Lyapunov function.

$$\begin{aligned}
V(x_{i+1}) - V(x_i) &= x_i^T (A - BH)^T Q_f (A - BH)x_i - x_i Q_f x_i \\
&= x_i(-Q_f - H^T RH)x_i \\
&\leq -x_i^T Q_f x_i
\end{aligned}$$

Thus, $x_i \to 0$. This completes the proof. ∎

It is noted that controls in Theorems 6.3, 6.9, and 6.10 are assumed to exist at each time. In order to guarantee the feasibility on the infinite horizon, we introduce an ellipsoid invariant set in the next section.

When certain constraints are imposed on the given system, it is an important issue whether the optimization problem formulated has a solution for all times k.

Terminal Invariant Ellipsoid

From the optimization problem (6.97), we require additional constraint

$$\gamma_2 \leq \alpha \tag{6.107}$$

in order to improve the feasibility.

Here, we introduce an algorithm to solve the optimization problem (6.97). Now, we will use γ_2^* instead of α, which brings out the advantage that it is not necessary to make the final state driven to the specific ellipsoid. We can obtain the ellipsoid while optimizing the given performance criterion. The optimal cost $x^T Q_f x$ for the final state is equal to γ_2^*, so that we choose Q_f and γ_2^* to make the ellipsoid $x^T Q_f x \leq \gamma_2^*$ invariant. For these Q_f and γ_2^*, the final state is on the boundary of the ellipsoid $x^T Q_f x \leq \gamma_2^*$. Thus, we have only to replace X with $\gamma_2 Q_f^{-1}$ and find it so that the ellipsoid is invariant. The LMIs (6.71) and (6.73) should be changed to

$$\begin{bmatrix} (u_{lim}^j)^2 & E_j Y \\ (E_j Y)^T & \frac{1}{\gamma_2} X \end{bmatrix} > 0 \tag{6.108}$$

and

$$\begin{bmatrix} (g_{lim}^j)^2 & E_j G X \\ X^T (E_j G)^T & \gamma_2 X \end{bmatrix} > 0 \tag{6.109}$$

Lemma 6.11. *Suppose that there exist $X > 0$ and Y satisfying (6.103), (6.71), and (6.73). Then, the state feedback controller $u_i = Y X^{-1} x_i = H x_i$ exponentially stabilizes the closed-loop system for all $x_0 \in \mathcal{E}_{Q_f,\alpha}$ while satisfying the constraint (6.2). And the resultant state trajectory x_i always remains in the region $\mathcal{E}_{Q_f,\alpha}$.*

The following theorem summarizes the stabilizing receding horizon control.

Theorem 6.12. *Consider the optimization problem at time k as follows:*

$$\min_{\Lambda} \quad \gamma_1 + \gamma_2 \tag{6.110}$$

subject to (6.52), (6.53), (6.81),

$$\begin{bmatrix} \gamma_2 I & (A^N x_k + \bar{B} U_k)^T \\ A^N x_k + \bar{B} U_k & \hat{X} \end{bmatrix} \geq 0 \tag{6.111}$$

$$\begin{bmatrix} \hat{X} & (A\hat{X} - B\hat{Y})^T & (Q^{\frac{1}{2}}\hat{X})^T & (R^{\frac{1}{2}}\hat{Y})^T \\ A\hat{X} - B\hat{Y} & \hat{X} & 0 & 0 \\ Q^{\frac{1}{2}}\hat{X} & 0 & \gamma_2 I & 0 \\ R^{\frac{1}{2}}\hat{Y} & 0 & 0 & \gamma_2 I \end{bmatrix} \geq 0 \tag{6.112}$$

$$\begin{bmatrix} (u_{lim}^j)^2 & E_j\hat{Y} \\ E_j\hat{Y}^T & \hat{X} \end{bmatrix} > 0 \qquad (6.113)$$

$$\begin{bmatrix} (g_{lim}^j)^2 & E_jG\hat{X} \\ \hat{X}E_jG^T & \hat{X} \end{bmatrix} > 0 \qquad (6.114)$$

where $j = 1, 2, \cdots, m$, $\Lambda = \{\gamma_1, \gamma_2, \hat{X}, \hat{Y}, U_t\}$. Then, we can obtain $Q_f = \gamma_2\hat{X}$ and $H = \hat{Y}\hat{X}^{-1}$. If this optimization problem has a solution at the initial time, then we can obtain a stabilizing receding horizon control for all the time that satisfies the constraints on the input and the state.

Proof. In order to combine two variables γ_2 and X, we introduce $\hat{X} = \gamma_2X$.
From (6.98), we have

$$\begin{bmatrix} \gamma_2 I & (A^N x_k + \bar{B}U_k)^T \\ A^N x_k + \bar{B}U_k & X \end{bmatrix} \geq 0 \qquad (6.115)$$

Representing X in terms of $\hat{X}$ and scaling elements in the matrix on the left side of (6.111) yields

$$T\begin{bmatrix} \gamma_2 I & (A^N x_k + \bar{B}U_k)^T \\ A^N x_k + \bar{B}U_k & \gamma_2^{-1}\hat{X} \end{bmatrix} T^T \qquad (6.116)$$

$$= \begin{bmatrix} I & (A^N x_k + \bar{B}U_k)^T \\ A^N x_k + \bar{B}U_k & \hat{X} \end{bmatrix} \geq 0 \qquad (6.117)$$

where

$$T = \begin{bmatrix} \sqrt{\gamma_2}^{-1} & 0 \\ 0 & \sqrt{\gamma_2} \end{bmatrix} \qquad (6.118)$$

Thus, we obtain the LMI (6.111). By using $\hat{X}$ instead of X, the left side of the LMI (6.103) for the cost monotonicity condition can be changed into

$$\begin{bmatrix} \gamma_2^{-1}\hat{X} & \gamma_2^{-1}(A\hat{X} - B\hat{Y})^T & (Q^{\frac{1}{2}}\gamma_2^{-1}\hat{X})^T & (R^{\frac{1}{2}}\gamma_2^{-1}\hat{Y})^T \\ \gamma_2^{-1}(A\hat{X} - B\hat{Y}) & \gamma_2^{-1}\hat{X} & 0 & 0 \\ Q^{\frac{1}{2}}\gamma_2^{-1}\hat{X} & 0 & I & 0 \\ R^{\frac{1}{2}}\gamma_2^{-1}\hat{Y} & 0 & 0 & I \end{bmatrix} \qquad (6.119)$$

Multiplying (6.119) by T_2 given as

$$T_2 = \text{diag}\{\sqrt{\gamma_2}, \sqrt{\gamma_2}, \sqrt{\gamma_2}, \sqrt{\gamma_2}\}$$

yields (6.112). The input and the state constraints (6.108) and (6.109) in an ellipsoid can be represented in a form of the LMIs as

$$\begin{bmatrix} I & 0 \\ 0 & \gamma_2 I \end{bmatrix} \begin{bmatrix} (u_{lim}^j)^2 & E_jY \\ E_jY^T & \frac{1}{\gamma_2}X \end{bmatrix} \begin{bmatrix} I & 0 \\ 0 & \gamma_2 I \end{bmatrix} = \begin{bmatrix} (u_{lim}^j)^2 & E_j\hat{Y} \\ E_j\hat{Y}^T & \hat{X} \end{bmatrix} > 0 \qquad (6.120)$$

and

$$\begin{bmatrix} I & 0 \\ 0 & \gamma_2 I \end{bmatrix} \begin{bmatrix} (g_{lim}^j)^2 & E_j GX \\ XE_j G^T & \gamma_2^{-1} X \end{bmatrix} \begin{bmatrix} I & 0 \\ 0 & \gamma_2 I \end{bmatrix} = \begin{bmatrix} (g_{lim}^j)^2 & E_j G\hat{X} \\ \hat{X} E_j G^T & \hat{X} \end{bmatrix} > 0 \quad (6.121)$$

This completes the proof. ∎

The trajectory of the receding horizon control of Theorem 6.12 is depicted in Figure 6.5.

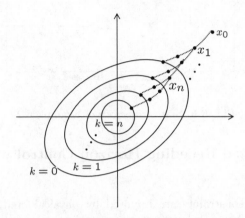

Fig. 6.5. State trajectory due to the proposed RHC in Theorem 6.12

Example 6.2

To illustrate the validity of the RHC, numerical examples are given for a linear discrete time-invariant system

$$x_{i+1} = \begin{bmatrix} 1 & 0.2212 \\ 0 & 0.7788 \end{bmatrix} x_i + \begin{bmatrix} 0.0288 \\ 0.2212 \end{bmatrix} u_i \quad (6.122)$$

where the initial state is set to $[1 \; 0.3]^T$, and u_i and x_i should be constrained in the following region:

$$-0.5 \leq u_i \leq 0.5 \quad (6.123)$$

$$\begin{bmatrix} -1.5 \\ -0.3 \end{bmatrix} \leq Gx_i = \begin{bmatrix} 1 & 0 & 1.5 \\ 0 & 1 & 1 \end{bmatrix} x_i \leq \begin{bmatrix} 1.5 \\ 0.3 \end{bmatrix} \quad (6.124)$$

Here, the performance criterion is taken as

$$\sum_{j=0}^{4} [\, x_{k+j}^T x_{k+j} + u_{k+j}^2 \,] + x_{k+5}^T Q_f x_{k+5} \quad (6.125)$$

where Q_f is determined each time. State trajectories are shown in Figure 6.6.

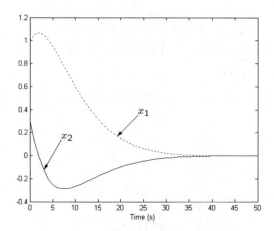

Fig. 6.6. State trajectory of Example 6.2

6.6 Constrained Receding Horizon Control with Mixed Constraints

Generally, input constraints are imposed by physical limitations of actuators, valves, pumps, etc., whereas state constraints cannot be satisfied all the time, and hence some violations are allowable. In particular, even if state constraints are satisfied in nominal operations, unexpected disturbances may put the states aside from the region where state constraints are satisfied. In this case, it may happen that some violations of state constraints are unavoidable, while input constraints can be still satisfied. We introduce so-called mixed constraints, which consist of a hard input constraint and a soft state constraint, and apply these constraints to the RHC.

The mixed constraints are given by

$$-u_{lim} \leq u_i \leq u_{lim} \tag{6.126}$$

$$-g - \epsilon_i \leq Gx_i \leq g + \epsilon_i \tag{6.127}$$

where $i = 0, 1, \cdots, \infty$ and $\epsilon_i \geq 0$ denotes tolerance for violation of state constraints.

In order to deal with the mixed constraints (6.126), we modify the performance criterion and the corresponding optimization problem as

$$\underset{u_k, \cdots, u_{k+N-1}, \epsilon_k, \cdots, \epsilon_{k+N-1}}{\text{Minimize}} J_m(x_k, k) \tag{6.128}$$

where

$$J_m(x_k, k) = J(x_k, k) + \sum_{j=0}^{N-1} \epsilon_{k+j}^T S\epsilon_{k+j} \tag{6.129}$$

subject to

$$-u_{lim} \le u_{k+j} \le u_{lim}, \quad j = 0, 1, \cdots, N-1 \tag{6.130}$$

$$-g - \epsilon_{k+j} \le Gx_{k+j} \le g + \epsilon_{k+j}, \quad j = 0, 1, \cdots, N-1 \tag{6.131}$$

where $J(x_k, k)$ is given in (6.94) and $S > 0$ is a weighting matrix for violation on state constraints. The constraints (6.126) and (6.130) are hard constraints, whereas the constraints (6.127) and (6.131) are soft constraints. They together are called mixed constraints if both exist. Note that the additional cost is included so as to penalize a measure of the state constraints violation. We require that $u_{k+j} = Hx_{k+j}$ is applied for $j \ge N$ and the state trajectory satisfies

$$-u_{lim} \le Hx_{k+j} \le u_{lim} \tag{6.132}$$

$$-g \le Gx_{k+j} \le g \tag{6.133}$$

for $j = N, \cdots, \infty$.

It is shown that the cost monotonicity condition for a hard constraint still holds under a specific condition for (6.126).

u_i^1 and ϵ_i^1 are optimal for $J(x_\tau, \tau, \sigma + 1)$, and u_i^2 and ϵ_i^2 are optimal for $J(x_\tau, \tau, \sigma)$. If we replace u_i^1 and ϵ_i^1 by u_i^2 and ϵ_i^2 up to $\sigma - 1$, and we use $u_\sigma^1 = -Hx_\sigma$ at time σ, then by the optimal principle we have

$$J^*(x_\tau, \sigma + 1) = \sum_{i=\tau}^{\sigma} [x_i^{1T}Qx_i^1 + u_i^{1T}Ru_i^1 + \epsilon_i^{1T}S\epsilon_i^1] + x_{\sigma+1}^{1T}Q_f x_{\sigma+1}^1$$

$$\le \sum_{i=\tau}^{\sigma-1} [x_i^{2T}Qx_i^2 + u_i^{2T}Ru_i^2 + \epsilon_i^{2T}S\epsilon_i^2] + x_\sigma^{2T}Qx_\sigma^2 + x_\sigma^{2T}H^TRHx_\sigma^2$$

$$+ x_\sigma^{2T}(A - BH)Q_f(A - BH)x_\sigma^2 \tag{6.134}$$

where $x_{\sigma+1} = (A - BH)x_\sigma$.

If x_σ satisfies the condition (6.133), then the optimal value of ϵ_σ must be zero. How to make x_σ satisfy (6.133) will be discussed later. It follows from (6.134) that

$$\delta J^*(x_\tau, \sigma) = J^*(x_\tau, \sigma + 1) - J^*(x_\tau, \sigma) \tag{6.135}$$

$$= \sum_{i=\tau}^{\sigma} [x_i^{1T}Qx_i^1 + u_i^{1T}Ru_i^1 + \epsilon_i^{1T}S\epsilon_i^1] + x_{\sigma+1}^{1T}Q_f x_{\sigma+1}^1$$

$$- \sum_{i=\tau}^{\sigma-1} [x_i^{2T}Qx_i^2 + u_i^{2T}Ru_i^2 + \epsilon_i^{1T}S\epsilon_i^1] - x_\sigma^{2T}Q_f x_\sigma^2$$

$$\le x_\sigma^{2T}Qx_\sigma^2 + x_\sigma^{2T}H^TRHx_\sigma^2 + x_\sigma^{2T}(A - BH)Q_f(A - BH)x_\sigma^2$$

$$- x_\sigma^{2T}Q_f x_\sigma^2$$

$$= x_\sigma^{2T}\{Q + H^TRH + (A - BH)^TQ_f(A - BH) - Q_f\}x_\sigma^2 \le 0 \tag{6.136}$$

Finally, we have the cost monotonicity condition:

$$Q + H^T R H + (A - BH)^T Q_f (A - BH) - Q_f \leq 0 \qquad (6.137)$$

In order to consider the cost function of a state constraint violation, we introduce the variable γ_3 as

$$E_k^T S_N E_k < \gamma_3 \qquad (6.138)$$

$$\begin{bmatrix} \gamma_3 & E_k^T \\ E_k & S_N^{-1} \end{bmatrix} > 0 \qquad (6.139)$$

where $E_k = [\epsilon_k, \cdots, \epsilon_{k+N-1}]^T$ and $S_N = \mathrm{diag}\{\underbrace{S, \cdots, S}_{N}\}$.

Finally, we have the following LMI problem:

$$\mathrm{Min}_{\gamma_1, \gamma_2, \gamma_3, U_k, \epsilon} \gamma_1 + \gamma_2 + \gamma_3 \qquad (6.140)$$

subject to (6.81), (6.98), (6.111), (6.139), (6.112), (6.113), (6.114).

If there is a solution to the optimization problem (6.140), then it satisfies all the conditions (6.126), (6.132), (6.133), and (6.137).

6.7 Constrained Output Feedback Receding Horizon Control

Constrained Output Feedback Receding Horizon Control with a Disturbance Invariant Set

We will show that an output feedback RHC guarantees the closed-loop stability for a constrained system under some conditions. Since the system with constraints is nonlinear, we cannot apply the separation principle to prove the stability of the system.

We consider the following system with input, systems disturbance, and measurement disturbance constraints:

$$x_{k+1} = Ax_k + Bu_k + Gw_k \qquad (6.141)$$
$$y_k = Cx_k + v_k \qquad (6.142)$$

where

$$\begin{aligned} -u_{lim} &\leq u_k \leq u_{lim} \\ -v_{lim} &\leq v_k \leq v_{lim} \\ -w_{lim} &\leq w_k \leq w_{lim} \end{aligned} \qquad (6.143)$$

The state is not accessible and an observer of the plant

$$\hat{x}_{k+1} = A\hat{x}_k + Bu_k + L(y_k - C\hat{x}_k) \qquad (6.144)$$

is used to provide estimated states instead. The dynamics of the state estimation error $e_k = x_k - \hat{x}_k$ can be derived as

$$e_{k+1} = (A - LC)e_k + Gw_k - Lv_k \tag{6.145}$$

It is assumed that the bounds on the initial estimation error are given as

$$|e_0| \le \bar{e}_0 \tag{6.146}$$

By combining (6.144) and (6.145), it is possible to build the following augmented system:

$$x_{k+1}^a = \begin{bmatrix} A & LC \\ 0 & A - LC \end{bmatrix} x_k^a + \begin{bmatrix} B \\ 0 \end{bmatrix} u_k + \begin{bmatrix} 0 & L \\ G & -L \end{bmatrix} d_k^a \tag{6.147}$$

where

$$x_k^a = \begin{bmatrix} \hat{x}_k^T & e_k^T \end{bmatrix}^T, \quad d_k^a = \begin{bmatrix} w_k^T & v_k^T \end{bmatrix}^T \tag{6.148}$$

It is assumed that the control u_k is represented as

$$u_k = F\hat{x}_k + \nu_k \tag{6.149}$$

where ν_k is an additional control which will be explained later. F in (6.149) and L in (6.145) are chosen so that $A - BF$ and $A - LC$ respectively are Hurwitz and meet a certain performance. Then, (6.147) can be rewritten as

$$\begin{aligned} x_{k+1}^a &= \begin{bmatrix} \Phi_F & LC \\ 0 & \Phi_L \end{bmatrix} x_k^a + \begin{bmatrix} B \\ 0 \end{bmatrix} c_k + \begin{bmatrix} 0 & L \\ G & -L \end{bmatrix} d_k^a \\ &= \Phi x_k^a + \begin{bmatrix} B \\ 0 \end{bmatrix} c_k + \Xi d_k^a \end{aligned} \tag{6.150}$$

where $\Phi_F = A + BF$, $\Phi_L = A - LC$, and

$$\Phi = \begin{bmatrix} \Phi_F & LC \\ 0 & \Phi_L \end{bmatrix}, \quad \Xi = \begin{bmatrix} 0 & L \\ G & -L \end{bmatrix} \tag{6.151}$$

Now, we set ν_k to zero and introduce a disturbance invariant set where all trajectories satisfy the constraint once the initial state comes inside.

For the matrix $A \in \Re^{m \times n}$, $|A|$ is defined as a matrix each of whose elements is $|A_{ij}|$. The following disturbance invariance set is considered:

$$\mathcal{E}_{W,\alpha} = \{x_k^a \in \Re^{2n} | \; |Wx_k^a| \le \alpha, \; 0 < \alpha \in \Re^{2n}\} \tag{6.152}$$

with a weighting symmetric matrix W that is a block diagonal matrix such as

$$W = \begin{bmatrix} W_1 & 0 \\ 0 & W_2 \end{bmatrix} \tag{6.153}$$

It is noted that $\mathcal{E}_{W,\alpha}$ in (6.152) is of a polyhedral form, not an ellipsoid.

From the inequality

$$|Wx_{k+1}^a| = |W(\Phi x_k^a + \Xi \begin{bmatrix} w_k \\ v_k \end{bmatrix})| = |W(\Phi W^{-1}Wx_k^a + \Xi \begin{bmatrix} w_k \\ v_k \end{bmatrix})|$$

$$\leq |W\Phi W^{-1}| \, |Wx_k^a| + |W\Xi| \, | \begin{bmatrix} w_k \\ v_k \end{bmatrix}|$$

$$\leq |W\Phi W^{-1}|\alpha + |W\Xi| \begin{bmatrix} w_{lim} \\ v_{lim} \end{bmatrix}$$

the invariance condition of the set $\mathcal{E}_{W,\alpha}$ is given as

$$|W\Phi W^{-1}|\alpha + |W\Xi| \begin{bmatrix} v_{lim} \\ w_{lim} \end{bmatrix} \leq \alpha \qquad (6.154)$$

After some obvious manipulation, the equation can be divided as follows:

$$|W_1\Phi_F W_1^{-1}|\alpha_1 + |W_1 L C W_2^{-1}|\alpha_2 + |W_1 L|w_{lim} \leq \alpha_1 \qquad (6.155)$$
$$|W_2\Phi_L W_2^{-1}|\alpha_2 + |W_2 G|v_{lim} + |W_2 L|w_{lim} \leq \alpha_2 \qquad (6.156)$$

where $\alpha = [\alpha_1^T \ \alpha_2^T]^T$. It is easy to see that the state feedback control $u_k = F\hat{x}_k$ satisfies the input constraints for any $x_{k+N}^a \in \mathcal{E}_{W,\alpha}$ provided that

$$|FW_1^{-1}|\alpha_1 \leq u_{lim} \qquad (6.157)$$

which comes from the following fact:

$$|F\hat{x}_k| = |FW_1^{-1}W_1\hat{x}_k| \leq |FW_1^{-1}||W_1\hat{x}_k| \leq |FW_1^{-1}|\alpha_1 \qquad (6.158)$$

Theorem 6.13. *If (6.155), (6.156), and (6.157) are satisfied, then $u_k = F\hat{x}_k$ stabilizes the system (6.141) and (6.142) for any initial estimation $\hat{x}_0$ and initial estimation error e_0 such $x_0^a \in \mathcal{E}_{W,\alpha}$.*

It is noted that ν_k is set to zero inside the invariant set. Now, we force the state into the disturbance invariant set by using nonzero ν_k and a receding horizon scheme.

A control algorithm of the constrained output feedback RHC for linear constrained systems is as follows:

1. At time instant k, determine N additional controls $[\nu_k, \nu_{k+1}, \cdots, \nu_{k+N-1}]$ for (6.150) to satisfy the input constraints in (6.143) and $x_{k+N}^a \in \mathcal{E}_{W,\alpha}$, and minimize the following performance criterion:

$$J([\nu_k, \nu_{k+1}, \cdots, \nu_{k+N-1}]) = \sum_{i=0}^{N-1} \nu_{k+i}^T \nu_{k+i} \qquad (6.159)$$

2. Apply $u_k = F\hat{x}_k + \nu_k$ to the plant (6.141) and (6.142).
3. Repeat this procedure at the next time instant $k + 1$.

It is noted that the above approach provides the stabilizing output feedback control and tends not to much deviate from the trajectory generated from the control (6.149) with the observer (6.145).

The disturbance invariant set can be designed once or repeatedly each time. In the first case, as in dual-mode controls, the control is switched to the linear output feedback control when the state enters the disturbance invariant set. In the second case, the disturbance invariance set is designed repeatedly for a better performance, which will be given as a problem in Section 6.9.

Constrained Linear Quadratic Finite Memory Control

It is shown that LQFMC can easily handle input constraints just by adding LMIs. Assume that the system matrix A is nonsingular and the following LMI problem is feasible:

$$\min_{F,W} \ \mathrm{tr}(W)$$

subject to

$$\begin{bmatrix} W & \hat{R}^{\frac{1}{2}}(FM^T + H_0) \\ (FM^T + H_0)^T \hat{R}^{\frac{1}{2}} & (\bar{G}_{N_f} Q_{N_f} \bar{G}_{N_f}^T + R_{N_f})^{-1} \end{bmatrix} > 0 \qquad (6.160)$$

$$-u_{lim} \leq HY_{k-1} - H\bar{B}_{N_f}U_{k-1} \leq u_{lim} \qquad (6.161)$$

where $H_0 = -\mathcal{K}_1(\bar{C}_{N_f}^T \bar{C}_{N_f})^{-1}\bar{C}_{N_f}$ and the columns of M^T consist of the basis of the null space of $\bar{C}_{N_f}^T$. Then the gain matrices of the constrained LQFMC is given by

$$H = FM^T + H_0, \quad L = -H\bar{B}_{N_f}$$

It is noted that inequalities (6.161) are of an LMI form, which decreases the feasibility.

6.8 References

Reachable sets and maximal output admissible sets in Section 6.2 are mostly based on [BGFB94] and [GT91] respectively. In particular, constrained LQ controls in Section 6.2 using a maximal admissible set appeared in [KG88] [SR98].

The cost monotonicity for the fixed terminal case $(x_{k+N} = 0)$ in Theorem 6.2 and the stability of the RHC in Theorem 6.3 hold for nonlinear and constrained systems, which can be easily proved from well-known results without the help of any references. The stability in the case of a zero terminal constraint in Theorem 6.3 appeared in [MM93] for constrained nonlinear systems.

The invariant ellipsoid for the linear state feedback control in Lemma 6.4 and the feasibility of a solution in Theorem 6.5 appeared in [LKC98]. The finite arriving time to the invariant set in Theorem 6.6 is a discrete-time version of [MM93].

The constrained receding horizon control in Theorem 6.7 appeared in [LKC98]. A cost monotonicity condition for constrained systems in Theorem 6.8 and its stability in Theorem 6.9 appeared in [LKC98]. The stability of a linear feedback control for linear unconstrained systems in Theorem 6.10 appeared in [LKC98] without a proof. The property of an invariant ellipsoid in Theorem 6.11 and an LMI representation for the constrained RHC in Theorems 6.12 appeared in [LKC98]. The RHC with a mixed constraint in Section 6.6 is based on [ZM95].

The constrained output feedback RHC in Theorem 6.13 appeared in [LK01]. The disturbance invariant set in Section 6.7 is introduced in [KG98] [Bla91], and how to find it for the dynamic equation (6.147) is dealt with in [LK99].

6.9 Problems

6.1. Consider the system (6.1) with $\sum_{i=0}^{N-1} u_i^T u_i \leq 1$.

(1) Show that $(x_N - A^N x_0)^T (WW^T)^{-1}(x_N - A^N x_0) \leq 1$, where $W = [A^{N-1}B \cdots B]$.

(2) If A and B are given by

$$A = \begin{bmatrix} 0.9746 & 0.1958 \\ -0.1953 & 0.5839 \end{bmatrix}, \quad B = \begin{bmatrix} 0.8831 \\ -0.8811 \end{bmatrix} \tag{6.162}$$

compute P satisfying the LMI (6.14).

(3) The ellipsoid (6.15) made by P contains the reachable set. Check whether $WW^T \leq P^{-1}$ for the system (6.162).

6.2. Show that the maximal output admissible set is positively invariant.

6.3. Suppose that a system and an off-origin ellipsoid are given by

$$x_{k+1} = -\frac{1}{2}x_k + u_k$$

$$(x_k - 0.5)^2 \leq \alpha$$

Find α such that the ellipsoid is invariant with a linear stable feedback control. Additionally, compute the corresponding state feedback controller inside the ellipsoid.

6.4. Consider polytopic uncertain systems [BGFB94] described by

$$x_{k+1} = \widetilde{A}x_k + \widetilde{B}u_k, \quad -\bar{u}_i \leq u_k \leq \bar{u}_i \tag{6.163}$$

where $x \in R^n$, $u \in R^m$, and

$$(\widetilde{A}, \widetilde{B}) = \sum_{i=1}^{n_p} \eta_i (A_i, B_i), \quad \eta_i \geq 0, \quad \sum_{i=1}^{n_p} \eta_i = 1$$

If we use a state feedback controller $u = Fx$ and a state transformation $x = Vz$, we get

$$z_{k+1} = \widetilde{\Phi} z_k, \quad \widetilde{\Phi} = V^{-1}(\widetilde{A} + \widetilde{B}F)V \tag{6.164}$$

(1) Denote that $|z| = \{|z_i|\}$ and α is a positive column vector. Derive a necessary and sufficient condition to make the set

$$\{x \in R^n | \ |z| \leq \alpha\} \tag{6.165}$$

be invariant with respect to the closed-loop dynamics of (6.164).

(2) Derive a condition to make F robustly stabilizing.

6.5. Considering the following system with constraints on the state and input:

$$x_{i+1} = \begin{bmatrix} 1 & 1 \\ 0 & 1 \end{bmatrix} x_i + \begin{bmatrix} 0 \\ 1 \end{bmatrix} u_i \tag{6.166}$$

where

$$\begin{bmatrix} -1 \\ -1 \end{bmatrix} \leq x_i \leq \begin{bmatrix} 1 \\ 1 \end{bmatrix}, \quad -1 \leq u_i \leq 1 \tag{6.167}$$

Find the invariant set $x^T P x \leq 1$ and compute the corresponding state feedback control which makes the state stay inside the terminal ellipsoid forever.

6.6. Consider the systems (2.287) and (2.288) in the second problem of Problem 2.2. Obtain an RHC to minimize the performance criterion

$$J = \sum_{j=0}^{3} 0.5[x_{1,k+j}^2 + x_{2,k+j}^2 + u_{k+j}^2] \tag{6.168}$$

subject to $x_{1,k+4} = x_{2,k+4} = 0$. The control and states are constrained by (2.290) and (2.291).

6.7. Consider an RHC design problem with the minimum energy performance criterion given by

$$J = \sum_{i=k}^{k+N-1} u_i^T R u_i \tag{6.169}$$

subject to the system $x_{k+1} = A x_k + B u_k$ and the equality constraint $x_{k+N} = 0$. Find the RHC u_k satisfying the constraint $|u_k| \leq \bar{u}_k$.

6.8. Consider a system represented as follows:

$$x_{i+1} = \begin{bmatrix} 0.8676 & -0.3764 \\ -0.0252 & 0.8029 \end{bmatrix} x_i + \begin{bmatrix} -0.3764 \\ -0.8029 \end{bmatrix} u_i \qquad (6.170)$$

where the initial state x_0 is set to $[1.2 \ \ 0.3]^T$, and u_i and x_i should be constrained in the following region:

$$-1 \leq u_i \leq 1 \qquad (6.171)$$

$$\begin{bmatrix} -1.5 \\ -0.3 \end{bmatrix} \leq Gx_i = \begin{bmatrix} 1 & 0 \\ 0 & 1 \end{bmatrix} x_i \leq \begin{bmatrix} 1.5 \\ 0.3 \end{bmatrix} \qquad (6.172)$$

Here, the following performance criterion is considered:

$$J(x_k, k) \triangleq \sum_{j=0}^{4} \{x_{k+j}^T x_{k+j} + u_{k+j}^2\} \qquad (6.173)$$

(1) Find P such that

$$\mathcal{E}_{P,1} = \{x \in \Re^n | x^T P x \leq 1\} \qquad (6.174)$$

is an invariant set with respect to some linear state feedback control.
(2) Compute the constrained receding horizon dual-mode control for the terminal set constraint (6.174).

6.9. Consider an RHC design problem with the following performance index:

$$J = \sum_{i=0}^{4} (x_{1,k+i}^2 + x_{2,k+i}^2 + u_{k+i}^2) + x_{k+5}^T Q_f x_{k+5}$$

with initial state $x_0 = [1.5 \ 0.3]^T$ subject to the systems and the constraints

$$x_{1,k+1} = -x_{1,k} - 0.5x_{2,k}$$
$$x_{2,k+1} = -1.2x_{2,k} - x_{1,k} + u_k$$
$$-2 \leq x_{1,k} \leq 2$$
$$-2 \leq x_{2,k} \leq 2$$
$$-2 \leq u_k \leq 2$$

(1) Find the condition on Q_f satisfying the cost monotonicity.
(2) Represent the condition in an LMI form such that the ellipsoid $\{x_k | x_k^T Q_f x_k \leq 1\}$ is a positive invariance set with a linear state feedback controller $u_k = Hx_k$.
(3) Obtain an RHC in an LMI form.

6.10. Under the conditions of Problem 2.6, consider an RHC design problem for the system (2.302) with the following performance criterion:

$$J = \beta x_{k+N} + \sum_{j=0}^{N-1} (x_{k+j} - u_{k+j}) \tag{6.175}$$

(1) Derive the condition on β satisfying the cost monotonicity.
(2) Check the stability of the closed-loop system controlled by the RHC.

6.11. (1) Find a condition on Q_f satisfying the cost monotonicity for the following system and performance criterion:

$$x_{k+1} = \frac{1}{2}x_k + u_k$$

$$J = \sum_{i=0}^{N-1} (x_{k+i}^2 + u_{k+i}^2) + x_{k+N}^T Q_f x_{k+N}$$

Design an RHC in an LMI form under the following constraints:

$$-1 \leq x_k \leq 1$$
$$-1 \leq u_k \leq 1$$

(2) Find a condition on Q_f satisfying the cost monotonicity for the following system and performance criterion:

$$x_{1,k+1} = x_{1,k} + \frac{1}{2}x_{2,k}$$
$$x_{2,k+1} = -ax_{1,k} - bx_{2,k} + cu_k$$
$$J = \sum_{i=0}^{N-1} (x_{1,k+i}^2 + x_{2,k+i}^2 + u_{k+i}^2) + x_{k+N}^T Q_f x_{k+N}$$

Obtain an RHC in an LMI form under the following constraints:

$$-1 \leq x_{1,k} \leq 1$$
$$-1.5 \leq x_{2,k} \leq 1.5$$
$$-2 \leq u_k \leq 2$$

6.12. Consider an RHC design problem for the system

$$x_{k+1} = Ax_k + Bu_k$$

with the following performance criterion:

$$J(x_k, u_k) = \sum_{i=k}^{k+N-1} \begin{bmatrix} x_i \\ u_i \end{bmatrix}^T \begin{bmatrix} Q & S \\ S^T & R \end{bmatrix} \begin{bmatrix} x_i \\ u_i \end{bmatrix} + x_{k+N}^T Q_f x_{k+N}$$

(1) Derive a condition on Q_f satisfying the cost monotonicity.

(2) Design the RHC in an LMI form under the following constraints:

$$-\bar{x} \leq x_k \leq \bar{x}$$
$$-\bar{u} \leq u_k \leq \bar{u}$$

where $\bar{x}$ and $\bar{u}$ are given vectors.

6.13. Consider the following system:

$$x_{k+1} = \begin{bmatrix} 0.9347 & 0.5194 \\ 0.3835 & 0.8310 \end{bmatrix} x_k + \begin{bmatrix} -1.4462 \\ -0.7012 \end{bmatrix} u_k + \begin{bmatrix} 1 \\ 0 \end{bmatrix} w_k$$
$$y_k = \begin{bmatrix} 0.5 & 0.5 \end{bmatrix} x_k + v_k$$

Assume that input and disturbances are constrained to

$$-1 \leq u_k \leq 1$$
$$-0.05 \leq w_k \leq 0.05$$
$$-0.1 \leq v_k \leq 0.1$$

and the bound on the initial estimation error is $[0.2\ 0.2]^T$.

(1) When the state feedback law F and filter gain L are chosen to be LQ and Kalman filter gains, obtain F and L.

(2) Find W and α to determine the disturbance invariance set.

(3) Design a constrained output feedback RHC and perform the simulation.

6.14. Consider the following system:

$$x_{k+1} = Ax_k + Bu_k + Gw_k$$

and the following min-max problem:

$$J^*(x_{k_0}, k_0, k_1) = \min_{u_k} \max_{w_k} J(x_{k_0}, k_0, k_1)$$

where

$$J(x_{k_0}, k_0, k_1) = \sum_{k=k_0}^{k_1} [x_k^T Q x_k + u_k^T R u_k - \gamma^2 w_k^T w_k] + x_{k_1}^T Q_f x_{k_1}$$

(1) Derive a condition on Q_f to satisfy $J^*(x_{k_0}, k_0, k_1+1) - J^*(x_{k_0}, k_0, k_1) \leq 0$.

(2) Let $V(x_{k_0}, N) = J^*(x_{k_0}, k_0, k_1)$, where $N = k_1 - k_0$. Show that $V(x_{k_0}, N) \geq 0$ for all x_{k_0}.

(3) Show that $V(0, N) = 0$ under the condition of 1).

(4) Assume that $w_k \in W$, $x_k \in X$, and $u_k \in U$, where

$$W = \{w_k \in R^n | \ w_k^T D w_k \leq 1 \ \}$$
$$X = \{x_k \in R^n | \ x^- \leq x_k \leq x^+ \ \}$$
$$U = \{u_k \in R^m | \ u^- \leq u_k \leq u^+ \ \}$$

Consider the following ellipsoid set:

$$E = \{x_k \in R^n | x_k^T Q_f x_k \leq 1\}$$

Derive a condition on Q_f to satisfy $x_{k+1}^T Q_f x_{k+1} \leq x_k^T Q_f x_k$ for all $w_k \in W$ and all $x_k \in \partial E$, the boundary of E.

(5) Show that the constrained RHC under conditions obtained in (1) and (4) guarantee the H_∞ performance.

(6) Show that, in the case of $w_k = 0$, the RHC scheme of (5) guarantees asymptotic stability.

7

Nonlinear Receding Horizon Controls

7.1 Introduction

In real plants, control systems are often represented by nonlinear dynamic systems. Inputs and some states may be unconstrained or constrained depending on the system characteristics. Generally, the finite-time optimal controls can be obtained more easily than infinite-time optimal controls. Since the receding horizon controls are obtained repeatedly over the finite horizon, they may be easier to obtain than the infinite-time optimal controls. The optimization problems over the finite horizon, on which the RHC is based, can be applied to a broad class of systems, including nonlinear systems and time-delayed systems. Thus, the RHC has the same broad applications even for nonlinear systems.

Often, the optimal controls are given as open-loop controls. Since the receding horizon control uses the first control on a horizon, it is automatically a closed-loop control, i.e. a function of the the current state. The cost monotonicity conditions for nonlinear systems can be easily obtained as in linear systems. The terminal cost function becomes a control Lyapunov function (CLF) under the cost monotonicity condition. In this chapter, cost monotonicity conditions for the nonlinear RHC will be investigated according to a terminal equality constraint, a free terminal cost and a terminal invariant set with or without a free terminal cost. And it is also shown that the stability can be guaranteed from this cost monotonicity condition. The receding horizon control provides a systematic way to design a stabilizing control. Receding horizon controls for nonlinear systems are possible not only for minimization criteria, but also for minimaximization criteria.

The organization of this chapter is as follows. In Section 7.2, the RHC with a terminal equality constraint is introduced for nonlinear systems. In particular, a cost monotonicity condition is presented in order to guarantee the closed-loop stability. In Section 7.3, the nonlinear RHC with a fixed terminal set is introduced. In Section 7.4, the nonlinear RHC with a free terminal cost is introduced for unconstrained and constrained systems. We shall also show that the nonlinear RHC can be obtained under a quadratic cost function. In Section, 7.5, the nonlinear RHC with an infinite cost horizon is introduced. In Section 7.6, the minimax RHC is presented for the H_∞ performance criterion.

7.2 Nonlinear Receding Horizon Control with Terminal Equality Constraint

We consider the following discrete time-invariant nonlinear system:

$$x_{i+1} = f(x_i, u_i) \tag{7.1}$$

where the input and the state belong to

$$u_i \in \mathcal{U} \subset \Re^m \tag{7.2}$$

$$x_i \in \mathcal{X} \subset \Re^n \tag{7.3}$$

$f(\cdot, \cdot) \in C^2$ is a nonlinear function describing the system dynamics, and $f(0,0) = 0$. If $\mathcal{U} = \Re^m$ and $\mathcal{X} = \Re^n$, the system (7.1) is called an "unconstrained system". In the case of $\mathcal{U} \subsetneq \Re^m$ or $\mathcal{X} \subsetneq \Re^n$, the system (7.1) with constraints (7.2) and (7.3) is called a "constrained system". For constrained systems, $\mathcal{U}$ is often taken as a compact set including the origin.

We try to minimize the performance criterion such as

$$J(x_k, k, k+N) = \sum_{j=0}^{N-1} g(x_{k+j}, u_{k+j}) \tag{7.4}$$

subject to the terminal equality constraint

$$x_{k+N} = 0 \tag{7.5}$$

where $g(x_k, u_k)$ denotes an intermediate cost. We assume that

$$c_1(\|(x_k, u_k)\|) \le g(x_k, u_k) \le c_2(\|(x_k, u_k)\|) \tag{7.6}$$

where $c_1, c_2 : R^+ \to R^+$ are all strictly increasing functions with $c_1(0) = c_2(0) = 0$. Note that $g(0,0) = 0$. x_{k+j} and u_{k+j} are the predicted state and control at the current time k. x_{k+j} and u_{k+j} satisfy $x_{k+j+1} = f(x_{k+j}, u_{k+j})$ with constraints

$$u_{k+j} \in \mathcal{U}, \quad j = 0, 1, \cdots, N-1 \tag{7.7}$$

$$x_{k+j} \in \mathcal{X}, \quad j = 0, 1, \cdots, N \tag{7.8}$$

The optimal control which minimizes the performance criterion (7.4) is assumed to exist for every initial state $x_k \in \mathcal{X} \subset \Re^n$, which is defined as follows:

$$U_k^* = [u_k^{*T}, u_{k+1}^{*T}, \cdots, u_{k+N-1}^{*T}]^T \qquad (7.9)$$

Then, the receding horizon control law is represented by

$$u_k^* = [1\ 0 \cdots 0]\, U_k^* \qquad (7.10)$$

All controls u_{k+j} on the horizon are obtained as the open-loop controls depending on x_k. Since the receding horizon control uses the first control on the horizon, it is automatically a closed-loop control because it is a function of the current state.

If there is a feasible solution for the system (7.1) with constraints (7.5), (7.7) and (7.8) at the initial time 0 and some horizon N^*, then the next solutions are guaranteed to exist, since at least one feasible solution $u_{[k+1,k+N]} = [u_{[k+1,k+N-1]}\ 0]$ exists. We have the following general result on feasibility.

Lemma 7.1. *If there is a feasible solution on $[k_0,\ k_0 + N^*]$, then there exists a feasible solution on $[k,\ k + N]$ for all $k \geq k_0$ and all $N \geq N^*$.*

Theorem 7.2. *If it is assumed that there is a feasible solution on $[k,\ \sigma]$ for the system (7.1) with constraints (7.5), (7.7), and (7.8), then the optimal performance criterion $J^*(x_k, k, \sigma)$ does not increase monotonically as σ increases.*

$$J^*(x_k, k, \sigma + 1) \leq J^*(x_k, k, \sigma) \qquad (7.11)$$

Proof. This can be proved by contradiction. Assume that u_i^1 and u_i^2 are optimal controls to minimize $J(x_\tau, \tau, \sigma + 1)$ and $J(x_\tau, \tau, \sigma)$ respectively. If (7.11) does not hold, then

$$J^*(x_\tau, \tau, \sigma + 1) > J^*(x_\tau, \tau, \sigma)$$

Replace u_i^1 by u_i^2 up to $\sigma - 1$ and then $u_i^1 = 0$ at $i = \sigma$. In this case, $x_\sigma^1 = 0$, $u_\sigma^1 = 0$, and thus $x_{\sigma+1}^1 = 0$ due to $f(0,0) = 0$. Therefore, the cost for this control is $\bar{J}(x_\tau, \tau, \sigma + 1) = J^*(x_\tau, \tau, \sigma)$. Since this control may not be optimal for $J(x_\tau, \tau, \sigma + 1)$, we have $\bar{J}(x_\tau, \tau, \sigma + 1) \geq J^*(x_\tau, \tau, \sigma + 1)$, which implies that

$$J^*(x_\tau, \tau, \sigma) \geq J^*(x_\tau, \tau, \sigma + 1) \qquad (7.12)$$

This is a contradiction. It is noted that if $J^*(x_k, k, \sigma)$ exists, then $J^*(x_k, k, \sigma + 1)$ exists according to Lemma 7.1. This completes the proof. ∎

Theorem 7.3. *If it is assumed that there is a feasible solution for the system (7.1) with constraints (7.5), (7.7), and (7.8) at the initial time and some horizon N^*, then the receding horizon control (7.10) is asymptotically stable for all horizons $N \geq N^*$.*

Proof. The optimal performance criteria for the current time and the next time are related as

$$J^*(x_k, k, k+N) = g(x_k^1, u_k^1) + J^*(x_{k+1}, k+1, k+N)$$
$$\geq g(x_k^1, u_k^1) + J^*(x_{k+1}, k+1, k+N+1) \quad (7.13)$$

from which we have

$$J^*(x_k, k, k+N) - J^*(x_{k+1}, k+1, k+N+1) \geq l(x_k^1, u_k^1) \quad (7.14)$$

Since the optimal cost $J^*(x_k, k, k+N)$ does not increase monotonically as time increases and is lower bounded, $J^*(x_k, k, k+N)$ approaches a constant value, the limit value. $g(x_k, u_k)$ is always positive, except for $x_k = 0$ and $u_k = 0$. Since the left-hand side in (7.14) approaches zero, $g(x_k, u_k)$ should also approach zero. Therefore, the closed-loop system is asymptotically stable. This completes the proof. ∎

7.3 Nonlinear Receding Horizon Control with Terminal Set Constraints

Consider a receding horizon control of the time-invariant nonlinear systems described by (7.1). An RHC at the current state x_k is obtained by solving the following finite-horizon optimal control problem $J(x_k, k, u_{k+.}, N_k)$ defined by

$$J(x_k, k, u_{k+.}, N_k) = \sum_{j=0}^{N_k} g(x_{k+j}, u_{k+j}) \quad (7.15)$$

subject to the control constraint

$$u_{k+j} \in \mathcal{U}, \ x_{k+j} \in \mathcal{X} \quad (7.16)$$

where $j = 0, 1, 2, \cdots$, and $\mathcal{U} \subset \Re^m$ and $\mathcal{X} \subset \Re^n$ are compact subsets containing the origin in its interior. It is noted that the horizon size N_k at time k depends on the current time k, whereas the horizon size was fixed in previous sections. In this section, we include the control u_k as an argument of the cost function for a better understanding.

Since a solution to the finite horizon optimal control problem $J(x_k, k, u_{k+.}, N_k)$ with a terminal equality constraint $x_{k+N_k} = 0$ rarely exists due to constraints, we can relax this constraint to a fixed terminal set of the form

$$x_{k+N_k} \in \mathcal{E} \quad (7.17)$$

where $\mathcal{E}$ is a terminal set that is some neighborhood of the origin. The constraint (7.17) is a terminal set constraint. If the terminal equality constraint is relaxed in this way, then the receding horizon control is more feasible, but must consider its stabilizing properties inside $\mathcal{E}$. To handle the terminal set constraint, we introduce a dual-mode receding horizon control scheme which uses a locally stabilizing linear control law inside $\mathcal{E}$ and a receding horizon controller outside $\mathcal{E}$. We assume that a locally stabilizing linear feedback control exists in $\mathcal{E}$, which will be obtained later.

The dual-mode receding horizon controller is given below. Suppose that one has an admissible control horizon pair $(u_{k+\cdot}, N_k)$ for $J(x_k, k, u_{k+\cdot}, N_k)$, so that $u_{k+\cdot}$ for $0 \leq j \leq N_k - 1$ steers the state x_k to $\mathcal{E}$ in time N_k. The control $u'_{k+1+\cdot}$, which is defined to be the control $u_{k+\cdot}$ restricted to the interval $[k+1 \ k+N_k]$, steers the system from x_{k+1} to $\mathcal{E}$ in a time $N'_{k+1} = N_k - 1$. This is an admissible control pair $(u_{k+1+\cdot}, N_{k+1})$ for $J(x_{k+1}, k+1, u_{k+1+\cdot}, N_{k+1})$.

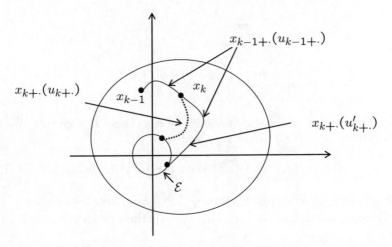

Fig. 7.1. Dual-mode receding horizon control

Procedures to obtain a dual mode receding horizon control

At time $k = 0$, if $x_0 \in \mathcal{E}$, switch to a local linear control, employ the linear feedback law $u_k = \mathcal{K} x_k$ thereafter such that $x_k \in \mathcal{E}$. Otherwise, compute a control horizon pair $(u_{0+\cdot}, N_0)$ for the problem $J(x_0, 0, u_{0+\cdot}, N_0)$. Apply the control u_0 to the real system.

Assume that the control pair $(u_{k-1+\cdot}, N_{k-1})$ is found. At the time k, if $x_k \in \mathcal{E}$, switch to a local linear control, employ the linear feedback control law $\mathcal{K} x_k$ thereafter such that $x_k \in \mathcal{E}$. Otherwise, we compute another admissible control horizon pair $(u_{k+\cdot}, N_k)$, which is better than the preceding control horizon pair in the sense that

$$J(x_k, k, u_{k+\cdot}, N_k) \leq J(x_k, k, u'_{k+\cdot}, N_{k-1} - 1) \tag{7.18}$$

where $u'_{k+.}$ is equal to the restriction of $u_{k-1+.}$ to the interval $[k, \ k-1+N_{k-1}]$ and is an admissible control for $J(x_k, k, u_{k+.}, N_k)$. Finally, we apply the control u_k to the real system.

(7.18) is necessary for a good performance. This is similar to the basic concept of the existing receding horizon controllers which require the repeated solution of an optimal control problem.

Let $X \in \Re^n$ denote the set of initial states which can be steered to $\mathcal{E}$ by a control in $\mathcal{U}$. The above procedure is depicted in Figure 7.1.

The following procedure shows how to construct $\mathcal{W}$. If the linearized system of (7.1) near the origin is stabilizable, then it can be shown that $\mathcal{W}$ is always constructed. From the nonlinear system (7.1), we have the following linear system:

$$x_{i+1} = Ax_i + Bu_i \tag{7.19}$$

where

$$A = \left.\frac{\partial f}{\partial x}\right|_{x=0, u=0} \qquad B = \left.\frac{\partial f}{\partial u}\right|_{x=0, u=0} \tag{7.20}$$

By the assumption of the stabilizability, we can choose H such that $A - BH$ is Hurwitz. $f(x_i, -Hx_i)$ can be represented as

$$f(x_i, -Hx_i) = (A - BH)x_i + \phi(x_i) = A_H x_i + \phi(x_i) \tag{7.21}$$

Clearly, $\phi(x_i)$ satisfies $\phi(0) = 0$ and $\frac{\phi(x_i)}{x_i} \to 0$ as x_i approaches zero.

Thus, for any $0 < \kappa < 1 - \rho^2(A_H)$, $\frac{A_H}{\sqrt{1-\kappa}}$ is Hurwitz and thus there exists a $\hat{P} > 0$ such that

$$\left(\frac{A_H}{\sqrt{1-\kappa}}\right)^T \hat{P}\left(\frac{A_H}{\sqrt{1-\kappa}}\right) - \hat{P} = -M \tag{7.22}$$

for some positive definite matrix M. Letting $P = \frac{\hat{P}}{1-\kappa}$, it follows that

$$A_H^T P A_H - (1-\kappa)P = -M \tag{7.23}$$

Additionally, there exists α_1 such that

$$2x_i^T P\phi + \phi^T P\phi \leq \kappa x_i^T P x_i \tag{7.24}$$

for all $x_i \in \mathcal{E}_{P,\alpha_1}$, where $\mathcal{E}_{P,\alpha_1} = \{x | x^T P x \leq \alpha_1, \ \alpha_1 > 0\}$ since $\frac{\phi(x_i)}{x_i} \to 0$.

It is noted that input or state constraints must be satisfied in $\mathcal{E}_{P,\alpha_1}$. Assume that there exists a region $\mathcal{E}_{P,\alpha_2} = \{x | x^T P x \leq \alpha_2, \ \alpha_2 > 0\}$ such that $-Hx \in \mathcal{U}$ and $x \in \mathcal{X}$ for all $x \in \mathcal{E}_{P,\alpha_2}$. The terminal region $\mathcal{E}_{P,\alpha}$ should be inside $\mathcal{E}_{P,\alpha_1} \cap \mathcal{E}_{P,\alpha_2}$. This can be achieved numerically by adjusting α_1 such that $\alpha_1 \leq \alpha_2$.

Now, we show that $\mathcal{E}_{P,\alpha}$ is an invariant set as follows:

$$
\begin{aligned}
x_{i+1}^T P x_{i+1} - x_i^T P x_i &= f(x_i, -Hx_i)^T P f(x_i, -Hx_i) - x_i^T P x_i \\
&= (A_H x_i + \phi)^T P (A_H x_i + \phi) - x_i^T P x_i \\
&= x_i^T A_H^T P A_H x_i + 2 x_i^T P \phi + \phi^T P \phi - x_i^T P x_i \\
&\leq x_i^T A_H^T P A_H x_i - (1 - \kappa) x_i^T P x_i \\
&= -x_i^T M x_i
\end{aligned}
\tag{7.25}
$$

where $\phi(x_i)$ is shortened to ϕ for simplicity. Finally, we have

$$
x_{i+1}^T P x_{i+1} - x_i^T P x_i \leq -x_i M x_i < 0
\tag{7.26}
$$

for nonzero x_i. Thus, we can say that if x_i enters the set $\mathcal{E}_{P,\alpha} = \{x_i | x_i^T P x_i < \alpha\}$, then the state stays there forever with $u_i = -Hx_i$.

Letting $\beta = \max\limits_{x_i^T P x_i \leq \alpha} \|\phi(x_i)\| / \|x_i\|$, we have

$$
\begin{aligned}
|2 x^T P \phi + \phi^T P \phi| &\leq 2 \|x\| \|P\| \|\phi\| + \|\phi\| \|P\| \|\phi\| \\
&= \left(2 \frac{\|\phi\|}{\|x\|} + \frac{\|\phi\|^2}{\|x\|^2} \right) \|P\| \|x\|^2 \\
&\leq \left(2 \frac{\|\phi\|}{\|x\|} + \frac{\|\phi\|^2}{\|x\|^2} \right) \frac{\|P\|}{\lambda_{\min}(P)} x^T P x \\
&\leq (2\beta + \beta^2) \frac{\|P\|}{\lambda_{\min}(P)} x^T P x
\end{aligned}
\tag{7.27}
$$
$$
\tag{7.28}
$$

where κ can be taken as

$$
(2\beta + \beta^2) \frac{\|P\|}{\lambda_{min}(P)}
$$

What we have done so far is summarized in the following theorem.

Theorem 7.4. *Suppose that the system (7.1) is stabilizable near the origin. The invariant set that satisfies $-Hx \in \mathcal{U}$ and $x \in \mathcal{X}$ for all $x \in \mathcal{E}_{P,\alpha}$ is given by*

$$
\mathcal{E}_{P,\alpha} = \{x_i | x_i^T P x_i \leq \alpha, \ \alpha > 0\}
\tag{7.29}
$$

where P is obtained from (7.23) and α is chosen so that (7.24) is satisfied.

We are now in a position to prove the stability. Instead of (7.15), the following quadratic cost function is considered:

$$
J(x_k, k, u_{k+\cdot}, N_k) = \sum_{j=0}^{N_k} \left[x_{k+j}^T Q x_{k+j} + u_{k+j}^T R u_{k+j} \right]
\tag{7.30}
$$

Theorem 7.5. *Suppose that the linearized systems with respect to (7.1) is stabilizable. Then, for $N_k > 1$, there exists a control horizon pair (u_k, N_k) such that the following equation is satisfied:*

$$J(x_k, k, u_k, N_k) \leq J(x_k, k, u_k', N_{k-1} - 1) \tag{7.31}$$

and there exists constant $0 < \eta < \infty$ satisfying

$$J(x_{k+1}, k+1, u_{k+1}, N_{k+1}) \leq J(x_k, k, u_k, N_k) - \eta \tag{7.32}$$

for all k such that both x_k and x_{k+1} are in $\mathcal{E}_{P,\alpha}$.

Proof. The first statement is already explained in (7.18). Since the trajectory by (u_k, N_k) is the same as one by $(u_{k+1}, N_k - 1)$ on $[k+1, k+N_k]$, we have

$$J(x_k, k, u_k, N_k) - J(x_{k+1}, k+1, u_{k+1}, N_{k+1}) \tag{7.33}$$
$$\geq J(x_k, k, u_k, N_k) - J(x_{k+1}, k+1, u_{k+1}', N_k - 1) \tag{7.34}$$
$$= x_k^T Q x_k + u_k^T R u_k \tag{7.35}$$
$$\geq \frac{\lambda_{\min}(Q)}{\lambda_{\max}(P)} \alpha = \eta \tag{7.36}$$

for all k such that both x_k and x_{k+1} lie in $\mathcal{E}_{P,\alpha}$, as seen in Theorem 6.5. This completes the proof. ∎

Theorem 7.6. *The dual-mode receding horizon controller is asymptotically stabilizing with a region of attraction X. For all $x_0 \in X$, there exists a finite time N^* such that $x_{N^*} \in \mathcal{E}_{P,\alpha}$.*

Proof. The proof of this theorem is the same as Theorem 6.6 ∎

In this section, suboptimal controls satisfying (7.18) were considered. However, we can even use optimal controls for the following performance criterion:

$$\min_{u_i} \sum_{j=0}^{N-1} g(x_{k+j}, u_{k+j}) \tag{7.37}$$

instead of (7.18). This requires heavier computation compared with suboptimal controls, but provides better performance.

7.4 Nonlinear Receding Horizon Control with Free Terminal Cost

We consider the discrete time-invariant nonlinear systems (7.1) and minimize a cost function such as

$$J(x_k, k, k+N) = \sum_{j=0}^{N-1} g(x_{k+j}, u_{k+j}) + h(x_{k+N}) \qquad (7.38)$$

where $h(\cdot)$ is a terminal weighting function. We assume that

$$c_3(\|x\|) \leq h(x) \leq c_4(\|x\|)$$

where $c_3, c_4 : R^+ \to R^+$ are all strictly increasing functions with $c_3(0) = c_4(0) = 0$. Note that $h(0) = 0$.

Theorem 7.7. *Assume that there is a feasible solution for the system (7.1) with constraints (7.7) and (7.8). If the following inequality is satisfied:*

$$g(x_\sigma, \mathcal{K}(x_\sigma)) + h(f(x_\sigma, \mathcal{K}(x_\sigma))) - h(x_\sigma) \leq 0 \qquad (7.39)$$

where $x_\sigma \in \mathcal{X}$, $f(x_\sigma, \mathcal{K}(x_\sigma)) \in \mathcal{X}$, and $\mathcal{K}(x_\sigma) \in \mathcal{U}$, then the optimal performance criterion $J^*(x_k, k, \sigma)$ does not increase monotonically as σ increases,

$$J^*(x_k, k, \sigma + 1) \leq J^*(x_k, k, \sigma) \qquad (7.40)$$

Proof. In the same way as Theorem 3.2, we can prove the nonincreasing monotonicity of the optimal cost. u_i^1 and u_i^2 are optimal controls to minimize $J(x_k, k, \sigma + 1)$ and $J(x_k, k, \sigma)$ respectively. If we replace u_i^1 by u_i^2 up to $\sigma - 1$ and $u_\sigma^1 = \mathcal{K}(x_\sigma)$, then the cost for this control is given by

$$\bar{J}(x_k, \sigma + 1) \triangleq \sum_{j=k}^{\sigma-1} g(x_j^2, u_j^2) + g(x_\sigma^2, \mathcal{K}(x_\sigma^2)) + h(x_{\sigma+1}^2)$$

$$\geq J^*(x_k, \sigma + 1) \qquad (7.41)$$

where the last inequality comes from the fact that this control may not be optimal. The difference between the adjacent optimal costs is less than or equal to zero as

$$J^*(x_k, \sigma + 1) - J^*(x_k, \sigma) \leq \bar{J}(x_k, \sigma + 1) - J^*(x_k, \sigma)$$
$$= g(x_\sigma^2 \mathcal{K}(x_\sigma^2)) + h(x_{\sigma+1}^2) - h(x_\sigma^2) \leq 0 \quad (7.42)$$

where $J^*(x_k, \sigma)$ is given by

$$J^*(x_k, \sigma) = \sum_{j=k}^{\sigma-1} g(x_j^2, u_j^2) + h(x_\sigma^2) \qquad (7.43)$$

This completes the proof. ∎

It is noted that it is difficult to check the nonlinear matrix inequality (7.39) because of the terminal state, x_σ. Therefore, we can give a sufficient condition such as

$$g(x, \mathcal{K}(x)) + h(f(x, \mathcal{K}(x))) - h(x) \le 0 \qquad (7.44)$$

for any $x \in \mathcal{X}$ such that $f(x, \mathcal{K}(x)) \in \mathcal{X}$ and $\mathcal{K}(x) \in \mathcal{U}$.

In the following theorem, using Theorem 7.7, we consider the closed-loop stability of the proposed RHC for nonlinear discrete-time systems.

Theorem 7.8. *Assume that there is a feasible solution for the system (7.1) with constraints (7.7) and (7.8) for all k and some horizon N^*. If the terminal cost function $h(x)$ satisfies the inequality condition (7.44) for some $\mathcal{K}(x)$, then the system (7.1) driven by the receding horizon control obtained from (7.38) is asymptotically stable for $N \ge N^*$.*

Proof. All the proof procedures are the same as Theorem 3.6. This completes the proof. ∎

The cost monotonicity condition (7.39) satisfies a condition for the CLF, which is a proper and positive definite function $h(x_k)$ such that

$$\inf_{u_k} \left[h(x_{k+1}) - h(x_k) + g(x_k, u_k) \right] \le 0 \qquad (7.45)$$

for all $x_k \ne 0$ and $u_k = \mathcal{K}(x_k)$. It is well known that if $h(x_k)$ satisfies (7.45), then the asymptotic stability is achieved.

As in the CLF, $u_k = \mathcal{K}(x_k)$ from the cost monotonicity condition stabilizes the closed-loop systems as follows.

Theorem 7.9. *Assume that there is a feasible solution for the system (7.1) with constraints (7.7) and (7.8). If the terminal cost function $h(x_k)$ satisfies the inequality condition (7.44) for some $\mathcal{K}(x_k)$, then the system (7.1) driven by the control $u_k = \mathcal{K}(x_k)$ is asymptotically stable.*

Proof. A positive definite function $h(x_t)$ can be a Lyapunov function since $h(x_k)$ decreases at least by $-g(x_k, \mathcal{K}(x_k))$ as follows:

$$h(f(x_k, \mathcal{K}(x_k))) - h(x_k) = h(x_{k+1}) - h(x_k) \le -g(x_k, \mathcal{K}(x_k)) < 0 \ (7.46)$$

Since $h(x_k)$ decreases monotonically as time increases and is lower bounded, $h(x_k)$ approaches a constant value, the limit value. $g(x_k, \mathcal{K}(x_k))$ is always positive, except for $x_k = 0$ and $u_k = 0$. Since the left-hand side in (7.46) approaches zero, $g(x_k, u_k)$ should also approach zero. Therefore, the closed-loop system is asymptotically stable.

Thus, the asymptotic stability for the system (7.1) is guaranteed. This completes the proof. ∎

It is noted that the receding horizon control obtained from the cost monotonicity condition integrated with $\mathcal{K}(x_k)$ is asymptotically stabilizing if a feedback control law $u_k = \mathcal{K}(x_k)$ is also asymptotically stabilizing.

We now introduce a terminal invariant ellipsoid $\mathcal{W}_\alpha$ which is defined as

$$\mathcal{E}_{h,\alpha} = \{x_k \in \mathcal{X} \mid h(x_k) \leq \alpha \ \ \mathcal{K}(x_k) \in \mathcal{U}, \ f(x_k, \mathcal{K}(x_k)) \in \mathcal{X}, \ \alpha > 0\} \quad (7.47)$$

Once x_k enters the terminal region $\mathcal{E}_{h,\alpha}$, the state trajectory due to $u_k = \mathcal{K}(x_k)$ would remain in $\mathcal{E}_{h,\alpha}$ because $h(x_{k+N+1}^{\mathcal{K}}) \leq h(x_{k+N}) - g(x_{k+N}, \mathcal{K}(x_{k+N})) \leq \alpha$. Hence, the input and state constraints in (7.63) are all satisfied. We call $\mathcal{E}_{h,\alpha}$ an invariant terminal region. If there exists a control that steers the initial state to $\mathcal{E}_{h,\alpha}$, then there exist feasible controls for all times since the feasible control at the time k can be one of the candidates of the feasible controls at the next time $k + 1$. What we have done so far is summarized as follows.

Lemma 7.10. *If there exists a control that steers the initial state to $\mathcal{E}_{h,\alpha}$, then there exist feasible controls for all times.*

The optimization problem (7.38) can be transformed into

$$\underset{\gamma_1, \gamma_2, \mathcal{K}, h, U_k}{\text{Minimize}} \quad \gamma_1 + \gamma_2$$
$$\text{subject to}$$
$$\begin{cases} x_{k+j|k} \in \mathcal{X}, & j = 0, 1, \cdots, N - 1 \\ u_{k+j|k} \in \mathcal{U}, & j = 0, 1, \cdots, N - 1 \\ \sum_{j=0}^{N-1} g(x_{k+j}, u_{k+j}) \leq \gamma_1, \\ h(x_{k+N|k}) \leq \gamma_2 \end{cases} \quad (7.48)$$

where $h(\cdot)$ satisfies the inequality (7.44). Additionally, another constraint

$$\gamma_2 \leq \alpha \quad (7.49)$$

is required for introducing an invariant ellipsoid $\mathcal{E}_{h,\alpha}$ such as (7.47). It is noted that the problem (7.48) and (7.49) is slightly different from the original problem (7.38).

In the following theorem, we consider the closed-loop stability of the proposed RHC derived from (7.48) and (7.49) for nonlinear systems. According to Lemma 7.10, Theorem 7.8 is changed to the following.

Theorem 7.11. *If there exists a control that steers the initial state to $\mathcal{E}_{h,\alpha}$, then the system (7.1) with the receding horizon control obtained from (7.48) and (7.49) is asymptotically stable.*

In the case of constrained linear systems, the final weighting matrix Q_f can be obtained each time for good performance. However, in the case of nonlinear systems it is not easy to find a final weighting function $h(\cdot)$ each time. Thus, we find $h(\cdot)$ once and then may use it thereafter. The condition $\gamma_2 \leq \alpha$ is required so that $\mathcal{E}_{h,\gamma_2}$ has an invariance property, $\mathcal{E}_{h,\gamma_2} \subset \mathcal{E}_{h,\alpha}$.

If, at the first step, we can find a terminal region $\mathcal{E}_{h,\alpha}$ and a control sequence that steers the current state into $\mathcal{E}_{h,\alpha}$, then feasible solutions exist thereafter. However, in order to obtain good performance such as fast convergence to zero, we will try to find the $\mathcal{E}_{h,\alpha}$ each time. As $\mathcal{E}_{h,\alpha}$ shrinks more and more, the control makes the current state closer to the origin.

Figure 6.5 shows the state trajectory due to the RHC which stems from the optimization (7.48) conceptually. In the proposed RHC, the state at the end of the horizon falls on the boundary of the terminal region $\mathcal{E}_{h,\min(\alpha,\gamma_2)}$. This is because h, $\mathcal{K}$, and $\mathcal{E}_{h,\min(\alpha,\gamma_2)}$ are also free parameters for optimization. However, computing h, $\mathcal{K}$, and $\mathcal{E}_{h,\min(\alpha,\gamma_2)}$ at each time is not easy, especially for nonlinear systems, due to the computational burden or the lack of a suitable numerical algorithm. However, this can be done at least for constrained linear systems, using SDP as shown in Section 6.4.

Now, we consider the following quadratic cost function:

$$\sum_{j=0}^{N-1}\left[x_{k+j}^T Q x_{k+j} + u_{k+j}^T R u_{k+j}\right] + x_{k+N} Q_f x_{k+N} \tag{7.50}$$

for the nonlinear systems (7.1). In a quadratic cost function, the cost monotonicity condition (7.39) is changed to

$$f^T(x_k, -\mathcal{K}(x_k)) Q_f f(x_k, -\mathcal{K}(x_k)) - x_k^T Q_f x_k$$
$$\leq -x_k^T Q x_k - u_k^T R u_k \tag{7.51}$$

In the previous section it was shown that the terminal cost function $h(x)$ is chosen to be a CLF. However, to find $\mathcal{K}(x)$ and $h(x)$ satisfying (7.39) is not so easy, as mentioned before.

Here, the cost function is given in a quadratic form as (7.50) and $\mathcal{K}(x)$ is a local linear state-feedback $-Hx$ as shown in Section 6.5.

To obtain a local linear state-feedback control, we consider the Jacobian linearization (7.19) of the system (7.1) at the origin where A and B are given by (7.20). From A and B in (7.20), H is chosen such that $0 < \rho(A - BH) < 1$, where $\rho(\cdot)$ is the spectral radius of a matrix.

Since $h(x) = x^T Q_f x$, the terminal region is represented by $\mathcal{E}_{Q_f,\alpha} = \{x | x^T Q_f x \leq \alpha\}$. How to determine α is given later in this section.

For any $0 < \kappa < 1 - \rho^2(A_H)$, it is noted that there exist matrices H and Q_f such that

$$\bar{A}_H^T \bar{Q}_f \bar{A}_H - \bar{Q}_f = -[Q + H^T R H] \tag{7.52}$$

where $\bar{A}_H = \frac{A - BH}{\sqrt{1-\kappa}}$, $A_H = A - BH$, and $\bar{Q}_f = \frac{\bar{Q}_f}{\sqrt{1-\kappa}}$. Since $||\phi||/||x|| \to 0$, there exists α_1 such that

$$2x_k^T Q_f \phi + \phi^T Q_f \phi \leq \kappa x_k^T Q_f x_k \tag{7.53}$$

for all $x_k \in \mathcal{E}_{Q_f,\alpha_1}$, where $\mathcal{E}_{Q_f,\alpha_1} = \{x_k | x_k^T Q_f x_k \leq \alpha_1, \ \alpha_1 > 0\}$.

It is noted that input or state constraints must be satisfied in $\mathcal{E}_{Q_f,\alpha_1}$. Assume that there exists a region $\mathcal{E}_{Q_f,\alpha_2} = \{x_k | x_k^T Q_f x_k \leq \alpha_2, \ \alpha_2 > 0\}$ such that $-Hx_k \in \mathcal{U}$ and $x_k \in \mathcal{X}$ for all $x_k \in \mathcal{E}_{Q_f,\alpha_2}$. The terminal region $\mathcal{E}_{Q_f,\alpha}$ should be inside $(\mathcal{E}_{Q_f,\alpha_1} \cap \mathcal{E}_{Q_f,\alpha_2})$. This can be achieved numerically by adjusting α_1 such that $\alpha_1 \leq \alpha_2$.

Now we show that Q_f in (7.52) satisfies the inequality (7.51) as follows:

$$
\begin{aligned}
&f(x_k, -Hx_k)^T Q_f f(x_k, -Hx_k) - x_k^T Q_f x_k \\
&= (A_H x_k + \phi)^T Q_f (A_H x_k + \phi) - x_k^T Q_f x_k \\
&= x_k^T A_H^T Q_f A_H x_k + 2x_k^T Q_f \phi + \phi^T Q_f \phi - x_k^T Q_f x_k \\
&\leq x_k^T A_H^T Q_f A_H x_k - (1 - \kappa) x_k^T Q_f x_k \\
&= -x_k^T (Q + H^T R H) x_k
\end{aligned}
\tag{7.54}
$$

where $\phi(x_k)$ is shortened to ϕ for simplicity.

The cost monotonicity condition and the closed-loop stability of the nonlinear RHC is summarized in the following theorems.

Lemma 7.12. *If Q_f and $\mathcal{E}_{Q_f,\alpha}$ are found out from (7.52) and (7.53), then the cost monotonicity condition (7.51) holds. In that case, the state starting inside $\mathcal{E}_{Q_f,\alpha}$ remains in $\mathcal{E}_{Q_f,\alpha}$ continuously with $u_k = -Hx_k$ while satisfying the input and the state constraints.*

Theorem 7.13. *If Q_f and $\mathcal{E}_{Q_f,\alpha}$ are found out from (7.52) and (7.53), the receding horizon control based on the cost function (7.50) asymptotically stabilizes the system (7.1) if it is feasible at the initial time.*

Given Q_f, the terminal region $\mathcal{E}_{Q_f,\alpha}$ can be determined. The larger that α is, the smaller the control horizon N needed to stabilize the systems. Furthermore, a smaller horizon leads to a lesser computational burden. Therefore, a larger α is preferable.

In the following, we give a numerical algorithm to decide a larger α. The state constraint $\mathcal{X}$ and the input constraint $\mathcal{U}$ are assumed as

$$
\begin{cases}
-u_{lim} \leq u_i \leq u_{lim} \\
-g_{lim} \leq Gx_i \leq g_{lim}
\end{cases}
\tag{7.55}
$$

for $i = 0, 1, \cdots, \infty$. It is guaranteed that the input and state constraints are satisfied in the ellipsoid if the minimum values of the following optimization problems are all positive:

$$
\begin{cases}
\min \ (u_{lim} - Hx_i)^{(i)}, & i = 1, \cdots, m \\
\min \ (u_{lim} + Hx_i)^{(i)}, & i = 1, \cdots, m \\
\min \ (g_{lim} - Gx_i)^{(i)}, & i = 1, \cdots, n \\
\min \ (g_{lim} + Gx_i)^{(i)}, & i = 1, \cdots, n
\end{cases}
\tag{7.56}
$$

subject to $x_i \in \mathcal{E}_{Q_f,\alpha}$, where v^i is an ith component of a vector v and $\mathcal{E}_{Q_f,\alpha} = \{x_i | x_i^T Q_f x_i \leq \alpha, \ \alpha > 0\}$.

The inequality (7.53) is satisfied if the minimum value of the following optimization problem is positive:

$$\begin{cases} \min \ \kappa x_i^T Q_f x_i - 2x_i^T Q_f \phi_i - \phi_i^T Q_f \phi_i \\ \text{subject to} \ \ x_i \in \mathcal{E}_{Q_f,\alpha} \end{cases} \tag{7.57}$$

α is chosen so that the minimum values of optimization problems (7.56) and (7.57) are all positive simultaneously. If there is a negative minimum value, then we should decrease α and solve the optimization problem again. If the minimum values are all positive, then we increase α and solve the optimization problem to get a larger α.

Control Lyapunov Function-based Nonlinear Receding Horizon Contol:

A CLF can be combined with the receding horizon control. These control schemes retain the global stability due to a Lyapunov function and the improvement of the performance due to a receding horizon performance. The CLF can be used with a one-step RHC. In this case, a control u_k is obtained by solving the following optimization problem:

$$\min \ u_k^T u_k \tag{7.58}$$

subject to

$$h(x_{k+1}) - h(x_k) \leq -\sigma(x_k) \tag{7.59}$$

where $\sigma(x_k)$ is a positive definite function. This scheme minimizes the control energy while requiring that h be a Lyapunov function for the closed-loop system and decrease by at least $\sigma(x_k)$ at every point, guarantee the stability. The CLF-based one-step RHC is extended to a general RHC based on the CLF such as

$$\min \ \sum_{j=0}^{N-1} [l(x_{k+j}) + u_{k+j}^T u_{k+j}] \tag{7.60}$$

subject to

$$h(x_{k+1}) - h(x_k) \leq -\sigma(x_k) \tag{7.61}$$

Stability is guaranteed for any horizon N and the one-step RHC is recovered if the horizon size N becomes 1. It is noted that the inequality (7.61) may be imposed over the horizon $[k, \ k+N]$ as

$$h(x_{k+1+j}) - h(x_{k+j}) \leq -\sigma(x_{k+j}) \tag{7.62}$$

for $0 \leq j \leq N - 1$. A terminal ellipsoid can be introduced in the optimization problem (7.48) for better feasibility.

The optimization problem (7.60) and (7.61) may require heavier computation since the constraint (7.61) should be considered each time. However, the optimization problem (7.48) does not require any constraint once $h(\cdot)$ is chosen according to the inequality (7.44).

In the next section we will consider the closed-loop stability of the RHC for constrained nonlinear systems with an infinite cost horizon.

7.5 Nonlinear Receding Horizon Control with Infinite Cost Horizon

In this section a stabilizing RHC for nonlinear discrete-time systems with input and state constraints is proposed. The proposed RHC is based on a finite terminal cost function and an invariant terminal region denoted by Ω_α.

Consider the nonlinear discrete time-invariant system (7.1) with input and state constraints:

$$
\begin{aligned}
u_k &\in \mathcal{U}, \quad k = 0, 1, \cdots, \infty \\
x_k &\in \mathcal{X}, \quad k = 0, 1, \cdots, \infty
\end{aligned}
\tag{7.63}
$$

We assume that $\mathcal{U}$ and $\mathcal{X}$ are convex and compact sets including the origin.

So far, the control horizon has been the same as the cost horizon. However, we can introduce an infinite cost horizon along with a finite control horizon such as the following problem:

$$
\underset{u_k, \ldots, u_{k+N-1}}{\text{Minimize}} \sum_{j=0}^{\infty} g(x_{k+j}, u_{k+j})
\tag{7.64}
$$

subject to

$$
\left\{
\begin{array}{ll}
u_{k+j} \in \mathcal{U}, & j = 0, 1, \cdots, N - 1 \\
u_{k+j} = \mathcal{K}(x_{k+j}) \in \mathcal{U}, & j = N, N + 1, \cdots, \infty \\
x_{k+j} \in \mathcal{X}, & j = 0, 1, \cdots, \infty
\end{array}
\right.
\tag{7.65}
$$

for some $\mathcal{K}(x)$. It is noted that the control is optimized over $[k, \ k + N - 1]$. The controls after $k + N - 1$ are given as stabilizing state feedback controls. These kinds of controls have some advantages, as shown below.

Lemma 7.14. *If the optimization problem (7.65) is feasible at the initial time k, then it is feasible for all times.*

Proof. Suppose that a control law $u_{k+j} = u^*_{k+j}$ exists in the optimization problem (7.65) at time k. Then, at the next time $k + 1$, consider the following control sequence:

$$\left.\begin{array}{r} u_{k+1|k+1} = u_{k+1|k} \\ u_{k+1+j|k+1} = u^*_{k+1+j|k}, \quad j = 1, 2, \cdots, \infty \end{array}\right\} \qquad (7.66)$$

Then, the above control sequence gives a feasible solution for the optimization problem (7.65) at time $k + 1$. Hence, by induction, if the optimization problem (7.65) is feasible at the initial time $k = 0$, then we observe that the optimization problem is always feasible for all times k. This completes the proof. ∎

The cost function (7.64) over the infinite cost horizon is divided into two parts as follows:

$$\sum_{j=0}^{\infty} g(x_{k+j}, u_{k+j}) = \sum_{j=0}^{N-1} g(x_{k+j}, u_{k+j}) + \sum_{j=N}^{\infty} g(x_{k+j}, u_{k+j}) \qquad (7.67)$$

$$= \sum_{j=0}^{N-1} g(x_{k+j}, u_{k+j}) + h(x_{k+N}) \qquad (7.68)$$

where

$$h(x_{k+N}) = \sum_{j=N}^{\infty} g(x_{k+j}, \mathcal{K}(x_{k+j})) \qquad (7.69)$$

It is noted that x_i for $i \geq k + N$ is the trajectory starting from x_{k+N} with the stabilizing control $u_i = \mathcal{K}(x_i)$. As seen in Figure 7.2, the cost function for the final state represents one incurred when the stabilizing controller is applied for the rest of the time after the end of the horizon. From (7.69) we have

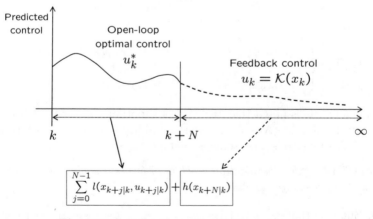

Fig. 7.2. Cost function with infinite cost horizon

$$h(x_{k+N+1}) - h(x_{k+N}) = \sum_{j=N+1}^{\infty} g(x_{k+j}, \mathcal{K}(x_{k+j})) - \sum_{j=N}^{\infty} g(x_{k+j}, \mathcal{K}(x_{k+j}))$$

$$= -g(x_{k+N}, \mathcal{K}(x_{k+N})) \tag{7.70}$$

It follows finally that

$$h(x_{k+N+1}) - h(x_{k+N}) + g(x_{k+N}, \mathcal{K}(x_{k+N})) = 0 \tag{7.71}$$

This is a special case of (7.44). Then, the RHC automatically stabilizes the system due to (7.71) only if input and state constraints are all satisfied. Results in previous sections can be applied to the nonlinear RHC for the infinite cost horizon owing to (7.68) and (7.71).

Calculation of $h(x_\sigma)$ in (7.69) requires the summation of the closed-loop system with the stabilizing controller $u_i = \mathcal{K}(x_i)$ over a possibly long period. For linear systems, this is easily obtained from the algebraic Lyapunov equation, but for nonlinear systems no easily computable method exists, which may further increase the burden of online computation of (7.69).

7.6 Nonlinear Receding Horizon Minimax Control with Free Terminal Cost

We consider the nonlinear system

$$x_{i+1} = f(x_i, u_i, w_i) \tag{7.72}$$

In this section, the input and the state constraints are not assumed. The finite horizon optimal differential game at time k consists of the minimization with respect to u_i, and the maximization with respect to w_i of the cost function

$$J(x_k, u_k, w_k, N) = \sum_{j=0}^{N-1} \left[l(x_{k+j}, u_{k+j}) - \gamma^2 q(w_{k+j}) \right] + h(x_{k+N}) \tag{7.73}$$

For any state $x_k \in \Re^n$, if a feedback saddle-point solution exists, then we denote the solution as $u^*(k+j, x_k)$ and $w^*(k+j, x_k)$. In the following, the optimal value will be denoted by $J^*(x_k, k, k+N)$. In the receding horizon control, at each time k, the resulting feedback control at the state x_k is obtained by optimizing (7.73) and setting

$$u_k = u^*(k, x_k) \tag{7.74}$$

Theorem 7.15. *If the following inequality is satisfied for some $\mathcal{K}(x_k)$ and all $x_k \neq 0$:*

$$l(x_k, \mathcal{K}(x_k)) - \gamma^2 q(w^*(x_k)) + h(f(x_k, \mathcal{K}(x_k), w_k^*)) - h(x_k) < 0 \tag{7.75}$$

then the optimal cost decreases monotonically as follows:

$$J^*(x_k, k, k+N+1) - J^*(x_k, k, k+N) \le 0 \qquad (7.76)$$

for all positive integer N.

Proof. Subtracting $J^*(x_k, k, k+N)$ from $J^*(x_k, k, k+N+1)$ yields

$$J^*(x_k, k, k+N+1) - J^*(x_k, k, k+N)$$
$$= \sum_{j=0}^{N} \left[l(x_{k+j}^1, u_{k+j}^1) - \gamma^2 q(w_{k+j}^1) \right] + h(x_{k+N+1}^1)$$
$$- \sum_{j=0}^{N-1} \left[l(x_{k+j}^2, u_{k+j}^2) - \gamma^2 q(w_{k+j}^2) \right] - h(x_{k+N}^2) \qquad (7.77)$$

where the pair u_{k+j}^1 and w_{k+j}^1 is a saddle-point solution for $J(x_k, k, k+N+1)$ and the pair u_{k+j}^2 and w_{k+j}^2 for $J(x_k, k, k+N)$. If we replace u_{k+j}^1 by u_{k+j}^2 and w_{k+j}^2 by w_{k+j}^1 on $[k \ \ k+N-1]$, then the following inequalities are obtained by $J(u^*, w^*) \le J(u, w^*)$:

$$\sum_{j=0}^{N} \left[l(x_{k+j}^1, u_{k+j}^1) - \gamma^2 q(w_{k+j}^1) \right] + h(x_{k+N+1}^1)$$
$$\le \sum_{j=0}^{N-1} \left[l(\tilde{x}_{k+j}, u_{k+j}^2) - \gamma^2 q(w_{k+j}^1) \right] + l(\tilde{x}_{k+N}, u_{k+N}^2)$$
$$- \gamma^2 q(w_{k+N}^1) + h(\tilde{x}_{k+N+1}) \qquad (7.78)$$

where $\tilde{x}_i$ is a trajectory associated with u_i^2 and w_i^1, $u_\sigma^1 = \mathcal{K}(\tilde{x}_\sigma)$, and $\tilde{x}_{k+N+1} = f(\tilde{x}_{k+N}, \mathcal{K}(\tilde{x}_{k+N}))$. By $J(u^*, w^*) \ge J(u^*, w)$, we have

$$\sum_{j=0}^{N-1} \left[l(x_{k+j}^2, u_{k+j}^2) - \gamma^2 q(w_{k+j}^2) \right] + h(x_{k+N+1}^2)$$
$$\ge \sum_{j=0}^{N-1} \left[l(\tilde{x}_{k+j}, u_{k+j}^2) - \gamma^2 q(w_{k+j}^1) \right] + h(\tilde{x}_{k+N+1}) \qquad (7.79)$$

By using the two inequalities (7.78) and (7.79), it follows from (7.77) that

$$J^*(x_k, k, k+N+1) - J^*(x_k, k, k+N)$$
$$\le l(\tilde{x}_{k+N}, \mathcal{K}(\tilde{x}_{k+N})) - \gamma^2 q(w_{k+N}^1(\tilde{x}_{k+N})) \le 0$$

where the last inequality comes from (7.75). This completes the proof.   ∎

Under the cost monotonicity condition, we can show that the closed-loop system without disturbances is asymptotically stabilized by RH H_∞ control.

Theorem 7.16. *If the cost monotonicity condition (7.75) is satisfied, then the system (7.72) with the RH H_∞ control (7.74) is asymptotically stabilized in the case that there is no disturbance.*

Proof. First, we show that the optimal cost decreases monotonically as follows:

$$
\begin{aligned}
J^*(x_k, k, k+N) &= l(x_k, u_k) - \gamma^2 q(w_k) + J^*(x(k+1; (x_k, k, u_k^*)), k+1, k+N) \\
&\geq l(x_k, u_k) + J^*(x^3(k+1; (x_k, k, u_k^*)), k+1, k+N) \\
&\geq l(x_k, u_k) + J^*(x^3(k+1; (x_k, k, u_k^*)), k+1, k+N+1)(7.80)
\end{aligned}
$$

where u_k is the receding horizon control at time k, x_{k+1}^3 is a state at time $k+1$ when $w_k = 0$, and the last inequality comes from the cost monotonicity condition (7.76). It can be easily seen that $J^*(x_k, k, k+N)$ is bounded below as

$$
\begin{aligned}
J^*(x_k, k, k+N) &= \sum_{j=0}^{N-1} \left[l(x_{k+j}, u_{k+j}) - \gamma^2 q(w_{k+j}) \right] + h(x_{k+N}) \\
&\geq \sum_{j=0}^{N-1} l(x_{k+j}^3, u_{k+j}^3) + h(x_{k+N}^3) \geq 0
\end{aligned}
$$

where x_{k+j}^3 and u_{k+j}^3 are the state and the input in the case that the disturbances on $[k\ ,\ k+N]$ are all zero. Therefore, $J(x_k, k, k+N)$ is bounded below, $J^*(x_k, k, k+N) \geq 0$. $J^*(x_k, k, k+N)$ approaches a constant number. It follows from (7.80) that

$$
l(x_k, u_k) \longrightarrow 0 \text{ as } k \longrightarrow 0 \tag{7.81}
$$

which implies that x_k goes to zero. This completes the proof. ∎

Under the cost monotonicity condition, we can show that the H_∞ performance bound is met.

Theorem 7.17. *If the cost monotonicity condition (7.75) is satisfied, then the nonlinear minimax RH control (7.74) for the system (7.72) satisfies*

$$
\frac{\sum_{k=0}^{\infty} l(x_k, u_k)}{\sum_{k=0}^{\infty} q(w_k)} \leq \gamma^2 \tag{7.82}
$$

Proof. Subtracting $J^*(x_k, k, k+N)$ from $J^*(x_{k+1}, k+1, k+N+1)$ yields

$$
\begin{aligned}
J^*(x_k, k, k+N) &= l(x_k, u_k) - \gamma^2 q(w_k) + J^*(x_{k+1}, k+1, k+N) \\
&\geq l(x_k, u_k) - \gamma^2 q(w_k) + J^*(x_{k+1}, k+1, k+N+1) \tag{7.83}
\end{aligned}
$$

where the last inequality comes from the cost monotonicity condition. Thus, we have

$$0 \leq J^*(x_\infty, \infty, \infty + N) - J^*(x_0, 0, N)$$
$$\leq \sum_{k=0}^{\infty} \left[-l(x_k, u_k) + \gamma^2 q(w_k) \right]$$

which implies the inequality (7.82). It is noted that x_0 is assumed to be zero. This completes the proof. ∎

7.7 References

The cost monotonicity condition for the terminal equality constraint in Theorem 7.2 is obtained directly from the optimal principle. The nonlinear RHC with the terminal equality constraint in Theorem 7.3 appeared in [KG88]. The invariant set for a nonlinear system in Theorem 7.4 appeared in [LHK04]. The dual-mode RHC in Theorem 7.5 and its stability in Theorem 7.6 appeared in [MM93]

The cost monotonicity for nonlinear systems in Theorem 7.7 and the stability of the nonlinear RHC under the cost monotonicity condition in Theorem 7.8 appeared in [NMS96a, MS97] for unconstrained systems and in [NMS96b] for constrained systems.

The nonlinear state feedback control in Theorem 7.9 appeared in [PND00]. The stability of the constrained RHC in Theorem 7.11 appeared in [LKC98].

The invariant set for a linearized system in Theorem 7.12 and the stability and the feasibility for the constrained nonlinear RHC in Theorem 7.13 are discrete versions of [LHK04].

The cost monotonicity of the saddle point in Theorem 7.15, the stability of the nonlinear H_∞ RHC in Theorem 7.16 and the preserved H_∞ norm of the nonlinear H_∞ RHC in Theorem 7.17 appeared in [Gyu02].

7.8 Problems

7.1. Find a final weighting functional $h(\cdot)$ satisfying the cost monotonicity condition (7.39) for the following nonlinear systems and the performance criterions:

(1)

$$x_{k+1} = \frac{1}{2}x_k + \frac{1}{4}\tanh(x_k) + u_k$$
$$J = \sum_{i=0}^{N-1} (x_{k+i}^2 + u_{k+i}^2) + h(x_{k+N})$$

(2)

$$x_{1,k+1} = x_{1,k} + \frac{1}{2}x_{2,k}$$

$$x_{2,k+1} = -a\sin(x_{1,k}) - bx_{2,k} + cu_k$$

$$J(x_k, k, k+N-1) = \sum_{i=0}^{N-1}(x_{1,k+i}^2 + x_{2,k+i}^2 + u_{k+i}^2) + h(x_{k+N})$$

7.2. Consider a nonlinear RHC with a terminal equality constraint, $x_{k+N} = 0$, for the following nonlinear system and the quadratic performance criterion:

$$x_{i+1} = f(x_i, u_i) = f_1(x_i) + f_2(x_i)u_i \tag{7.84}$$

and

$$J(x_k, k, k+N-1) = \sum_{i=0}^{N-1}(x_{k+i}^T Q x_{k+i}^T + u_{k+i}^T R u_{k+i}^T) \tag{7.85}$$

where $R > 0$ and $Q > 0$.

(1) Let $f_c(x) = f(x, k(x))$ where $k(x)$ is an RHC and $f_g(x) = f(x, \phi(k(x)))$ where $\phi(x)$ is a function $\phi : \Re^p \rightarrow \Re^p$ satisfying

$$\phi(x)^T R\phi(x) > (1+\alpha)(\phi(x) - x)^T R(\phi(x) - x), \ \forall x \neq 0 \tag{7.86}$$

and α is a real number such that

$$f_2^T(x)\frac{\partial^2 J^*(\beta f_c(x) + (1-\beta)f_g(x), k+1, k+N-1)}{\partial x^2}f_2(x) \leq \alpha R \tag{7.87}$$

for all x and $0 \leq \beta \leq 1$. Show that $J^*(f(x_k, \phi(k(x_k))), k+1, k+N-1) < J^*(x_k, k, k+N-1)$. Show that the closed-loop system with $\phi(\cdot)$ is asymptotically stable.

(2) Assume that the function $\psi : \Re^n \rightarrow \Re^p$ satisfies $x^T Q x > \frac{1}{2}(1+\alpha)\psi^T(x)R\psi(x)$ for $x \neq 0$. Show that $J^*(f(x_k, k(x_k) + \psi(x_k)), k+1, k+N-1) < J^*(x_k, k, k+N-1)$. Show that the closed-loop system with $\psi(\cdot)$ is asymptotically stable.

7.3. Consider the following nonlinear system:

$$x_{1,k+1} = x_{1,k} - 0.5x_{2,k} \tag{7.88}$$

$$x_{2,k+1} = -1.2x_{2,k} + x_{1,k} + x_{1,k}^2 x_{2,k} + u_k \tag{7.89}$$

(1) Find a linearized model $x_{k+1} = Ax_k + Bu_k$ around the origin and choose a feedback gain H such that $A - BH$ is Hurwitz.

(2) For some κ satisfying $0 < \kappa < 1 - \rho^2(A - BH)$, find P such that $(A - BH)^T P(A - BH) - (1 - \kappa)P = -I$.

(3) Find α such that (7.24) is satisfied in the positive invariant ellipsoid $\{x|x^T P x \leq \alpha\}$.

7.4. Suppose that a system and an off-origin ellipsoid are given by

$$x_{k+1} = -\frac{1}{2}x_k + \frac{1}{4}x_k^2 + u_k$$
$$(x_k - 0.5)^2 \leq \alpha$$

Find α such that the ellipsoid is invariant with a linear stable feedback control. Additionally, compute the corresponding state feedback controller inside the ellipsoid.

7.5. (1) For the ϕ_i defined in (7.21), show that there exists a Ψ such that

$$|\phi_i| \leq \Psi|x_i| \tag{7.90}$$

where $|v|$ of a vector v is $\{|v_i|\}$.
(2) For a given vector $\bar{x}$, show that the set $\{x \in \Re^n | |x_i| \leq \bar{x}\}$ is invariant if $(|A - BK| + \Psi)\bar{x} < \bar{x}$.
(3) For a given vector $\bar{u}$, consider the following input constraint:

$$|u| \leq \bar{u} \tag{7.91}$$

where $|u| = \{|u_i|\}$. Find the condition to guarantee the feasibility when there is an input constraint.

7.6. Consider an RHC design problem with a free terminal cost. We assume that a local stabilizing control law $\mathcal{K}$, an associated Lyapunov function F, and a level set $\mathcal{X}_F$ are known. More specifically, we assume that there exist $\mathcal{K}$, F, $\alpha > 0$ such that

$$\mathcal{X}_F = \{x \in \mathcal{X}|F(x) \leq \alpha\} \tag{7.92}$$
$$g(x_\sigma, \mathcal{K}(x_\sigma)) + F(f(x_\sigma, \mathcal{K}(x_\sigma))) - F(x_\sigma) \leq 0 \tag{7.93}$$

Define

$$\mathcal{X}_0 = \{x_k \in \mathcal{X}|x_{k+N} \in \mathcal{X}_F\}$$

The terminal cost is selected as

$$h(x_k) := \begin{cases} F(x_k), & x_k \in \mathcal{X}_F \\ \alpha, & x_k \notin \mathcal{X}_F \end{cases}$$

(1) For each $x_0 \in \mathcal{X}_0$, show that the closed-loop system by the control law from the above RHC scheme is asymptotically stable.
(2) Show that $\mathcal{X}_0$ is invariant and a domain of attraction for the origin.

(3) If the assumption

$$F(x_k) = \sum_{j=0}^{\infty} g(x_{k+j}, u_{k+j}), \quad \forall x_k \in \mathcal{X}$$

is satisfied for all $x_0 \in \mathcal{X}_N$, show that the RHC generates a control sequence that solves the optimal control problem with the infinite horizon performance index.

7.7. Consider a real system $x_{k+1} = f_r(x_k, u_k)$ and its model $x_{k+1} = f(x_k, u_k)$. Assume that f_r and f are twice continuously differentiable, $\|f_r(x_k, u_k) - f(x_k, u_k)\| \le \beta\|(x_k, u_k)\|$ is satisfied, and f is Lipschitz continuous. Consider the following linearized system of f with a local linear control law Kx:

$$x_{k+1} = Ax_k$$

where $A = \frac{\partial f}{\partial x_k}(0,0) + \frac{\partial f}{\partial u_k}(0,0)K$. We define P to satisfy

$$\frac{1}{2}x_k^T P x_k = \frac{1}{2}\sum_{k=0}^{\infty}[x_k^T Q x_k + x_k^T K^T R K x_k]$$

For a given $\beta_L > 0$, show that the set $W_\alpha = \{x_k \in R^n | x_k^T P x_k \le \alpha\}$ is invariant if there exist an $\alpha > 0$ and a K such that

$$x_k^T P[f(x_k, Kx_k) - Ax_k] - \frac{1}{2}x_k^T Q x_k + \frac{\lambda_{\min}(P^{-1/2}QP^{-1/2})}{2}x_k^T P x_k$$
$$+ \beta_L[1 + \|K\|_P]x_k^T P x_k \le 0$$

where $\|\cdot\|_P$ is a weighted norm.

7.8. Consider the following nonlinear system:

$$x_{k+1} = f(x_k, u_k)$$

with the performance index (7.38). The state and input vectors are subject to the constraints $x_k \in \mathcal{X}$ and $u_k \in \mathcal{U}$, where $\mathcal{X}$ and $\mathcal{U}$ are closed subsets of R^n and R^m respectively.

Assume

$$\lim_{\|(x_k, u_k)\| \to 0} \sup_{k \ge 0} \frac{\|\tilde{f}(x_k, u_k)\|}{\|(x_k, u_k)\|} = 0$$

where $\tilde{f}(x_k, u_k) = f(x_k, u_k) - [A_k x_k + B_k u_k]$ and

$$A = \left.\frac{\partial f(x_k, u_k)}{\partial x_k}\right|_{x_k = u_k = 0} \quad ; \quad B = \left.\frac{\partial f(x_k, u_k)}{\partial u_k}\right|_{x_k = u_k = 0}$$

(1) When a stable linear state feedback control law $u_k = Kx_k$ for linearized system is applied, the closed-loop system is given by

$$x_{k+1} = f_c(x_k) \tag{7.94}$$

where $f_c(x_k) = f(x_k, Kx_k)$. Let $\tilde{f}_c(x) = f_c(x_k) - [A_k + B_k K]x_k$. Show that

$$\lim_{\|x_k\| \to 0} \sup_{k \geq 0} \frac{\|\tilde{f}_c(x_k)\|}{\|x_k\|} = 0$$

(2) In (7.38), assume that terminal cost is selected as

$$h(x_k) = \begin{cases} \sum_{j=k}^{\infty} g(x_{j,k}^c, Kx_{j,k}^c), & x_k \in X(K) \\ \infty, & \text{otherwise} \end{cases}$$

where $x_{j,k}^c$ is the solution of (7.94) at time $j \geq k$ when x_k is given at time k and $X(K)$ is a stability region for the closed-loop system (7.94). Show that the optimal performance index satisfies the following relations:

$$J^*(x_k, k, k + N + 1) \leq J^*(x_k, k, k + N) \tag{7.95}$$
$$J^*(x_{k+1}, k + 1, k + 1 + N) \leq J^*(x_k, k, k + N) \tag{7.96}$$

(3) Show that the closed loop system controlled by the RHC is asymptotically stable.

7.9. Consider a class of nonlinear systems:

$$x_{k+1} = f(x_k, u_k, w_k) = a(x_k) + b(x_k)u_k + g(x_k)w_k \tag{7.97}$$
$$z_k = \begin{bmatrix} h(x_k) \\ u_k \end{bmatrix} \tag{7.98}$$

We design an RHC based on the following min-max difference game:

$$V(x_k, N) = \min_u \max_w J(x_k, N) \tag{7.99}$$

where

$$J(x_k, N) = \sum_{j=0}^{N-1} (\|z_{k+j}\|^2 - \gamma^2 \|w_{k+j}\|^2) + V_f(x_{k+N}) \tag{7.100}$$

(1) Derive a condition on V_f such that

$$V(x_k, N + 1) - V(x_k, N) \leq 0 \tag{7.101}$$

(2) Under the condition of (1), if $h(x_k) \in \mathcal{K}_\infty$, then show that the closed-loop system is input-to-state stable with respect to w_k, where $\mathcal{K}_\infty$ is a set of strictly increasing functions with zero initial condition which approach ∞.

(3) Consider the following nonlinear discrete-time switched systems:

$$x_{k+1} = f_p(x_k, u_k, w_k), \qquad f_p \in \mathcal{F} \tag{7.102}$$

where $\mathcal{F} = \{f_p | p \in \mathcal{P}\}$ indexed by the parameter p takes on values in the set of parameters $\mathcal{P}$ which is either a finite set or a compact subset of some finite-dimensional, normed linear space. For each piece-wise switching signal as $\sigma(\cdot) : [0, \infty) \to \mathcal{P}$, we can define the switched nonlinear systems $x_{k+1} = f_\sigma(x_k, u_k, w_k)$. Also, we define the set of switching times as $T = \{k_0, k_1, k_2, ...\}$ and assume $V_f \in \mathcal{K}_\infty$. Using the result of (2) for each p, derive the condition to make the closed-loop switched systems controlled by the RHC satisfy

$$\lim_{i \to \infty} ||x_{k_i}|| = 0 \tag{7.103}$$

(4) Using the results of (2) and (3) for each p, derive the condition to make the closed-loop switched systems controlled by the RHC satisfy

$$\lim_{i \to \infty} ||x_i|| = 0 \tag{7.104}$$

7.10. Consider the RHC design problem mentioned in Problem 7.9.

(1) Define $J_\infty^*(x_0)$ as the optimal cost of the infinite horizon difference game with the following performance index:

$$J_\infty(x_0) = \sum_{i=0}^{\infty} (||z_i||^2 - \gamma^2 ||w_i||^2)$$

Show that

$$\lim_{k \to \infty} V(x_k, N) = J_\infty^*(x_0)$$

(2) Derive a condition to make the RHC guarantee a global stability result.

7.11. Consider the following nonlinear system and the performance criterion:

$$x_{k+1} = \frac{1}{2}x_k + \frac{1}{4}\tanh(x_k) + w_k \tag{7.105}$$

$$J = \sum_{i=0}^{N-1} (x_{k+i}^2 + u_{k+i}^2 - w_{k+i}^2) + h(x_{k+N}) \tag{7.106}$$

Find a state feedback control $\mathcal{K}(x_k)$ satisfying (7.75).

7.12. If $h(x_k)$ satisfies the following inequality:

$$h(x_k) > \min_{u_k} \max_{w_k} \left[l(x_k)^T l(x_k) + u_k^T u_k - w_k^T w_k \right.$$

$$\left. + h(f(x_k, \mathcal{K}(x_k), w_k)) \right] \tag{7.107}$$

then show that $h(x_k)$ also satisfies the inequality (7.75).

7.13. Consider the input uncertainty in the nonlinear H_∞ control (7.74), as shown in Figure 7.3. Assume that input uncertainty Δ is given by

$$\widetilde{x}_{k+1} = \widetilde{a}(\widetilde{x}_k) + \widetilde{b}(\widetilde{x}_k)\widetilde{u}_k \tag{7.108}$$

$$\widetilde{y}_k = \widetilde{\phi}(\widetilde{x}_k) \tag{7.109}$$

The input $\widetilde{u}_k$ and output $\widetilde{y}_k$ of the uncertainty Δ satisfies

$$\mathcal{V}(\widetilde{x}_{k+1}) - \mathcal{V}(\widetilde{x}_k) \le \widetilde{u}_k^T \widetilde{y}_k - \rho \widetilde{u}_k^T \widetilde{u}_k$$

where $\mathcal{V}(x_k)$ is some nonnegative function (this is called the dissipative property) and $\rho \in R$. Consider the following feedback interconnection with the input uncertainty Δ:

$$\widetilde{u}_k = u_k^{RHC}$$
$$u_k = -\widetilde{y}_k$$

Find the condition on ρ so that the closed-loop system has an H_∞ norm bound γ for all dynamic input uncertainty Δ.

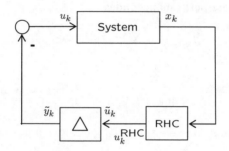

Fig. 7.3. Feedback interconnection of Problem 7.13

A

Matrix Equality and Matrix Calculus

A.1 Useful Inversion Formulae

The following relation is very useful even though it looks simple.

Theorem A.1. *If $(I + GK)$ is nonsingular, then the following relation is satisfied:*

$$(I + GK)^{-1}G = G(I + KG)^{-1} \tag{A.1}$$

Proof.

$$(I + GK)^{-1}G = G(I + KG)^{-1}$$
$$\longleftrightarrow \quad G(I + KG) = (I + GK)G$$
$$\longleftrightarrow \quad G + GKG = G + GKG$$

This completes the proof. ∎

Theorem A.2. *If it is assumed that A, C, and $(A + BCD)$ are nonsingular, then we have the formula as*

$$(A + BCD)^{-1} = A^{-1} - A^{-1}B(DA^{-1}B + C^{-1})^{-1}DA^{-1} \tag{A.2}$$

Proof. Multiplying the right side of (A.2) by $A + BCD$, we have

$$(A + BCD)[A^{-1} - A^{-1}B(DA^{-1}B + C^{-1})^{-1}DA^{-1}]$$
$$= I + BCDA^{-1} - B(DA^{-1}B + C^{-1})^{-1}DA^{-1}$$
$$\quad - BCDA^{-1}B(DA^{-1}B + C^{-1})^{-1}DA^{-1}$$
$$= I + BCDA^{-1} - B(I + CDA^{-1}B)(DA^{-1}B + C^{-1})^{-1}DA^{-1}$$
$$= I + BCDA^{-1} - BC(C^{-1} + DA^{-1}B)(DA^{-1}B + C^{-1})^{-1}DA^{-1} = I$$

Postmultiplying the right side of (A.2) by $A + BCD$, we have the same result. This completes the proof. ∎

It is noted that Theorem A.2 is often called the matrix inversion lemma.

- The following matrix equality is obtained as

$$(I + GK)^{-1} = I - G(KG + I)^{-1}K \tag{A.3}$$
$$= I - GK(GK + I)^{-1} \tag{A.4}$$

by replacing A, B, C, and D by I, G, I, and K respectively in Theorem A.2. It is noted that the equality (A.4) comes from (A.1). Additionally, if C in (A.2) is not invertible, then we can write

$$(A + BCD)^{-1} = A^{-1} - A^{-1}B(CDA^{-1}B + I)CDA^{-1} \tag{A.5}$$

- The inverse of some matrices such as $(I + KG)$ and $(C^{-1} + DA^{-1}B)$ is required on the right side of (A.1) and (A.2). It can be seen that the nonsingularity of $I + GK$ implies that of $I + KG$ in Theorem A.1. It is also shown in Theorem A.2 that the nonsingularity of $(A + BCD)$ implies that of $(C^{-1} + DA^{-1}B)$ under the assumption that A and C are nonsingular. The nonsingularity of these matrices is checked as follows:

$$\begin{bmatrix} I & 0 \\ -DA^{-1} & I \end{bmatrix} \begin{bmatrix} A & -B \\ D & C^{-1} \end{bmatrix} \begin{bmatrix} I & A^{-1}B \\ 0 & I \end{bmatrix} = \begin{bmatrix} A & 0 \\ 0 & C^{-1} + DA^{-1}B \end{bmatrix} \tag{A.6}$$

$$\begin{bmatrix} I & BC \\ 0 & I \end{bmatrix} \begin{bmatrix} A & -B \\ D & C^{-1} \end{bmatrix} \begin{bmatrix} I & 0 \\ -CD & I \end{bmatrix} = \begin{bmatrix} A + BCD & 0 \\ 0 & C^{-1} \end{bmatrix} \tag{A.7}$$

Using the product rule for determinants, the matrices decomposition in (A.6) and (A.7) give

$$\det \begin{bmatrix} A & -B \\ D & C^{-1} \end{bmatrix} = \det A \det(C^{-1} + DA^{-1}B) \tag{A.8}$$

$$= \det C^{-1} \det(A + BCD) \tag{A.9}$$

which tells us

$$\det(C^{-1} + DA^{-1}B) \neq 0 \iff \det(A + BCD) \neq 0 \tag{A.10}$$

under the assumption that A and C are invertible. By replacing A, B, C, and D by I, G, I, and K respectively, it is also shown that the nonsingularity of $I + GK$ implies that of $I + KG$ in Theorem A.1.

In the following theorem, it is shown that matrices satisfying the equality constraint can be parameterized.

Theorem A.3. *Suppose that $A \in \Re^{m \times n}$ is of full rank and $n \geq m$. Solutions X to the problem $AX = Y$ can be parameterized as $X = A^{\perp}Y + MV$ for arbitrary matrices V, where $A^{\perp}$ is the right inverse of A and M spans the null space of A and its columns are orthogonal to each other. The number of rows of V is equal to the dimension of the null space of A.*

A.2 Matrix Calculus

The differentiation involving vectors and matrices arises in dealing with control and estimation problems. Here, we introduce useful formulae and relations.

- The derivative of the row vector $x \in \Re^n$ with respect to the scalar α is defined as

$$\frac{\partial x}{\partial \alpha} = \begin{bmatrix} \dfrac{\partial x_1}{\partial \alpha} & \dfrac{\partial x_2}{\partial \alpha} & \cdots & \dfrac{\partial x_n}{\partial \alpha} \end{bmatrix}. \tag{A.11}$$

- The derivative of a scalar y with respect to the vector $x \in \Re^n$ is defined as

$$\frac{\partial y}{\partial x} = \begin{bmatrix} \dfrac{\partial y}{\partial x_1} \\ \dfrac{\partial y}{\partial x_2} \\ \vdots \\ \dfrac{\partial y}{\partial x_n} \end{bmatrix} \tag{A.12}$$

- For $x \in \Re^n$ and $y \in \Re^m$ vector, the derivative of y with respect to x is defined as

$$\frac{\partial y}{\partial x} = \begin{bmatrix} \dfrac{\partial y_1}{\partial x_1} & \dfrac{\partial y_2}{\partial x_1} & \cdots & \dfrac{\partial y_m}{\partial x_1} \\ \dfrac{\partial y_1}{\partial x_2} & \dfrac{\partial y_2}{\partial x_2} & \cdots & \dfrac{\partial y_m}{\partial x_2} \\ \vdots & \vdots & \ddots & \cdots \\ \dfrac{\partial y_1}{\partial x_n} & \dfrac{\partial y_2}{\partial x_n} & \cdots & \dfrac{\partial y_m}{\partial x_n} \end{bmatrix} \tag{A.13}$$

- Some useful formulae involving matrix and vector derivatives are:

$$\frac{\partial y^T x}{\partial x} = \frac{\partial x^T y}{\partial x} = y$$

$$\frac{\partial Ax}{\partial x} = \frac{\partial x^T A^T}{\partial x} = A^T$$

$$\frac{\partial y^T Ax}{\partial x} = \frac{\partial x^T A^T y}{\partial x} = A^T y$$

$$\frac{\partial y^T f(x)}{\partial x} = \frac{\partial f^T(x) y}{\partial x} = f_x^T y$$

$$\frac{\partial y^T(x) f(x)}{\partial x} = \frac{\partial f^T(x) y(x)}{\partial x} = f_x^T y + y_x^T f$$

$$\frac{\partial x^T Ax}{\partial x} = Ax + A^T x$$

$$\frac{\partial (x-y)^T A(x-y)}{\partial x} = A(x-y) + A^T(x-y) \tag{A.14}$$

If A is symmetrical, then simplified relations are obtained:

$$\frac{\partial x^T A x}{\partial x} = 2Ax$$

$$\frac{\partial^2 x^T A x}{\partial x^2} = 2A$$

$$\frac{\partial (x-y)^T A(x-y)}{\partial x} = 2A(x-y)$$

$$\frac{\partial^2 (x-y)^T A(x-y)}{\partial x^2} = 2A \tag{A.15}$$

- The Hessian of a scalar function α with respect to a vector x is defined as

$$\frac{\partial^2 \alpha}{\partial x^2} = \left[\frac{\partial^2 \alpha}{\partial x_i \partial x_j} \right] \tag{A.16}$$

- The Taylor series expansion of a scalar function $\alpha(x)$ about x_0 is

$$\alpha(x) = \alpha(x_0) + (\frac{\partial \alpha}{\partial x})^T (\alpha - \alpha_0) + \frac{1}{2}(\alpha - \alpha_0)^T (\frac{\partial^2 s}{\partial x^2})^T (\alpha - \alpha_0)$$
$$+ \text{ terms of higher order} \tag{A.17}$$

B

System Theory

B.1 Controllability and Observability

In this appendix we provide a brief overview of some concepts in system theory that are used in this book. We start by introducing the notions of controllability and observability. Consider the n-dimensional linear state-space model

$$x_{k+1} = Ax_k + Bu_k \tag{B.1}$$

Definition B.1. *The state-space model (B.1) is said to be controllable at time t if there exists a finite $N > 0$ such that for any x_k and any x^* there exists an input $u_{[k,k+N]}$ that will transfer the state x_k to the state x^* at time $k + N$.*

By definition, the controllability requires that the input u_k be capable of moving any state in the state-space to any other state in a finite time. Whether the system is controllable is easily checked according to the following well-known theorem.

Theorem B.2. *The system (B.1) is controllable if and only if there exists a finite N such that the following specific matrix, called the controllability Grammian, is nonsingular:*

$$W_c = \sum_{i=0}^{N-1} A^i BB^T A^{iT} \tag{B.2}$$

The smallest number N is called the controllability index, denoted by n_c in this book.

Now, in order to introduce the notion of observability, consider the n-dimensional state-space model:

$$\begin{aligned} x_{k+1} &= Ax_k + Bu_k \\ y_k &= Cx_k \end{aligned} \tag{B.3}$$

Definition B.3. *The system (B.3) is said to be observable at k if there exists a finite N such that the knowledge of the input $u_{[k,k+N]}$ and the output $y_{[k,k+N]}$ over the time interval $[k, k + N]$ uniquely determines the state x_k.*

We have the following well known theorem.

Theorem B.4. *The system (B.3) is observable at time k if and only if there exists a finite N such that the following matrix observability Grammian is nonsingular:*

$$W_o = \sum_{i=0}^{N-1} A^{iT} C^T C A^i \tag{B.4}$$

The smallest number N is called the observability index, denoted by n_o in this book.

Note that observability is dependent only on A and C. The following theorem shows the invariance of the observability.

Theorem B.5. *The observability of (A, C) implies that of $(A - BH, L)$ where $C^T C + H^T R H = L^T L$.*

Proof: From the observability condition,

$$\text{rank}\left(\begin{bmatrix} \lambda I - A \\ C \end{bmatrix} \right) = n$$

for any real λ, is satisfied. Since the addition of a row vector does not change the rank for full rank matrices, the rank of the following matrix is preserved:

$$\text{rank}\left(\begin{bmatrix} \lambda I - A \\ C \\ R^{\frac{1}{2}} H \end{bmatrix} \right) = n.$$

If the third-row block is added to the first-row block with a pre-multiplication by $BR^{-\frac{1}{2}}$, the above matrix is changed as

$$\text{rank}\left(\begin{bmatrix} \lambda I - A + BH \\ C \\ R^{\frac{1}{2}} H \end{bmatrix} \right) = n.$$

From the fact that $\text{rank}(V) = n$ with $V \in \Re^{n \times m}$ and $n \geq m$ implies $\text{rank}(V^T V) = n$,

$$\text{rank}\left(\begin{bmatrix} \lambda I - A + BH \\ C \\ R^{\frac{1}{2}} H \end{bmatrix}^T \begin{bmatrix} \lambda I - A + BH \\ C \\ R^{\frac{1}{2}} H \end{bmatrix} \right) = n,$$

$$\text{rank}\left(\begin{bmatrix} \lambda I - A + BH \\ L \end{bmatrix}^T \begin{bmatrix} \lambda I - A + BH \\ L \end{bmatrix} \right) = n,$$

$$\text{rank}\left(\begin{bmatrix} \lambda I - A + BH \\ L \end{bmatrix} \right) = n.$$

Therefore, $(A - BH, L)$ is observable. This completes the proof. ∎

The method of adjoint equations has been used in the study of terminal control problems. In addition, the duality between control and estimation can be precisely defined in terms of the adjoint system.

Suppose $G : S_1 \to S_2$ is a linear system and S_1 and S_2 are Hilbert spaces. The adjoint system is a linear system $G^* : S_2 \to S_1$ that has the property

$$< Gw, y >_{S_2} = < w, G^*y >_{S_1} \tag{B.5}$$

for all $w \in S_1$ and all $y \in S_2$. To determine the adjoint of G in the Hilbert space $l_2[0 \; k]$, consider inputs $w \in l_2[0 \; k]$ and represent the linear system $z = Gw$ by

$$z_k = \sum_{i=0}^{k} G_{k-i} w_i \tag{B.6}$$

For any $y \in l_2[0 \; k]$,

$$< Gw, y > = \sum_{j=0}^{k}\sum_{i=0}^{j} y_k G_{j-i} w_i = \sum_{j=0}^{k}\sum_{i=0}^{j} (G_{j-i}^T y_k)^T w_i = < w, \eta > \tag{B.7}$$

where

$$\eta_j = \sum_{i=0}^{j} G_{j-i}^T y_k \tag{B.8}$$

The $l_2[0 \; k]$ adjoint of G is therefore the system G^* defined by

$$(G^*y)_j = \sum_{i=0}^{j} G_{j-i}^T y_k \tag{B.9}$$

In the case of a state-space system,

$$G_{j-k} = CA^{j-k}B \tag{B.10}$$

and it is easily shown that $\eta_j = \sum_{i=0}^{j} G_{j-i}^T y_k$ satisfies

$$p_{i+1} = A^T p_i + C^T y_i \tag{B.11}$$

$$\eta_i = B^T p_i \tag{B.12}$$

which is therefore a state-space realization of the adjoint system G^*.

B.2 Stability Theory

Here, we give some definitions on stability, attractivity, asymptotic stability, uniformly asymptotic stability, and exponential stability.

Consider the following nonlinear discrete time-varying system:

$$x_{i+1} = f(x_i) \tag{B.13}$$

where $x_i \in \Re^n$ and $f(\cdot) : \Re^n \to \Re^n$ for all $i \geq i_0$. The solution evaluated at the ith time with the initial time i_0 and the initial state x_0 is denoted by $x(i; i_0, x_0)$.

A point $x_e \in \Re^n$ is called an *equilibrium* of the system (B.13), if

$$f(x_e) = x_e, \ \forall i \geq i_0$$

Definition B.6. *(Stability) The equilibrium x_e of the system (B.13) is stable if, for each $\epsilon > 0$ and each $i_0 \leq 0$, there exists a $\delta = \delta(\epsilon, i_0)$ such that*

$$||x_0|| < \delta(\epsilon, i_0) \to ||s(i, i_0, i_0)|| < \epsilon, \ \forall i \geq i_0$$

Definition B.7. *(Attractivity) The equilibrium x_e is attractive if, for each $k_0 \geq 0$, there exists an $\eta(k_0)$ such that*

$$||x_0|| < \eta(i_0) \to ||s(i, i_0, i_0)|| \to 0 \ as \ i \to \infty$$

Definition B.8. *(Asymptotic stability) The equilibrium x_e is asymptotically stable if it is stable and attractive.*

Definition B.9. *(Exponential stability) The equilibrium x_e is exponentially stable if there exist constants $\eta, a > 0$ and $p < 1$ such that*

$$||x_0|| < \eta, \ i_0 \geq 0 \to ||s(i_0 + i, i_0, i_0)|| \leq a||x_0||p^i, \ \forall i \geq 0$$

Theorem B.10. *(Lyapunov theorem) Suppose that $V_1, V_2: \Re^+ \to \Re^+$ are continuous nondecreasing functions such that $V_1(s)$ and $V_2(s)$ are positive for $s > 0$ and $V_1(0) = V_2(0) = 0$. If there exists a continuous function $V : \Re^n \to \Re^+$ satisfying*

$$V_1(||x||) \leq V(x) \leq V_2(||x||) \tag{B.14}$$

then the system is asymptotically stable.

Theorem B.11. *(LaSalle's theorem) Let $\Omega \subset D$ be a compact set that is positively invariant with respect to the system $x_{k+1} = f(x_k)$. Let $V : D \to R$ be a function such that $V_{k+1} - V_k \leq 0$ in Ω. Let E be the set of all points in Ω where $V_{k+1} - V_k = 0$. Let M be the largest invariant set in E. Then every solution starting in Ω approaches M as $t \to \infty$.*

B.3 Lyapunov and Riccati Matrix Equations

Lyapunov and Riccati equations arise in many contexts in linear systems theory. The characteristics of the two equations are summarized in this appendix.

Lyapunov Matrix Equation

We consider the following linear matrix equation, called a Lyapunov equation:

$$K - A^T K A = Q \tag{B.15}$$

where Q is symmetric. If K exists and is unique, then K should be symmetric.

Theorem B.12. *The Lyapunov equation (B.15) has the following properties:*

(1) A unique solution K exists if and only if $\lambda_i(A)\lambda_j(A) \neq 1$ for all i and j.
(2) If A is a stable matrix, then a unique solution K exists and can be expressed as

$$K = \sum_{i=0}^{\infty} A^{iT} Q A^i \tag{B.16}$$

(3) If A is a stable matrix and Q is positive (semi) definite, then the unique solution K is positive (semi) definite.
(4) If A is a stable matrix, $Q = C^T C$ is positive semidefinite, and the pair $\{A, C\}$ is observable, then the unique solution K is positive definite.
(5) Suppose that $Q = C^T C$ is positive semidefinite and the pair $\{A, C\}$ is observable. A is a stable matrix if and only if the unique solution K is positive definite.

Proof. We prove only the fifth statement.
($\Longleftarrow$)
Consider $V(x_i) = x_i^T K x_i$. Then we have

$$V(x_{i+1}) - V(x_i) \overset{\triangle}{=} x_{i+1}^T K x_{i+1} - x_i^T K x_i \tag{B.17}$$
$$= x_i^T (A^T K A - K) x_i = -x_i^T Q x_i \leq 0 \tag{B.18}$$

along any trajectory of $x_{i+1} = A x_i$.

We show that if Q is positive semidefinite and $\{A, C\}$ is observable, then $x_i^T Q x_i$ cannot be identically zero along any nontrivial trajectory of $x_{i+1} = A x_i$, which is proved as follows. Since

$$x_i^T Q x_i = x_0^T A^{iT} C^T C A^i x_0 = \|C A^i x_0\|^2 \tag{B.19}$$

and $\{A, C\}$ is observable, all rows of $C A^i$ are linearly independent and thus $C A^i x_0 = 0$ for all $i = 0, 1, 2, \cdots$ if and only if $x_0 = 0$. By Lasalle's theorem,

$x_{i+1} = Ax_i$ is stabilized.

$(\Longrightarrow)$

If the zero state of $x_{i+1} = Ax_i$ is asymptotically stable, then all the eigenvalues of A are inside the open unit disc. Consequently, for any Q, there exists a unique matrix K satisfying the Lyapunov equation. K can be expressed as

$$K = \sum_{i=0}^{\infty} A^{iT} Q A^i \tag{B.20}$$

Consider

$$x_0^T K x_0 = \sum_{i=0}^{\infty} x_0 A^{iT} C^T C A^i x_0 = \sum_{i=0}^{\infty} \|C A^i x_0\|^2 \tag{B.21}$$

Since $\{A, C\}$ are observable, we have $CA^i x_0 \neq 0$ for all i unless $x_0 = 0$. Hence, we conclude that $x_0^T K x_0$ for all x_0 and K is positive definite. This completes the proof.   ∎

Riccati Matrix Equations

Now we shall be concerned with the ARE

$$K = A^T K A - A^T K B [R + B^T K B]^{-1} B^T K A + Q \tag{B.22}$$

where $Q = C^T C$.

Theorem B.13. *Suppose (A, C) is observable. Then (A, B) is stabilizable if and only if*

(1) There is a unique positive definite solution to the ARE.
(2) The closed-loop system $A_{cl} := A - BL_\infty$ is asymptotically stable.

Proof.
$(\Longrightarrow)$

If (A, C) is observable, then

$$\left((A - BL), \begin{bmatrix} C \\ R^{\frac{1}{2}} L \end{bmatrix} \right)$$

is also observable for any L by Theorem B.5 in Appendix B.2. Now, stabilizability implies the existence of a feedback control $u_i = -Lx_i$, so that

$$x_{i+1} = (A - BL)x_i$$

is asymptotically stable. The cost of such a control on $[i, \infty]$ is

$$J_i = \frac{1}{2} x_i^T S x_i$$

where S is the limiting solution of

$$S_i = (A - BL)^T S_{i+1} (A - BL) + L^T RL + Q \tag{B.23}$$

The optimal closed-loop system has an associated cost on $[i, \infty]$ of

$$J_i^* = \frac{1}{2} \sum_{i=i_0}^{\infty} [x_i^T Q x_i + u_i^T R u_i] = \frac{1}{2} x_i^T K x_i \leq \frac{1}{2} x_i^T S^* x_i \tag{B.24}$$

where S^* is the limiting solution to Riccati Equation (B.23). Therefore, $C x_i \to 0$, and since $|R| \neq 0, u_i \to 0$. Then it follows that

$$\begin{bmatrix} C x_i \\ C x_{i+1} \\ \vdots \\ C x_{i+n-1} \end{bmatrix} = \begin{bmatrix} C \\ CA \\ \vdots \\ CA^{n-1} \end{bmatrix} x_i \longrightarrow 0$$

and so observability of (A, C) requires $x_i \to 0$. Hence, the optimal closed-loop system is asymptotically stable.

Write (B.22) as

$$K = (A - BL_\infty)^T K (A - BL_\infty) + \begin{bmatrix} C \\ R^{\frac{1}{2}} L_\infty \end{bmatrix}^T \begin{bmatrix} C \\ R^{\frac{1}{2}} L_\infty \end{bmatrix} \tag{B.25}$$

where $L_\infty = -R^{-1} B^T (I + KBR^{-1} B^T)^{-1} KA$ is the optimal feedback. Then, (B.25) is a Lyapunov equation with

$$\left((A - BK_\infty), \begin{bmatrix} C \\ R^{\frac{1}{2}} K_\infty \end{bmatrix} \right)$$

observable and $(A - BK_\infty)$ stable. Therefore, there is a unique positive definite solution S^* to (B.22).

$(\Longleftarrow)$

If $x_{i+1} = (A - BK_\infty) x_i$ is asymptotically stable, then (A, B) is certainly stabilizable. This completes the proof. ∎

Theorem B.14. *Consider Riccati Equation (B.22) with $\{A, B\}$ stabilizable and $\{A, C\}$ observable. Let $U \in \Re^n$ and $V \in \Re^n$ be any matrices that form a basis for the stable generalized eigenspace of the pair of matrices*

$$\begin{bmatrix} I & BR^{-1}B^T \\ 0 & A^T \end{bmatrix} \begin{bmatrix} U \\ V \end{bmatrix} \Lambda = \begin{bmatrix} A & 0 \\ -Q & I \end{bmatrix} \begin{bmatrix} U \\ V \end{bmatrix} \tag{B.26}$$

where Λ is an $n \times n$ matrix with all its eigenvalues inside the unit circle. Then

(i) V is invertible.

(ii) UV^{-1} is the unique stabilizing solution to (B.22).

C

Random Variables

C.1 Random Variables

The conditional probability density function (p.d.f.) can be written by using
the joint and marginal pdfs as follows:

$$p_{xy}(x|y) = \frac{p_{x,y}(x,y)}{p_y(y)} \tag{C.1}$$

where $x \in \Re^n$ and $y \in \Re^m$.

We introduce two important properties in random variables. Two random
variables x and y are said to be independent if $p_{x,y}(x,y) = p_x(x)p_y(y)$, i.e.
$p_{x,y}(x|y) = p_x(x)$ from (C.1). Two random variable x and y are said to be
uncorrelated if $E[xy] = E[x]E[y]$. Note that independent random variables
x and y are always uncorrelated, but the converse is not necessarily true. In
other words, it is possible for X and Y to be uncorrelated but not independent.

Suppose that a random variable x has a mean m_x and covariance P_x. Then
we can write, for a symmetric constant matrix $P_x \geq 0$,

$$
\begin{aligned}
E[x^T M x] &= E[(x - m_x + m_x)^T M (x - m_x + m_x)] \\
&= E[(x - m_x)^T M (x - m_x)] \\
&\quad + m_x^T M E[x - m_x] + E[(x - m_x)^T] M m_x + m_x^T M m_x \\
&= m_x^T M m_x + \operatorname{tr}(E[M(x - m_x)(x - m_x)^T]) \\
&= m_x^T M m_x + \operatorname{tr}(M P_x)
\end{aligned}
\tag{C.2}
$$

Note that $E[m_x - x] = 0$.

Theorem C.1. *Given measurement z, the means square estimation minimiz-
ing $E[(x - \hat{x}(z))^T(x - \hat{x}(z))|z]$ is given by*

$$\hat{x}(z) = E[x|z] \tag{C.3}$$

Proof. Expanding and rearranging $E[(x - \hat{x}(z))^T (x - \hat{x}(z))]$, we have

$$
\begin{aligned}
E[(x - \hat{x}(z))^T (x - \hat{x}(z))|z] &= E[x^T x - x^T \hat{x}(z) - \hat{x}^T(z)x + \hat{x}^T(z)\hat{x}(z)|z] \\
&= E[x^T x|z] - E[x|z]^T \hat{x}(z) - \hat{x}^T(z)E[x|z] \\
&\quad + \hat{x}^T(z)\hat{x}(z) \\
&= E[x^T x|z] + [\hat{x}(z) - E[x|z]]^T [\hat{x}(z) - E[x|z]] \\
&\quad - E[x|z]^T E[x|z]
\end{aligned}
\tag{C.4}
$$

Note that (C.4) is a function of z. It can be easily seen that the minimum is achieved at $\hat{x}(z) = E[x|z]$ in (C.4). This completes the proof. ∎

C.2 Gaussian Random Variable

In this appendix we collect several facts and formulas for a multivariable Gaussian density function that is used for deriving the Kalman filter and the LQG control.

The random variables z_1, z_2, $\cdots$, and z_n are said to be jointly Gaussian if their joint pdf is given by

$$
p_z(z) = \frac{\exp\{-\frac{1}{2}(z - m_z)^T P_z^{-1}(z - m_z)\}}{(2\pi)^{\frac{n}{2}}|\det(P_z)|^{\frac{1}{2}}}
\tag{C.5}
$$

where $z = [z_1 \, z_2 \, \cdots \, z_n]$, $m_z = E[z]$, $P_z = E[(z - m_z)(z - m_z)^T]$.

It is assumed that $\{z_i\}$ consists of x_1, x_2, $\cdots$, x_n, y_1, y_2, $\cdots$, and y_m. Thus, (C.5) can be given by

$$
p_{x,y}(x, y) = \frac{\exp\{-\frac{1}{2}(z - m_z)^T P_z^{-1}(z - m_z)\}}{(2\pi)^{\frac{n+m}{2}}|\det(P_z)|^{\frac{1}{2}}}
\tag{C.6}
$$

where

$$
z = \begin{bmatrix} x \\ y \end{bmatrix}, \quad m_z = \begin{bmatrix} m_x \\ m_y \end{bmatrix}, \quad P_z = \begin{bmatrix} P_x & P_{xy} \\ P_{yx} & P_y \end{bmatrix}
$$

In terms of P_x, P_y, P_{xy} and P_{yx}, P_z^{-1} can be represented as

$$
P_z^{-1} = \begin{bmatrix} P_{11} & P_{12} \\ P_{12}^T & P_{22} \end{bmatrix}
$$

where

$$
\begin{aligned}
P_{11} &= (P_x - P_{xy}P_y^{-1}P_{yx})^{-1} = P_x^{-1} + P_x^{-1}P_{xy}P_{22}P_{yx}P_x^{-1} \\
P_{12} &= -P_{11}P_{xy}P_y^{-1} = -P_x^{-1}P_{xy}P_{22} \\
P_{22} &= (P_y - P_{yx}P_x^{-1}P_{xy})^{-1} = P_y^{-1} + P_y^{-1}P_{yx}P_{11}P_{xy}P_y^{-1}
\end{aligned}
$$

As can be seen in (C.6), the pdf of jointly Gaussian random variables is completely determined by the mean and the covariance. Marginal pdfs generated from (C.6) are also Gaussian. If any n-dimensional jointly Gaussian random variables are transformed linearly, they are also Gaussian.

An estimation problem is often based on the conditional expectation. The following theorem is very useful for obtaining a conditional expectation of a joint Gaussian pdf.

Theorem C.2. *The conditional probability of x given y of a joint Gaussian pdf is written as*

$$p_{x|y}(x|y) = \frac{1}{\sqrt{(2\pi)^n |P_{x|y}|}} \exp[-\frac{1}{2}(x - m_{x|y})^T P_{x|y}^{-1}(x - m_{x|y})] \qquad (C.7)$$

where

$$m_{x|y} = E[x|y] = m_x + P_{xy}P_y^{-1}(y - m_y) \qquad (C.8)$$

$$P_{x|y} = P_x - P_{xy}P_y^{-1}P_{yx} \qquad (C.9)$$

Proof. Using the fact of $p_{xy}(x|y) = \frac{p_{x,y}(x,y)}{p_y(y)} = \frac{p_z(z)}{p_y(y)}$, we have

$$p_{x|y}(x|y) = \frac{1}{\sqrt{(2\pi)^n \frac{|P_z|}{|P_y|}}} \exp\left[-\frac{1}{2}(z - m_z)^T \begin{bmatrix} P_{11} & P_{12} \\ P_{12}^T & P_{22} - P_y^{-1} \end{bmatrix} (z - m_z)\right]$$

From facts $P_{12} = -P_x^{-1}P_{xy}P_{22}$ and $P_{22} = P_y^{-1} + P_y^{-1}P_{yx}P_{11}P_{xy}P_y^{-1}$, the terms inside the exponential function can be factorized as

$$(z - m_z)^T \begin{bmatrix} P_{11} & P_{12} \\ P_{12}^T & P_{22} - P_y^{-1} \end{bmatrix} (z - m_z)$$

$$= (x - m_x)^T P_{11}(x - m_x) + 2(x - m_x)^T P_{12}(y - m_y)$$

$$+ (y - m_y)^T (P_{22} - P_y^{-1})(y - m_y)$$

$$= \left[(x - m_x) - P_{xy}P_y^{-1}(y - m_y)\right]^T P_{11}\left[(x - m_x) - P_{xy}P_y^{-1}(y - m_y)\right]$$

where it can be seen that

$$m_{x|y} = E[x|y] = m_x + P_{xy}P_y^{-1}(y - m_y)$$

$$P_{x|y} = P_{11}^{-1} = P_x - P_{xy}P_y^{-1}P_{yx}$$

What remains to be proved is $|P_{z|y}| = |P_z|/|P_y|$. We try to prove $|P_z| = |P_{z|y}||P_y|$ as follows:

$$\det(P_z) = \det(\begin{bmatrix} P_x & P_{xy} \\ P_{yx} & P_y \end{bmatrix}) = \det(\begin{bmatrix} I & P_{xy}P_y^{-1} \\ 0 & I \end{bmatrix} \begin{bmatrix} P_x & P_{xy} \\ P_{yx} & P_y \end{bmatrix})$$

$$= \det(\begin{bmatrix} P_x - P_{xy}P_y^{-1}P_{yx} & 0 \\ P_{yx} & P_y \end{bmatrix}) = \det(P_x - P_{xy}P_y^{-1}P_{yx})\det(P_y)$$

$$= \det(P_{x|y})\det(P_y)$$

This completes the proof.   ∎

Note that uncorrelatedness implies independence in the case of Gaussian random variables.

Theorem C.3. *Let x, y, and z be joint Gaussian random variables. If y and z are independent, then $E(x|y,z) = E(x|y) + E(x|z) - m_x$ holds. For general random variables y and z, the relation*

$$E(x|y,z) = E(x|y,\tilde{z}) = E(x|y) + E(x|\tilde{z}) - m_x$$

holds, where $\tilde{z} = z - E(z|y)$.

Proof. In the case that y and z are independent, augmenting y and z into one vector $\xi = \begin{bmatrix} y^T & z^T \end{bmatrix}^T$ yields

$$E(x|y,z) = E(x|\xi) = m_x + P_{x\xi}P_\xi^{-1}(\xi - m_\xi)$$
$$P_{x\xi} = E(x\xi^T) = \begin{bmatrix} P_{xy} & P_{xz} \end{bmatrix}$$
$$P_\xi = \begin{bmatrix} P_y & 0 \\ 0 & P_z \end{bmatrix}$$

from which we have

$$E(x|\xi) = m_x + P_{xy}P_y^{-1}(y - m_y) + P_{xz}P_z^{-1}(z - m_z)$$
$$= E(x|y) + E(x|z) - m_x$$

Now we consider the case that y and z are general, i.e. the independence of two random variables is not required. From Theorem C.2, $\tilde{z}$ can be written as

$$\tilde{z} = z - E\{z|y\} = z - m_z - P_{zy}P_y^{-1}(y - m_y) \tag{C.10}$$

which means that x, y and $\tilde{z}$ are jointly Gaussian. The mean of $\tilde{z}$, i.e. $m_{\tilde{z}}$, is

$$m_{\tilde{z}} = E\{z - E\{z|y\}\} = E\{z\} - E\{E\{z|y\}\} = 0 \tag{C.11}$$

It can be seen that y and $\tilde{z}$ are uncorrelated as follows:

$$E\{(y - m_y)(\tilde{z} - m_{\tilde{z}})^T\} = E\{(y - m_y)\tilde{z}^T\} = E\{y\tilde{z}^T\}$$
$$= E\{yz^T\} - E\{yE\{z^T|y\}\},$$
$$= E\{yz^T\} - E\{yz^T\} = 0$$

It follows then that $E(x|y,\tilde{z})$ is written as

$$E(x|y,\tilde{z}) = m_x + P_{xy}P_y^{-1}(y - m_y) + P_{x\tilde{z}}P_{\tilde{z}}^{-1}\tilde{z} \tag{C.12}$$
$$= E(x|y) + E(x|\tilde{z}) - m_x$$

where

$$P_{x\tilde{z}} = P_{xz} - P_{xy}P_y^{-1}P_{yz}$$
$$P_{\tilde{z}} = P_z - P_{zy}P_y^{-1}P_{yz}$$

Through a trivial and tedious calculation for C.12, we obtain

$$E(x|y,\tilde{z}) = E(x|y,z)$$

This completes the proof.   ∎

C.3 Random Process

Lemma C.4. *Suppose that $u_i \in \Re^m$ is the wide sense stationary (WSS) input signal with power spectral density $U(e^{jw})$ and $y_i \in \Re^p$ is the output signal coming through the system $H(e^{jw})$. The power of the output signal can be represented as*

$$E[y_i^T y_i] = \text{tr}(E[y_i y_i^T]) = \frac{1}{2\pi} \int_0^{2\pi} \text{tr}(H(e^{j\omega})U(e^{j\omega})H^*(e^{j\omega}))\ d\omega \quad (C.13)$$

It is noted that

$$E\left[\lim_{N \to \infty} \frac{1}{N} \sum_{i=i_0}^{i_0+N} y_i^T y_i \right] = E[y_i^T y_i] \quad (C.14)$$

D

Linear Matrix Inequalities and Semidefinite Programming

D.1 Linear Matrix Inequalities

Now, we give a brief introduction to LMIs. The basic results of this appendix are from [BLEGB94].

We call $F(x)$ an LMI if it has the following form:

$$F(x) = F_0 + \sum_{i=1}^{m} x_i F_i > 0 \qquad (\text{D.1})$$

where the variable $x = [x_1 \cdots x_m]^T \in \mathcal{R}^m$ and the symmetric matrices $F_i = F_i^T \in \mathcal{R}^{n \times n}$, $0, \cdots, m$ are given. The inequality in (D.1) means that $F(x)$ is positive definite. It is noted that $F(x)$ depends affinely on x.

The LMI (D.1) is convex on x, i.e. the set $\{x | F(x) > 0\}$ is convex. Even though the LMI (D.1) may seem to have a specialized form, it can represent several nonlinear matrix inequalities, such as quadratic inequalities, including Riccati inequalities, matrix norm inequalities, and constraints that arise in control theory.

Multiple LMIs $F^1(x) > 0, \cdots, F^p(x) > 0$ can be expressed as the following single LMI:

$$\text{diag}(F^1(x), \cdots, F^p(x)) > 0$$

Therefore, it is usual to make no distinction between a set of LMIs and a single LMI, i.e. "the LMIs $F^1(x) > 0, \cdots, F^p(x) > 0$" also means "the LMI $\text{diag}(F^1(x), \cdots, F^p(x)) > 0$".

Nonlinear matrix inequalities can be represented as LMIs, which is possible using the Schur complement. Consider the following LMI:

$$\begin{bmatrix} Q(x) & S(x) \\ S(x)^T & R(x) \end{bmatrix} > 0 \qquad (\text{D.2})$$

where $Q(x)$ and $R(x)$ are symmetric. Note that $Q(x)$, $R(x)$, and $S(x)$ depend affinely on x. This LMI is equivalent to the following matrix inequalities:

$$R(x) > 0, \quad Q(x) - S(x)R(x)^{-1}S^T(x) > 0 \tag{D.3}$$

or equivalently:

$$Q(x) > 0, \quad R(x) - S^T(x)Q(x)^{-1}S(x) > 0 \tag{D.4}$$

The above equivalences can be easily proved by using the following matrix decomposition:

$$\begin{bmatrix} Q & S \\ S^T & R \end{bmatrix} = \begin{bmatrix} I & 0 \\ S^TQ^{-1} & I \end{bmatrix} \begin{bmatrix} Q & 0 \\ 0 & R - S^TQ^{-1}S \end{bmatrix} \begin{bmatrix} I & Q^{-1}S \\ 0 & I \end{bmatrix}$$
$$= \begin{bmatrix} I & SR^{-1} \\ 0 & I \end{bmatrix} \begin{bmatrix} Q - SR^{-1}S^T & 0 \\ 0 & R \end{bmatrix} \begin{bmatrix} I & 0 \\ R^{-1}S^T & I \end{bmatrix} \tag{D.5}$$

In other words, the set of nonlinear inequalities (D.3) or (D.4) can be represented as the LMI (D.2). If $R(x)$ and $Q(x)$ are set to unit matrices, we have the useful formula

$$I - SS^T > 0 \iff I - S^TS > 0 \tag{D.6}$$

If $R(x)$ and $S(x)$ are set to a unit matrix and a vector x, we have

$$xx^T < Q \iff x^TQ^{-1}x < I \tag{D.7}$$

where Q is assumed to be nonsingular.

The matrix norm constraint $\|Z(x)\|_2 < 1$, where $Z(x) \in \mathcal{R}^{p \times q}$ and depends affinely on x, is represented as the LMI

$$\begin{bmatrix} I & Z(x) \\ Z(x)^T & I \end{bmatrix} > 0 \tag{D.8}$$

from the fact that $\lambda_{max}(Z(x)^TZ(x)) < 1$ implies $y^TZ(x)^TZ(x)y - y^Ty < 0$ for all y, i.e. $Z(x)^TZ(x) < I$.

The constraint $c(x)^TP(x)^{-1}c(x) < 1$, $P(x) > 0$, where $c(x) \in \mathcal{R}^n$ and $P(x) = P(x)^T \in \mathcal{R}^{n \times n}$ depend affinely on x, is expressed as the LMI

$$\begin{bmatrix} P(x) & c(x) \\ c(x)^T & 1 \end{bmatrix} > 0 \tag{D.9}$$

More generally, the constraint

$$\mathrm{tr}(S(x)^TP(x)^{-1}S(x)) < 1, \ P(x) > 0 \tag{D.10}$$

where $P(x) = P(x)^T \in \mathcal{R}^{n \times n}$ depend affinely on x, is handled by introducing a new matrix variable $X = X^T \in \mathcal{R}^{p \times p}$, and the LMI (in x and X)

$$\mathrm{tr}(X) < 1, \ \begin{bmatrix} X & S(x)^T \\ S(x) & P(x) \end{bmatrix} > 0 \tag{D.11}$$

Even though the variable x is given by a vector in the LMI (D.1), matrix variables can be used. Consider the following continuous version of the Lyapunov inequality

$$A^T P + PA < 0 \tag{D.12}$$

where $A \in \mathcal{R}^{n \times n}$ is given and $P = P'$ is the variable which should be found. In this case, instead of representing the LMI explicitly in the form $F(x) > 0$, we will refer to (D.12) as the LMI where the matrix P is the variable. Of course, the Lyapunov inequality (D.12) can be expressed as the form (D.1) by letting $P_1, \cdots, P_m$ be a basis for symmetric $n \times n$ matrices and taking $F_0 = 0$ and $F_i = -A^T P_i - P_i A$.

As another example, consider the algebraic Riccati inequality

$$A^T P + PA + PBR^{-1}B^T P + Q < 0, \;\; R > 0 \tag{D.13}$$

where A, B, $Q = Q^T$, $R = R^T$ are given matrices of appropriate size, and $P = P^T$ is the variable. Inequality (D.13) can be expressed as the LMI

$$\begin{bmatrix} A^T P + PA + Q & PB \\ B^T P & -R \end{bmatrix} < 0 \tag{D.14}$$

We are often interested in whether a solution exists or not in an LMI.

- Feasibility problems
  Given LMI $F(x) > 0$, find a feasible solution x^{feas} such that $F(x^{feas}) > 0$ or prove that a solution does not exist.

D.2 Semidefinite Programming

Now, we give a brief introduction to SDP which can effectively solve many other optimization problems involving LMIs. The basic results are from [BLEGB94, VB96].

SDP is an optimization problem of minimizing a linear function of a variable $x \in \mathcal{R}^m$ subject to a matrix inequality:

$$\begin{aligned} & \text{minimize} \;\; c^T x \\ & \text{subject to } F(x) > 0 \end{aligned} \tag{D.15}$$

where

$$F(x) = F_0 + \sum_{i=1}^{m} x_i F_i$$

This SDP is a convex optimization problem, since its objective and constraint are convex: if $F(x) > 0$ and $F(y) > 0$, then, for all λ, $0 \le \lambda \le 1$,

$$F(\lambda x + (1 - \lambda y)) \leq \lambda F(x) + (1 - \lambda y)F(y)$$

Even though the SDP (D.15) may seem to be quite specialized, it can represent many important optimization problems, such as a linear programming (LP),

$$\begin{aligned} \text{minimize}\ \ & c^T x \\ \text{subject to}\ & Gx < h \end{aligned}$$

$$Gx < h \quad \Longleftrightarrow \quad \text{diag}(h_1\ h_2\ \cdots\ h_n) - \sum_{i=1}^{n} x_i\ \text{diag}(g_{1,i}\ g_{2,i}\ \cdots\ g_{n,i}) \geq 0$$

a matrix norm minimization,

$$\|A\|_2^2 < t^2 \quad \Longleftrightarrow \quad \lambda_{\max}(A^T A) < tI \quad \Longleftrightarrow \quad A^T A < tI$$

$$\Longleftrightarrow \quad \min\ t \text{ subject to } \begin{bmatrix} tI & A \\ A^T & I \end{bmatrix} > 0$$

a maximum eigenvalue minimization,

$$\begin{aligned} \text{minimize}\ \ & \lambda \\ \text{subject to}\ & \lambda I - A(x) > 0 \end{aligned} \qquad\qquad (\text{D.16})$$

where A is symmetric, and a quadratic programming (QP)

$$\begin{aligned} \text{minimize}\ \ & \tfrac{1}{2}x^T P x + q^T x + r \\ \text{subject to}\ & Gx < h \end{aligned} \quad \Longleftrightarrow \quad \begin{aligned} \text{minimize}\ \ & t \\ \text{subject to}\ & \begin{bmatrix} t - r - q^T x & x^T \\ x & 2P^{-1} \end{bmatrix} > 0 \\ & Gx < h \end{aligned}$$

Note that the quadratic programming problem involves minimization of a quadratic function subject to linear constraints.

In particular, it can represent a quadratic constrained QP (QCQP), which is needed to solve the optimization problems. Now we will show how QCQP can be cast as a special case of SDP. The convex quadratic constraint $(Ax + b)^T(Ax + b) - c^T x - d < 0$ can be expressed as

$$\begin{bmatrix} I & Ax + b \\ (Ax + b)^T & c^T x + d \end{bmatrix} < 0$$

The left-hand side depends affinely on the vector x, and it can be also expressed as

$$F(x) = F_0 + \sum_{i=1}^{m} x_i F_i$$

with

$$F_0 = \begin{bmatrix} I & b \\ b^T & d \end{bmatrix}, \ F_i = \begin{bmatrix} 0 & a_i \\ a_i^T & c_i \end{bmatrix}, \ i = 1, \cdots, m$$

where $A = [a_1 \ \cdots \ a_m]$. Therefore, a general quadratically constrained quadratic programming

$$\begin{aligned} \text{minimize} \ \ & f_0(x) \\ \text{subject to} \ \ & f_i(x) < 0, \ i = 1, \cdots, L \end{aligned}$$

where each f_i is a convex quadratic function $f_i(x) = (A_i x + b)^T (A_i x + b) - c_i^T x - d_i$ can be written as

$$\begin{aligned} \text{minimize} \ \ & t \\ \text{subject to} \ \ & \begin{bmatrix} I & A_0 x + b_0 \\ (A_0 x + b_0)^T & c_0^T x + d_0 + t \end{bmatrix} > 0 \end{aligned}$$

$$\begin{bmatrix} I & A_i x + b_i \\ (A_i x + b_i)^T & c_i^T x + d_i \end{bmatrix} > 0, \ i = 1, \cdots, L$$

which is an SDP with variables $x \in \mathcal{R}^m$ and $t \in \mathcal{R}$. The relationship among LP, QP, QCQP, and SDP is given as

$$\text{LP} \subset \text{QP} \subset \text{QCQP} \subset \text{SDP} \tag{D.17}$$

E

Survey on Applications

There are many papers that deal with applications of receding horizon controls. The terminology "MPC" is more popular in applications, so that we use it instead of "RHC".

Most survey papers [RRTP78, GPM89, RMM94, Kwo94, May95, LC97, May97, CA98, ML99, MRRS00, KHA04] and books [BGW90, Soe92, Mos95, MSR96, AZ00, KC00, Mac02, Ros03, HKH02, CB04] also report real applications of the RHC. There are a few survey papers on applications of industrial MPC technologies, particularly for process controls [OOH95] [QB97] [QB00] [QB03].

We can consider two different kinds of applications. One is a software product for general uses and the other is a specific process where RHCs are applied.

There are many commercial software products, such as Aspen Technology's DMCPlus (Dynamic Matrix Control) and SMCA (Setpoint Multivariable Contr. Arch.), Adersa's IDCOM (identification and command), HIECON (Hierarchical Constraint Control), and PFC (Predictive Functional Control), Honeywell's Profit Control, RMPCT (Robust MPC Technology), and PCT (Predictive Control Technology), Pavilion Technology's Process Perfector, SCAP Europa's APCS (Adaptive Predictive Control System), IPCOS's INCA (IPCOS Novel Control Architecture), Simulation Sciences' Connoisseur, and ABB's 3dMPC. This list may not be complete. For additional information, refer to [Mac02] [QB97][QB03].

Applied processes include refining, petrochemicals, chemicals, pulp and paper, air and gas, utility, mining/metallurgy, food processing, polymer, furnace, aerospace/defence, automotive, ammonia synthesis, and so on. It is not so easy to list all applied processes. The surveys on the applied processes can be done by vendors [QB03] and manufacturers [OOH95]. In particular, applications of MPC are widely used in refining and petrochemical processes [QB03].

In the beginning, linear RHCs were applied. Recent years have shown rapid progress in the development and application of industrial nonlinear MPC.

When the operating points change frequently and span a sufficiently wide range of nonlinear process dynamics, the nonlinear MPC is very efficient compared with linear MPC. While applications of linear MPC are concentrated in refining, those of nonlinear MPC cover a much broader range of chemicals, such as pH control, polymer manufacturing, and ammonia synthesis. However, it has been observed that the size of nonlinear MPC applications is typically much smaller than that of linear MPC [MJ98]. This is due to the computational complexity of nonlinear MPC.

F

MATLAB® Programs

Example 3.2: Receding Horizon Linear Quadratic Tracking Control

```
M = 1; bb = 0.01; Ac = [0 1 0 0;0 -bb/M 0 0;0 0 0 1;0 0 0 -bb/M];
Bc = [0 0 ;1/M 0 ; 0 0;0 1/M]; Cc = [1 0 0 0 ;0 0 1 0]; Dc = [0
0;0 0]; Ts = 0.055; [A B C D] = c2dm(Ac,Bc,Cc,Dc,Ts,'zoh');

% design parameter in cost function
% - state weighting matrix: Q
% - control weighting matrix: R
% - initial state: x0
Q = eye(2); R = eye(2); x0 = [1;0;1;0];

% arbitrary future references available over [i,i+N]
y1r_1 = 1:-0.01:0.01; y1r_2 = zeros(1,100); y1r_3 = 0:0.01:0.99;
y1r_4 = ones(1,100); y2r_1 = ones(1,100); y2r_2 = 1:-0.01:0.01;
y2r_3 = zeros(1,100); y2r_4 = 0:0.01:0.99; y1r = [y1r_1 y1r_2
y1r_3 y1r_4 ones(1,100)]; y2r = [y2r_1 y2r_2 y2r_3 y2r_4
ones(1,100)]; yr = [y1r;y2r];

% simulation step
is = 440;

% Discrete-time LQTC for Unconstrained Systems
[x,y,u] = dlqtc(x0,A,B,C,Q,R,yr,is);
```

Subroutine: | drhtc.m |

```
function  [x,y,u ] = drhtc(x0,A,B,C,Q,R,Qf,N,yr,is);

%DRHTC Discrete-time RHTC for Unconstrained Systems
```

```
% convert reference vector to row vector
[s1,s2] = size(yr); if (s2 == 1)
   yr = yr';
end

% check future reference length for simulation
if (length(yr) < (is + N))
   disp('The future reference is of too small length');
   return
end

% Riccati solution K(N-1) for RHTC
[K_N_1 K_history_vec] = drde2(A,B,C,Q,R,Qf,N);

% initialization of history variables
xi = x0; % state of plant x(i)
ui_history = []; % history of rhc u(i)
xi_history = x0; % history of state of plant x(i)
yi_history = C*x0; % history of output of plant y(i)


for i=1:is
   % time-varying feed feedward gain g for RHTC
   [g_1 g_history] = dvde(A,B,C,Q,R,Qf,N,K_history_vec,yr,i);

   % receding horizon tracking controller u(i)
   ui = -inv(R+B'*K_N_1*B)*B'*(K_N_1*A*xi+g_1);

   % plant is controlled by rhtc u(i) at time [i,i+1]
   xi = A*xi + B*ui;
   yi = C*xi;

   % history u(i), x(i), y(i)
   ui_history = [ui_history ui];
   xi_history = [xi_history xi];
   yi_history = [yi_history yi];
end

x = xi_history; % state trajectory x
y = yi_history; % output trajectory y
u = ui_history; % RHTC history vector u

return
```

Subroutine: | drde2.m |

```
function [K_N_1,K_history_vec] = drde2(A,B,C,Q,R,Qf,N);

%DRDE2  Discrete-time Riccati Difference Equation Solver
```

```
%        This RDE appears in LQ Tracking Problem

% system dimension
n = size(A,1);

% boundary condition
K_0 = C'*Qf*C; K_0_vec = mtx2vec(K_0); K_history_vec = K_0_vec;
K_i = K_0;

% solve Riccati Differential Equation 2
for i=1:N-1
   K_i = A'*K_i*inv(eye(n)+B*inv(R)*B'*K_i)*A + C'*Q*C;
   K_i_vec = mtx2vec(K_i);
   K_history_vec = [K_history_vec K_i_vec];
end

% constant feedback gain K(N-1) for RHTC
[s1,s2] = size(K_history_vec); K_N_1 =
vec2mtx(K_history_vec(:,s2));

return
```

Subroutine: `dvde.m`

```
function  [g_1,g_history ]  =
dvde(A,B,C,R,Q,Qf,N,K_history_vec,yr,i );

%DVDE Discrete-time Vector Differential Equation Solver

% system dimension
n = size(A,1);

% boundary condition
g_N = -C'*Qf*yr(:,i+N); g_history = g_N; g_j = g_N;

% solve Vector Difference Equation
for j=(N-1):-1:1
    K_j = vec2mtx(K_history_vec(:,(N-1)-j+1));
    g_j = A'*inv(eye(n)+K_j*B*inv(R)*B')*g_j - C'*Q*yr(:,i+j);
    g_history = [g_history g_j];
end

% time-varying feed-forward gain g(1) for RHTC
[m,n] = size(g_history); g_1 = g_history(:,n);

return;
```

Example 3.3: Receding Horizon H_∞ Tracking Control

```
clear;
% state-space model of original system and its discretization

M = 1; bb = 0.01; Ac = [0 1 0 0;0 -bb/M 0 0;0 0 0 1;0 0 0 -bb/M];
Bc = [0 0 ;1/M 0 ; 0 0;0 1/M]; Cc = [1 0 0 0 ;0 0 1 0]; Dc = [0
0;0 0]; Ts = 0.055;

[A B C D] = c2dm(Ac,Bc,Cc,Dc,Ts,'zoh'); Bw =  [ 0 0    ; 0 0.03  ;
0.02 0 ; 0 0.05]


% design parameter in cost function

% - state weighting matrix: Q
% - control weighting matrix: R
% - disturbance weighting matrix: Rw
% - terminal weighting matrix: Qf
% - prediction horizon: N
system_order = size ( A , 1) ; input_order = size ( B , 2 ) ;
dist_order = size ( Bw , 2 ) ;

Q = eye(system_order) ; R = eye(input_order) ; Rw = 1.2
*eye(dist_order) ; N = 5 ; gamma_2 = 5.0 ;

% arbitrary future references available over [i,i+N]

y1r_1 = 1:-0.01:0.01; y1r_2 = zeros(1,100); y1r_3 = 0:0.01:0.99;
y1r_4 = ones(1,100); y2r_1 = ones(1,100); y2r_2 = 1:-0.01:0.01;
y2r_3 = zeros(1,100); y2r_4 = 0:0.01:0.99; y1r = [y1r_1 y1r_2
y1r_3 y1r_4 ones(1,100)]; y2r = [y2r_1 y2r_2 y2r_3 y2r_4
ones(1,100)]; yr = [y1r;y2r];

% simulation step
is = 440;

% Discrete-time RHTC for Unconstrained Systems
% initial state
x0 = [ 1 ; 0 ; 1 ; 0 ]; Qf = 100*eye(2); Q = eye(2) ;

gamma_2=1.5; Bw =  [ 0.016  0.002   ; 0.01 0.009  ;  0.008 0 ; 0
0.0005] ;

[x_hinf,y_hinf,u_hinf , w] = drhtc_hinf ( x0 , A , B , Bw , C , Q
, R , Rw , Qf , gamma_2 , N , yr , is);

[x,y,u] = drhtc(x0,A,B,C,Q,R,Qf,N,yr,is, Bw,w);
```

```
for k = 0: is
    y_rhtc_ws ( k+1 ,: ) = [ k*0.01 y(:,k+1)' ];
end

for k = 0: is
    r_rhtc_ws ( k+1 ,: ) = [k*0.01 yr(:,k+1)'];
end

for k = 0: is
    y_rhtc_ws_hinf ( k+1 ,: ) = [k*0.01 y_hinf(:,k+1)'];
end

plot( r_rhtc_ws(:,2) ,  r_rhtc_ws(:,3) , y_rhtc_ws(:,2) ,
y_rhtc_ws(:,3) ,'-.' , y_rhtc_ws_hinf(:,2) , y_rhtc_ws_hinf(:,3)
);

legend(' Reference','H infinity RHTC' , 'LQ RHTC') ;
```

Subroutine: | drhtc_hinf.m |

```
function  [ x , y , u , w] = drhtc_hinf ( x0 , A , B , Bw , C , Q
, R , Rw , Qf , gamma_2, N , yr , is);



% convert reference vector to row vector
[s1,s2] = size(yr); if (s2 == 1)
    yr = yr';
end

% check future reference length for simulation
if (length(yr) < (is + N))
    disp('The future reference is of too small length');
    return
end

% Riccati solution K(N-1) for RHTC
[K_N_1 K_history_vec] = drde2_hinf( A , B , Bw , C , Q , R , Rw ,
Qf , gamma_2 , N);

% initialization of history variables
xi = x0; % state of plant x(i)
ui_history = []; % history of rhc u(i)
xi_history = x0; % history of state of plant x(i)
yi_history = C*x0; % history of output of plant y(i)

wi_history = [];

system_order = size(A,1);
```

```
for i=1:is
   % time-varying feed feedward gain g for RHTC
   [g_1 g_history] = dvde_hinf( A , B , Bw , C , Q , R , Rw , Qf ,...
   gamma_2, N , K_history_vec , yr , i);
   % receding horizon tracking controller u(i)
   Lambda = eye(system_order) + K_N_1 * ( B*inv(R)*B' ...
   - 1 / gamma_2 * Bw*inv(Rw)*Bw') ;

   ui = -inv(R) * B' * inv( Lambda )*( K_N_1 * A * xi + g_1 ) ;
   wi = 1/gamma_2* inv(Rw) * Bw' * inv ( Lambda ) ...
   * ( K_N_1 * A * xi + g_1 ) ;

   % plant is controlled by rhtc u(i) at time [i,i+1]

   xi = A*xi + B*ui + Bw*wi;
   yi = C*xi;

   ui_history = [ui_history ui];
   xi_history = [xi_history xi];
   yi_history = [yi_history yi];
   wi_history = [wi_history wi];

end

x = xi_history; % state trajectory x
y = yi_history; % output trajectory y
u = ui_history; % RHTC history vector u

w = wi_history; % RHTC history vector u

return
```

Subroutine: | drde2_hinf.m |

```
function [K_N_1,K_history_vec] = drde2_hinf (A , B , Bw , C , Q
, R , Rw , Qf , gamma_2 , N );


% system dimension
n = size(A,1);

% boundary condition
K_0 = C'*Qf*C;

K_0_vec = mtx2vec(K_0); K_history_vec = K_0_vec; K_i = K_0;

% solve Riccati Differential Equation
```

```
for i=1:N-1
    if  min ( real ( eig ( Rw - 1 / gamma_2 *Bw' * K_i * Bw) ) ) < 0
        error ('error');
    end
    Lambda = eye(n) + K_i * ( B*inv(R)*B' - 1 / gamma_2 ...
    * Bw * Rw^(-1) * Bw' ) ;
    K_i = A' * inv(Lambda) * K_i *A + C'*Q*C;
    K_i_vec = mtx2vec(K_i);
    K_history_vec = [K_history_vec K_i_vec];
end

% constant feedback gain K(N-1) for RHTC

[s1,s2] = size(K_history_vec); K_N_1 =
vec2mtx(K_history_vec(:,s2)); return
```

Subroutine: dvde_hinf.m

```
function  [g_1,g_history ]  = dvde_hinf( A , B , Bw , C , Q , R ,
Rw , Qf , gamma_2, N , K_history_vec , yr , i);

n = size(A,1);

g_N = -C'*Qf*yr(:,i+N); g_history = g_N; g_j = g_N;

for j = (N-1): -1: 1
    K_j = vec2mtx(K_history_vec(:,(N-1)-j+1));
        Lambda = eye(n) + K_j * ( B*inv(R)*B' ...
        - 1/gamma_2 * Bw * inv(Rw) * Bw' ) ;
        %g_j = A'*inv(eye(n)+K_j*B*inv(R)*B')*g_j - C'*Q*yr(:,i+j);
        if ( Rw -
    g_j = A'*inv(Lambda)*g_j - C'*Q*yr(:,i+j);

    g_history = [g_history g_j];
end

% time-varying feed-forward gain g(1) for RHTC
[m,n] = size(g_history); g_1 = g_history(:,n); return;
```

Example 3.4: Cost Monotonicity Condition for Receding Horizon Control

```
A = [ 0.6831    0.0353 ; 0.0928    0.6124 ] ; B = [0.6085;0.0158];
Q =eye(2); R =3;
```

```
SystemDim = size(A,1); InputDim = size(B,2);

% Getting started LMISYS description
setlmis([]);

% Defining LMI variables
Y=lmivar(2,[InputDim SystemDim]); X=lmivar(1,[SystemDim 1]);

% Defining LMI
lmiterm([-2 1 1 X],1,1);        % LMI (1,1): X
lmiterm([-2 2 1 Y],B,1);        % LMI (2,1): B*Y
lmiterm([-2 2 1 X],A,1);        % LMI (2,1): A*X
lmiterm([-2 2 2 X],1,1);        % LMI (2,2): X
lmiterm([-2 3 1 X],sqrt(Q),1);  % LMI (3,1): Q^(1/2)*X
lmiterm([-2 3 3 0],1);          % LMI (3,3): 1
lmiterm([-2 4 1 Y],sqrt(R),1);  % LMI (4,1): R^(1/2)*Y
lmiterm([-2 4 4 0],1);          % LMI (4,4): 1

LMISYS = getlmis; [tmin,xfeas]=feasp(LMISYS);
X=dec2mat(LMISYS,xfeas,X);
```

Example 4.1: Robustness of Minimum Variance Finite Impulse Response filters

```
% Systems and paramteres %
A = [ 0.9305 0 0.1107; 0.0077 0.9802 -0.0173; 0.0142 0 0.8953];
B = [0.0217 0.2510; 0.0192 -0.0051; 0.0247 0.0030];
C = [ 1 0 0;0 1 0 ]; G = [1 1 1]'; Q = 0.02^2; R = 0.04^2*eye(2);
D_1 = 0; D_2 = 0;

N_sample = 250 ; N_order =  size(A,1); N_horizon = 10;

% Making big matrices for FIR filters
[B_bar, C_bar, G_bar, Xi] = MakeBigMatrices (A,B,C,G,Q,R,10);
H = inv(C_bar'*inv(Xi) * C_bar) * C_bar'*inv(Xi);

% Parameter initialize
intial_state= [0 0 0 ]'; x = intial_state;
IIR_x_hat = [0 0 0]'; FIR_x_hat = IIR_x_hat;
P = 0.0*eye(N_order); real_state = zeros(N_order , N_sample);
estimated_state = zeros(N_order, N_sample); real_state(:,1)=x;
estimated_state(:,1) = IIR_x_hat;
FIR_estimated_state = zeros(N_order, N_sample);
measurements = zeros(2*N_sample);
```

```
delta_A = 0.1*[1 0 0;0 1 0;0 0 0.1 ];
delta_C = 0.1*[0.1  0 0;0 0.1   0 ];

% main procedure
for i = 1: N_sample-1

   if( i > 50 & i < 101 )
      x = (A+delta_A)*x + G*randn(1)*0.02;
      y =   (C+delta_C)*x + randn(2,1)*0.04;
    else
      x = A*x + G*randn(1)*0.02;
      y = C*x + randn(2,1)*0.04;
   end

   real_state(:,i+1)=x;

   % IIR_filter: one step predicted estimate

   IIR_x_hat = A * IIR_x_hat +  A * P *C' * inv( R + C * P * C')
             *( y - C*IIR_x_hat);
   P = A * inv( eye(N_order) + P * C' * inv(R) * C) * P * A'
      + G * Q * G';

   estimated_state(:,i+1)=IIR_x_hat;
   measurements(2*i-1:2*i) = y;

   % FIR filter

   if i>10
        FIR_x_hat = H * (measurements(2*i-19:2*i))';
   end

   FIR_estimated_state(:,i+1) = FIR_x_hat;

end

% Plot
plot( 1:N_sample , real_state(2,:)-estimated_state(2,:),
1:N_sample , real_state(2,:)-FIR_estimated_state(2,:))
```

Subroutine: | MakeBigMatrices.m |

```
function [B_bar , C_bar , G_bar , Xi] =
MakeBigMatrices(A,B,C,G,Q,R,horizon);

% Error check
A_inv = inv(A); IsInput = 1;

% initialization
```

```
if B == 0
   IsInput = 0;
   B_bar=0
else
    B_bar = C * A_inv * B;
end

C_bar = C * A_inv; G_bar = -C * A_inv * G;
Q_stack = Q ; R_stack = R ; A_inv_i = A_inv;

% parameter setting
N_input = size(B,2); N_output = size(C,1);
N_system_noise = size(G,2); N_order = size(A,1);

% main procedure
for i = 2: horizon
   A_inv_i = A_inv_i * A_inv ;

 if IsInput == 1
  B_bar = [B_bar -C_bar*A_inv*B; zeros(N_output, N_input*(i-1))
           -C*A_inv*B ];
 end
  G_bar = [G_bar -C_bar*A_inv*G; zeros(N_output, N_system_noise*(i-1))
           -C*A_inv*G ] ;
  C_bar = [C * A_inv_i ;C_bar];
  Q_stack = daug(Q_stack,Q); R_stack = daug(R_stack,R);
end

Xi = G_bar * Q_stack *G_bar' + R_stack ;
```

Example 5.1: Linear Quadratic Finite Memory Control

```
% Systems and paramteres
A = [ 0.9305   0   0.1107; 0.0077 0.9802  -0.0173 ; 0.0142 0
0.8953 ]; B = [0.0217 0.2510 ;
   0.0192 -0.0051 ;
   0.0247 0.0030 ] ;
G = [1 1 1]'; C = [ 1 0 0;0 1 0 ]; Q = 0.02^2; R = 0.04^2*eye(2);

N=zeros(size(A,1),size(B,2));

D_1 = 0; D_2 = 0;

[K,S,E] = dlqr(A,B,eye(3),eye(2),N)

horizon_size =10;
```

```
% Making big matrices
[ B_bar , C_bar , G_bar , Xi] =  MakeBigMatrices
(A,B,C,G,Q,R,horizon_size) ; H=inv(C_bar'*inv(Xi) * C_bar) *
C_bar'*inv(Xi);

N_sample = 800 ;
N_order =  size(A,1); N_horizon = 10;
intial_state= [-1 1 1]' ; x= intial_state ;
x_1=x; x_2=x;
IIR_x_hat = [0 0 0]' ; FIR_x_hat = IIR_x_hat ;

P = 0.5*eye(N_order) ; real_state = zeros(N_order , N_sample);
IIR_state = zeros(N_order, N_sample); real_state(:,1)=x;
IIR_state(:,1)=IIR_x_hat; measurements = zeros ( 2*N_sample);
FIR_state = zeros(N_order, N_sample);

% Temporary uncertainty
delta_A = 0.05*[1 0 0;0 1 0;0 0 0.1 ]; delta_C = 0.05*[0.1  0  0;0
0.1   0 ];

% Start time and stop time of temporary uncertainty
dist_start = 300; dist_stop = 501;

for i = 1: N_sample-1

   % FIR
   sys_noise=randn(1)*0.01;
   mea_noise=randn(2,1)*0.01;
   if( i > dist_start & i < dist_stop )
    x_1 = (A+delta_A)*x_1 - B*K*FIR_x_hat + G*sys_noise;
    y_1 =   (C+delta_C)*x_1 + mea_noise;
   else
    x_1 = A*x_1 - B*K*FIR_x_hat +G*sys_noise;
    y_1 = C*x_1 + mea_noise;
   end
   FIR_state(:,i+1) = x_1;
   measurements(2*i-1:2*i) = y_1;
   if i>10
      FIR_x_hat = H * (measurements(2*i-(2*horizon_size-1):2*i))';
   else
      FIR_x_hat = zeros(N_order,1);
   end

   % IIR
   if( i > dist_start & i < dist_stop )
    x_2 = (A+delta_A)*x_2 - B*K*IIR_x_hat + G*sys_noise;
    y_2 =   (C+delta_C)*x_2 + mea_noise;
    else
```

```
         x_2 = A*x_2 - B*K*IIR_x_hat +G*sys_noise;
         y_2 = C*x_2 + mea_noise;
    end
    IIR_state(:,i+1) = x_2;
    IIR_x_hat = A * IIR_x_hat +  A* P *C' *
        inv( R + C * P * C')*( y_2 - C*IIR_x_hat);
    P = A * inv( eye(N_order) + P * C' * inv(R) * C) * P * A' + G * Q * G';
end

% plot
time = 1:N_sample; time = time*0.05;

plot( time ,-IIR_state(1,1:N_sample),'r',time,-FIR_state(1,1:N_sample),'b' );
title('First state trajectory'); xlabel('Time'); figure;
plot( time ,-IIR_state(2,1:N_sample),'r' ,time ,-FIR_state(2,1:N_sample),'b');
title('Second state trajectory'); xlabel('Time'); figure;
plot( time ,-IIR_state(3,1:N_sample),'r' ,time,-FIR_state(3,1:N_sample),'b' );
title('Third state trajectory'); xlabel('Time');
```

Subroutine: | MakeBigMatrices.m |

```
    function [B_bar , C_bar , G_bar , Xi ] =
        MakeBigMatrices(A,B,C,G,Q,R,horizon);

    A_inv = inv(A);
    IsInput = 1;
    % initialization
    if B == 0
        IsInput = 0;
        B_bar=0
    else
        B_bar = C * A_inv * B;
    end
    C_bar = C * A_inv;
    G_bar = -C * A_inv * G;
    Q_stack = Q ;
    R_stack = R ;
    A_inv_i = A_inv

    % parameter setting
    N_input = size(B,2);
    N_output = size(C,1);
    N_system_noise = size(G,2);
    N_order = size(A,1);

    % main procedure
    for i = 2: horizon
        i
        A_inv_i = A_inv_i * A_inv ;
        if IsInput == 1
         B_bar = [ B_bar  -C_bar*A_inv*B ;
            zeros(N_output , N_input*(i-1))  -C*A_inv*B ];
```

```
      end
      G_bar = [ G_bar  -C_bar*A_inv*G ; zeros(N_output ,
         N_system_noise*(i-1))  -C*A_inv*G ] ;
      C_bar = [ C * A_inv_i ;C_bar];
      Q_stack = daug(Q_stack,Q);
      R_stack = daug(R_stack,R);
   end

   Xi = G_bar * Q_stack *G_bar' + R_stack ;
```

Example 6.2: Constrained Receding Horizon Control

```
%  Demonstration of RHC using LMI
clear all;
%% System State Equation
A= [1       0.2212;
    0       0.7788];
B = [0.0288;0.2212]; G =[1 0; 0 1; 1.5 1];

%% Initial State
x0 = [1;0.3];

[n,n]=size(A); [n,m]=size(B);

%  Let's get predictor
N = input('horizon N =' ); length = input('total simu time?=');

%  Input Constraint
u_lim=0.5; ubar_lim = []; for i=1:N,
   ubar_lim=[ubar_lim;u_lim];
end

%  State Constraint
x_lim = [1.5;0.3]; G =[1   0;
    0  1];
g_lim = [1.5; 0.3]; gbar_lim = []; for i=1:N,
   gbar_lim=[gbar_lim;g_lim];
end


[ng_r, ng_c] = size(G);

%  Weighting Matrix
Q = 1*eye(size(A)); R = 1;
```

```
A_hat=[]; B_hat=[]; for i=1:N-1,
   A_hat = daug(A_hat, A);
   B_hat = daug(B_hat, B);
end A_hat = [zeros(n,n*N);
        A_hat zeros(n*(N-1),n)];
B_hat = [zeros(n,m*N);
        B_hat zeros(n*(N-1),m)];
W_hat = inv(eye(n*N)-A_hat)*B_hat; V0_hat =
inv(eye(n*N)-A_hat)*[x0;zeros(n*(N-1),1)];

B_bar=[]; for i=1:N,
   B_bar = [B_bar A^(N-i)*B];
end

% Let's get stacked weighting matrix
[m,n]=size(Q); [p,q] = size(R); Q_hat = []; R_hat = []; for i=1:N,
   Q_hat = daug(Q_hat, Q);
   R_hat = daug(R_hat, R);
end

% Let's get constraint matrix
G_bar = []; for i=1:N
   G_bar = daug(G_bar, G);
end

W=W_hat'*Q_hat*W_hat + R_hat; W=(W+W')/2;

%  Simulation starts !
t= []; State = []; U = []; summed_cost=0; Summed_cost=[];
Cost_at_k=[]; x = x0;

R1=[]; R2=[]; for i=0:(length-1),
   i
   V0_hat = inv(eye(n*N)-A_hat)*[x;zeros(n*(N-1),1)];
   V = 2*W_hat'*Q_hat*V0_hat;
   V0 = V0_hat'*Q_hat*V0_hat;

   %% Solve LMI
   [X,Y,r1,r2,opt_u] =
   rhc_lmi(x,N,A,B,Q,R,W_hat,V0_hat,Q_hat,B_bar,...
   W,V,V0,G,G_bar,ubar_lim,gbar_lim,u_lim,g_lim);
   P = r2*inv(X);

   if (i==0),
       boundary0=[];
       for th=0:0.01:2*pi,
           z  = sqrt(r2)*inv(P^0.5)*[cos(th);sin(th)];
           boundary0=[boundary0  z];
       end
```

```
    elseif(i==1),
        boundary1=[];
        for th=0:0.01:2*pi,
            z = sqrt(r2)*inv(P^0.5)*[cos(th);sin(th)];
            boundary1=[boundary1  z];
        end
    elseif(i==2),
        boundary2=[];
        for th=0:0.01:2*pi,
            z = sqrt(r2)*inv(P^0.5)*[cos(th);sin(th)];
            boundary2=[boundary2  z];
        end
    elseif(i==3),
        boundary3=[];
        for th=0:0.01:2*pi,
            z = sqrt(r2)*inv(P^0.5)*[cos(th);sin(th)];
            boundary3=[boundary3  z];
        end
    elseif(i==4),
        boundary4=[];
        for th=0:0.01:2*pi,
            z = sqrt(r2)*inv(P^0.5)*[cos(th);sin(th)];
            boundary4=[boundary4  z];
        end
    end

    K = Y*inv(X);
    u = opt_u(1);
    State = [State x];
    U = [U u]; %% Control Input

    cost_at_k = r1+r2;
    real_cost = x'*Q*x + u'*R*u;
    R1=[R1 r1];
    R2=[R2 r2];
    summed_cost = summed_cost + real_cost;
    Cost_at_k = [Cost_at_k cost_at_k];
    Summed_cost = [Summed_cost summed_cost];

    %  State Update
    x = A*x + B*u;
    t= [t i+1];
    home;
end

Cost_at_k=[Cost_at_k Cost_at_k(length)]; Summed_cost=[Summed_cost
Summed_cost(length)]; U=[U u]; State=[State x]; t=[ 0 t];R2=[R2
R2(length)];R1=[R1 R1(length)];
```

```
% Let's take a look at the simulation results
figure; plot(t,State(1,:),'r:',t,State(2,:),'b');
legend('x1','x2',0); xlabel('time(sec)');grid; title('states');

if (i>5), figure;
    plot(State(1,:),State(2,:),'o',boundary0(1,:),boundary0(2,:),'b',
    boundary1(1,:),boundary1(2,:),'r',...
    boundary2(1,:),boundary2(2,:),'k',boundary3(1,:),boundary3(2,:),...
    'r',boundary4(1,:),boundary4(2,:),'r'); axis([-1.7 1.7 -0.35
    0.35]);
end

figure; stairs(t,U); xlabel('time(sec)');grid; title('control
input');

figure; stairs(t,Cost_at_k,'r'); xlabel('time(sec)');grid;
title('Expected cost at time k');

figure; stairs(t,Summed_cost,'b');grid; xlabel('time(sec)');
title('Summed cost');

figure; plot(t,R1,'r',t,R2,'b');
```

Subroutine: | dvde_rhclmi.m |

```
% Function: rhc_lmi.m
%
function [X_opt,Y_opt,r1_opt,r2_opt,u_opt] =
rhc_lmi(x,N,A,B,Q,R,W_hat,V0_hat,Q_hat,B_bar,W,V,V0,G,G_bar,...
ubar_lim,gbar_lim,u_lim,g_lim)

% LMI Variable  = X, Y, r1, r2, U

[n,n]=size(A); [n,nu]=size(B); [nG_r, nG_c]= size(G); [nglim_r,
nglim_c]= size(g_lim);

if (~isempty(G_bar)),
    GW = G_bar*W_hat;
    gg1 = gbar_lim + G_bar*V0_hat;
    gg2 = gbar_lim-G_bar*V0_hat;
end [nGW_r,nGW_c] = size(GW);

setlmis([]);

% LMI Variable
r1 = lmivar(1,[1 0]); r2 = lmivar(1,[1 0]); X = lmivar(1,[n 1]); Y
= lmivar(2,[nu n]); Z = lmivar(1,[nu 1]); U = lmivar(2,[N*nu 1]);
VV = lmivar(1,[nglim_r 1]);
```

```
lmiterm([-1 1 1 r1], 1, 1); lmiterm([-1 1 1 U], 0.5*V',1,'s');
lmiterm([-1 1 1 0], -V0);            %% (1,1) = r1-V'*U -V0
lmiterm([-1 2 1 U], W^0.5, 1);       %% (2,1) = W^0.5*U
lmiterm([-1 2 2 0], 1);              %% (2,2) = I


lmiterm([-2 1 1 0],1);               %% (1,1) = 1
lmiterm([-2 2 1 0],A^N*x);           %% (2,1) = A^N*x
lmiterm([-2 2 1 U], B_bar,1);        %% (2,1) = A^N*x + B_bar*U
lmiterm([-2 2 2 X],1,1);             %% (2,2) = X


lmiterm([-3 1 1 X], 1, 1);           %% (1,1) = X
lmiterm([-3 2 2 X], 1, 1);           %% (2,2) = X
lmiterm([-3 3 3 r2], 1,1);           %% (3,3) = r2*I
lmiterm([-3 4 4 r2], 1, 1);          %% (4,4) = r2*I
lmiterm([-3 2 1 X], A, 1);
lmiterm([-3 2 1 Y], B, 1);           %% (2,1) = A*X + B*Y
lmiterm([-3 3 1 X], real(Q^0.5), 1); %% (3,1) = Q^0.5*X
lmiterm([-3 4 1 Y], R^0.5, 1);       %% (4,1) = R^0.5*X


% Constraint on u during the horizon from 0 to N-1
Ucon_LMI_1 = newlmi; for i=1:N*nu,
    tmp = zeros(nu*N,1);
    tmp(i,1) = 1;
    %% U(i) >= -ubar_lim(i)
    lmiterm([-(Ucon_LMI_1+i-1) 1 1 U], 0.5*tmp',1,'s');
    lmiterm([-(Ucon_LMI_1+i-1) 1 1 0], ubar_lim(i));
    %% U(i) <= ubar_lim(i)
    lmiterm([(Ucon_LMI_1+N*nu+i-1) 1 1 U], 0.5*tmp', 1,'s');
    lmiterm([(Ucon_LMI_1+N*nu+i-1) 1 1 0], -ubar_lim(i));
end

% Constraint on u after the horizon N
Ucon_LMI_2 = newlmi;
lmiterm([-Ucon_LMI_2 1 1 Z], 1,1);   %% (1,1) = Z
lmiterm([-Ucon_LMI_2 1 2 Y], 1,1);   %% (1,2) = Y
lmiterm([-Ucon_LMI_2 2 2 X], 1,1);   %% (2,2) = X

Ucon_LMI_3 = newlmi; for i=1:nu,
    tmp = zeros(nu,1);
    tmp(i,1) = 1;
    lmiterm([(Ucon_LMI_3+i-1) 1 1 Z], tmp', tmp);
    %% (1,1) Z(i,i) <= u_lim(i)^2
    lmiterm([(Ucon_LMI_3+i-1) 1 1 0], -u_lim(i)^2);
end

% Constraint on x during the horizon from 0 to N-1
```

```
if (~isempty(G_bar)),
   Xcon_LMI1 = newlmi;
   for i=1:nGW_r,
      %% (1,1) GW*U + gg1 > = 0
      lmiterm([-(Xcon_LMI1+i-1)  1 1 U],  0.5*GW(i,:),1,'s');
      lmiterm([-(Xcon_LMI1+i-1)  1 1 0],  gg1(i));
      %% (1,1) -GW*U + gg2 >= 0
      lmiterm([-(Xcon_LMI1+nGW_r+i-1)  1 1 U], -0.5*GW(i,:),1,'s')
      lmiterm([-(Xcon_LMI1+nGW_r+i-1)  1 1 0], gg2(i));
   end
end

% Constraint on x after the horizon N
if (~isempty(G)),
   X_con_LMI2 = newlmi;
   lmiterm([X_con_LMI2  1 1 X], G, G');
   lmiterm([X_con_LMI2  1 1 VV], -1, 1); %% (1,1) G*X*G' <= VV
   for i=1:nglim_r,
      tmp = zeros(nglim_r,1);
      tmp(i,1) = 1;
      lmiterm([(X_con_LMI2+i)  1 1 VV], tmp',tmp);
      %% (1,1) VV(i,i) <= g_lim(i)^2
      lmiterm([(X_con_LMI2+i) 1 1 0], -g_lim(i)^2);
   end
end

rhclmi = getlmis;

% Now we're ready to solve LMI
n_lmi = decnbr(rhclmi); c = zeros(n_lmi,1); for j=1:n_lmi,
   [r1j, r2j]= defcx(rhclmi,j,r1,r2);
   c(j) = r1j + r2j;
end

[copt, Uopt]= mincx(rhclmi, c', [1e-6 300 0 0 0]);

u_opt = dec2mat(rhclmi, Uopt, U); r1_opt = dec2mat(rhclmi, Uopt,
r1); r2_opt = dec2mat(rhclmi, Uopt, r2); X_opt = dec2mat(rhclmi,
Uopt, X); Y_opt = dec2mat(rhclmi, Uopt, Y);
```

References

[AHK04]   C. K. Ahn, S. Han, and W. H. Kwon. Minimax finite memory controls for discrete-time state-space systems. *Seoul Nat'l Univ., School of EE & CS Tech. Report No. SNU-EE-TR-2004-3*, 2004-3, Nov. 2004.

[AM80]   B. D. O. Anderson and J. B. Moore. Coping with singular transition matrices in estimation and control stability theory. *Int. J. Contr.*, 31:571–586, 1980.

[AM89]   D. O. Anderson and J. B. Moore. *Optimal Control: Linear Quadratic Methods*. Prentice-Hall International Editions, Englewood Cliffs, NJ 07632, 1989.

[Ath71]   M. A. Athans. Special issue on the LQG problem. *IEEE Trans. Automat. Contr.*, 16(6):527–869, 1971.

[ÅW73]   K. J. Åström and B. Wittenmark. On self-tuning regulators. *Automatica*, 9(2):185–199, 1973.

[AZ99]   F. Allgower and A. Zheng. Nonlinear predictive control: assessment and future directions. In *Proc. of International Workshop on Model Predictive Control*, Ascona 1998, Springer, Berlin, 1999.

[AZ00]   Frank Allgower and Alex Zhen. *Nonlinear Model Predictive Control (Progress in Systems and Control Theory*. Birkhauser,Boston, 2000.

[BB91]   T. Basar and P. Bernhard. *H_∞ – Optimal Control and Related Minimax Design Problems: A Dynamic Game Approach*. Birkhauser, Boston, 1991.

[BD62]   R. E. Bellman and S. E. Dreyfus. *Applied Dynamic Programming*. Princeton University Press, Princeton, NJ, 1962.

[Bel57]   R. E. Bellman. *Dynamic Programming*. Princeton University Press, Princeton, NJ, 1957.

[BGFB94]   S. Boyd, L. E. Ghaoui, E. Feron, and V. Balakrishnan. *Linear Matrix Inequalities in System and Control Theory*, volume 15. SIAM, Philadelphia, PA, 1994.

[BGP85]   R. R. Bitmead, M. Gevers, and I. R. Petersen. Monotonicity and stabilizability properties of solutions of the Riccati difference equation: propositions, lemmas, theorems, fallacious conjecture and counterexamples. *Systems and Control Letters*, 5:309–315, 1985.

[BGW90]   R. R. Bitmead, M. Gevers, and V. Wertz. *Adaptive Optimal Control: The Thinking Man's GPC*. Prentice-Hall, 1990.

[BH75]   A. D. Bryson and Y. C. Ho. *Applied Optimal Control*. John Wiley & Sons, Inc., Hemisphere, New York, 1975.

[BH89]   D. S. Berstein and W. Haddad. Steady-state kalman filtering with h_∞ error bound. *Systems and Control Letters*, 12:9–16, 1989.

[Bie75]   G. J. Bierman. Fixed memory least squares filtering. *IEEE Trans. Automat. Contr.*, 20:690–692, 1975.

[BK65]   R. E. Bellman and R. E. Kalaba. *Dynamic Programming and Modern Control Theory*. Academic Press, New York, 1965.

[BK85]   A. M. Bruckstein and T. Kailath. Recursive limited memory filtering and scattering theory. *IEEE Trans. Inform. Theory*, 31:440–443, 1985.

[Bla91]   F. Blanchini. Constrained control for uncertain linear systems. *Journal of Optimizaton Theory and Applications*, pages 465–484, 1991.

[BLEGB94]   S. Boyd, E. Feron L. E. Ghaoui, and V. Balakrishnan. Linear matrix inequalities in system and control theory. *System and Control Theory*, 1994.

[BLW91]   Sergio Bittanti, Alan J. Laub, and Jan C. Willems. *The Riccati Equation*. Springer-Verlag, 1991.

[BO82]   T. B. and G. J. Olsde. *Dynamic Noncooperative Game Theory*. Academic Press, London, New York, 1982.

[Bur98]   Jeffrey B. Burl. *Linear Optimal Control*. Addison-Wesley, California, 1998.

[Bux74]   P. J. Buxbaum. Fixed-memory recursive filters. *IEEE Trans. Inform. Theory*, 20:113–115, 1974.

[CA98]   H. Chen and F. Allgower. Nonlinear model predictive control schemes with guaranteed stabilty. In *NATO ASI on Nonlinear Model Based Process Control*, 465–494, 1998.

[CB04]   E. F. Camacho and C. Bordons. *Model Predictive Control (Advanced Textbooks in Control and Signal Processing)*. Springer Verlag, 2004.

[CG79]   D. W. Clarke and P. J. Gawthrop. Self-tuning control. *IEE Proc. Pt. D*, 126(6):633–640, 1979.

[Cla94]   D. W. Clarke. Advances in model-based predictive control. *Advances in Model-Based Predictive Control*, pages 2–21, 1994.

[CMT87]   D. W. Clarke, C. Mohtadi, and P. S. Tuffs. Generalized predictive control– part I. the basic algorithm. *Automatica*, 23(2):137–148, 1987.

[CR80]   C. R. Cutler and B. L. Ramaker. Dynamic matrix control: a computer control algorithm. In *Joint Automatic Control Conf.*, San Francisco, U.S.A, 1980.

[DC93]   H. Demircioglu and D. W. Clarke. Generalised predictive control with end point weighting. *IEE Proc. Pt. D*, 140(4):275–282, 1993.

[DX94]   Liu Danyang and Liu Xuanhuang. Optimal state estimation without the requirement of a priori statistics information of the initial state. *IEEE Trans. Automat. Contr.*, 39(10):2087–2091, 1994.

[EWMR94]   R. W. Eustace, B. A. Woodyatt, G. L. Merrington, and A. Runacres. Fault signatures obtained from fault implant tests on an F404 engine. *ASME Trans. J. of Engine, Gas Turbines, and Power*, 116(1):178–183, 1994.

[GL95]   M. Green and D. J. N. Limebeer. *Linear Robust Control*. Prentice-Hall, Englewood Cliffs, New Jersey, 1995.

[GNLC95]   P. Gahinet, A. Nemirovski, A. J. Laub, and M. Chilali. *LMI Control Toolbox*. The MATHWORKS Inc., USA, 1995.

[GPM89]   C. E. García, D. M. Prett, and M. Morari. Model predictive control: Theory and practice–a survey. *Automatica*, 25(3):335–348, 1989.

[Gri02]   M. J. Grimble. Workshop on nonlinear predictive control. Oxford University, 18–19 July, 2002.

[GT91]   Elmer G. Gilbert and Kok Tin Tan.   Linear systems with state and control constraints: the theory and application of maximal output admissible sets . *IEEE Trans. on Auto. Control*, 36(9):1008–1020, 1991.

[Gyu02]   E. Gyurkovics. Receding horizon h_∞ control for nonlinear discrete-time systems. *IEE Proc.-Control Theory Appl.*, 149(6):540–546, 2002.

[HK04]   S. Han and W. H. Kwon. Receding horizon finite memory controls for output feedback controls of disrete-time state space models. In *Proc. of the 30th Annual Conference of the IEEE Industrial Electronics Society*, Busan, Korea, 2004.

[HKH02]   Sunan Huang, Tan Kok Kiong, and Lee Tong Heng. *Applied Predictive Control*. Springer Verlag, 2002.

[HKK99]   S. Han, P. S. Kim, and W. H. Kwon. Receding horizon FIR filter with estimated horizon initial state and its application to aircraft engine systems . In *Proc. IEEE Int. Conf. Contr. Applications*, pages 33–38, Hawaii, USA, 1999.

[HKK01]   S. Han, W. H. Kwon, and Pyung Soo Kim. Receding horizon unbiased FIR filters for continuous-time state space models without a priori initial state information. *IEEE Trans. Automat. Contr.*, 46(5):766–770, 2001.

[HKK02]   S. Han, W. H. Kwon, and Pyung Soo Kim. Quasi-deadbeat minimax filters for deterministic state-space models. *IEEE Trans. Automat. Contr.*, 47(11):1904–1908, 2002.

[HSK99]   Babak Hassibi, Ali H. Sayed, and Thomas Kailath. *Indefinite Quadratic Estimation and Control*. SIAM studies in applied and numerical mathematics, 1999.

[Jaz68]   A. H. Jazwinski. Limited memory optimal filtering. *IEEE Trans. Automat. Contr.*, 13:558–563, 1968.

[JHK04]   J. W. Jun, S. Han, and W. H. Kwon. Fir filters and dual iir filters. *Seoul Nat'l Univ., School of EE & CS Tech. Report No. SNU-EE-TR-2004-6*, 2004-6, Nov. 2004.

[Kal63]   R. E. Kalman. Lyapunov functions for the problem of lur'e in automatic control. *Proc. Nat. Acad. Sci.*, 49:201–205, 1963.

[KB60]   R. E. Kalman and R. S. Bucy. A new approach to linear filtering and prediction problems. *Trans. ASME J. of Basic Engr.*, 82:35–45, 1960.

[KB61]   R. E. Kalman and R. S. Bucy. New results in linear filtering and prediction theory. *Trans. ASME J. of Basic Engr.*, 83:95–108, 1961.

[KB89]   W. H. Kwon and D. G. Byun. Receding horizon tracking control as a predictive control and its stability properties. *Int. J. Contr.*, 50:1807–1824, 1989.

[KBK83]   W. H. Kwon, A. M. Bruckstein, and T. Kailath.   Stabilizing state-feedback design via the moving horizon method. *Int. J. Contr.*, 37:631–643, 1983.

[KBM96]   M. V. Kothare, V. Balakrishnan, and M. Morari. Robust constrained model predictive control using linear matrix inequalities. *Automatica*, 32:1361–1379, 1996.

[KC00]   Basil Kourvaritakis and Mark Cannon. *Nonlinear Predictive Control: Theory & Practice*. IEE Publishing, 2000.

[KG88]   S. S. Keerthi and E. G. Gilbert. Optimal infinite-horizon feedback laws for a general class of constrained discrete-time systems: stability and moving-horizon approximation. *J. of Optimization Theory and Applications*, 57:265–293, 1988.

[KG98]   I. Kolmanovsky and E. C. Gilbert. Theory and computation of disturbance invariant sets for discrete-time linear systems. *Math. Prob. Eng.*, pages 117–123, 1998.

[KGC97]   J. C. Kantor, C. E. Garcia, and B. Carnahan. In *In Fifth International Conference on Chemical Process Control, CACHE, A.I.Ch.E.*, 1997.

[KH04]   W. H. Kwon and S. Han. Receding horizon finite memory controls for output feedback controls of state space models. *IEEE Trans. Automat. Contr.*, 49(11):1905–1915, 2004.

[KHA04]   W. H. Kwon, S. Han, and C. K. Ahn. Advances in nonlinear predictive control: A survey on stability and optimality. *International Journal of Control, Automations, and Systems*, 2(1):15–22, 2004.

[KIF93]   Hans W. Knobloch, Alberto Isisori, and Dietrich Flockerzi. *Topics in Control Theory*. Birkhauser Verlag, 1993.

[Kir70]   D. E. Kirk. *Optimal Control Theory*. Prentice-Hall, Englewood Cliffs, New Jersey, 1970.

[KK00]   W. H. Kwon and K. B. Kim. On stabilizing receding horizon controls for linear continuous time-invariant systems. *IEEE Trans. Automat. Contr.*, 45(7):1329–1334, 2000.

[KKH02]   W. H. Kwon, Pyung Soo Kim, and S. Han. A receding horizon unbiased FIR filter for discrete-time state space models. *Automatica*, 38(3):545–551, 2002.

[KKP99]   W. H. Kwon, P. S. Kim, and P. Park. A receding horizon Kalman FIR filter for discrete time-invariant systems. *IEEE Trans. Automat. Contr.*, 44(9):1787–1791, 1999.

[Kle70]   D. L. Kleinman. An easy way to stabilize a linear constant system. *IEEE Trans. Automat. Contr.*, 15:692, 1970.

[Kle74]   D. L. Kleinman. Stabilizing a discrete, constant, linear system with application to iterative methods for solving the Riccati equation. *IEEE Trans. Automat. Contr.*, 19:252–254, 1974.

[KP77a]   W. H. Kwon and A. E. Pearson. A modified quadratic cost problem and feedback stabilization of a linear system. *IEEE Trans. Automat. Contr.*, 22(5):838–842, 1977.

[KP77b]   W. H. Kwon and A. E. Pearson. A note on the algebraic matrix Riccati equation. *IEEE Trans. on Automat. Contr.*, 22:143–144, 1977.

[KP77c]   W. H. Kwon and A. E. Pearson. A modified quadratic cost problem and feedback stabilization of a linear discrete time system. *Brown Univ., Div. Eng. Tech. Rep.*, AFOSR-75-2793C/1, Sept. 1977.

[KP78]   W. H. Kwon and A. E. Pearson. On feedback stabilization of time-varying discrete linear systems. *IEEE Trans. Automat. Contr.*, 23(3):479–481, 1978.

[KRC92]   B. Kouvaritakis, J. A. Rossiter, and A. O. T. Chang. Stable generalised predictive control: an algorithm with guaranteed stability. *IEE Proc. Pt. D*, 139(4):349–362, 1992.

[KS72]   H. Kwakernaak and R. Sivan. *Linear Optimal Control Systems*. Wiley Interscience, 1972.

[KSH00]   Thomas Kailath, Ali H. Sayed, and Babak Hassibi. *Linear Estimation.*
           Prentice Hall, 2000.

[Kuc91]   V. Kucera. *Analysis and Design of Discrete Linear Control Systems.*
           Prentice Hall, 1991.

[Kwo94]   W. H. Kwon. Advances in predictive control: theory and application.
           In Asian Control Conference, Tokyo, 1994.

[KYK01]   K. B. Kim, Tae-Woong Yoon, and W. H. Kwon. On stabilizing receding
           horizon H_∞ controls for linear continuous time-varying systems. *IEEE
           Trans. Automat. Contr.,* 46(8):1273–1279, 2001.

[LAKG92]   D. J. J. Limbeer, B. D. O. Anderson, P. P. Khargonekar, and M. Green.
           A game theoretic approach to H_∞ control for time varing systems. *SIAM
           Journal of Control and Optimization,* 30(2):262–283, 1992.

[LC97]   J. H. Lee and B. Cooley. Recent advances in model predictive control
           and other related areas. *In Fifth International Conference on Chemical
           Process Control, CACHE, AIChE,* pages 201–216, 1997.

[Lew86a]   F. L. Lewis. *Optimal Control.* John Wiley & Sons, Inc., 1986.

[Lew86b]   F. L. Lewis. *Optimal estimation.* John Wiley & Sons, Inc., 1986.

[LG94]   S. Lall and K. Glover. A game theoretic approach to moving horizon
           control. *Advances in Model-Based Predictive Control,* pages 131–144,
           1994.

[LHK04]   Y. S. Lee, S. Han, and W. H. Kwon. Nonlinear receding horizon controls
           with quadratic cost functions. *Seoul Nat'l Univ., School of EE & CS
           Tech. Report No. SNU-EE-TR-2004-7,* 2004-7, Nov. 2004.

[LK99]   Y. I. Lee and B. Kouvaritakis. Constrained receding horizon predictive
           control for systems with disturbances. *Int. J. Control,* 72(11):1027–1032,
           1999.

[LK01]   Y. I. Lee and B. Kouvaritakis. Receding horizon output feedback control
           for linear systems with input saturation. *IEE Proc.-Control Theory
           Appl.,* 148(2):109–115, 2001.

[LKC98]   J. W. Lee, W. H. Kwon, and J. H. Choi. On stability of constrained
           receding horizon control with finite terminal weighting matrix. *Auto-
           matica,* 34(12):1607–1612, 1998.

[LKL99]   J. W. Lee, W. H. Kwon, and J. H. Lee. Receding horizon H_∞ tracking
           control for time-varying discrete linear systems. *Int. J. Contr.,* 68:385–
           399, 1999.

[LL99]   K. V. Ling and K. W. Lim. Receding horizon recursive state estimation.
           IEEE Trans. Automat. Contr., 44(9):1750–1753, 1999.

[Mac02]   J. M. Maciejowski. *Predictive Control with Constraints.* Prentice Hall,
           2002.

[May95]   D. Q. Mayne. Optimization in model based control. In *Proc. of the IFAC
           Symposium on Dynamics and Control Chemical Reactors and Batch
           Processes,* 229–242, 1995.

[May97]   D. Q. Mayne. Nonlinear model predictive control: an assessment. *In
           Fifth International Conference on Chemical Process Control, CACHE,
           A.I.Ch.E,* pages 217–231, 1997.

[MG71]   J. M. Mendel and D. L. Gieseking. Bibliography on the linear quadratic
           Gaussian problem. *IEEE Trans. Automat. Contr.,* 16(6):847–869, 1971.

[MJ98]   G. Martin and D. Johnston. Continuous model-based optimization. In
           Proc. of Process Optimization Conference, Texas, 1998.

[ML99]   M. Morari and J. H. Lee. Model predictive control: past, present, and future. *Computers and Chemical Engineering*, 23:667–682, 1999.

[MM90]   D. Q. Mayne and H. Michalska. Receding horizon control of nonlinar systems. *IEEE Trans. Automat. Contr.*, 35(7):814–824, 1990.

[MM93]   H. Michalska and D. Q. Mayne. Robust receding horizon control of constrained nonlinear systems. *IEEE Trans. Automat. Contr.*, 38(11):1623–1633, 1993.

[Mos95]   Edoardo Mosca. *Optimal, Predictive and Adaptive Control*. Prentice Hall, 1995.

[MRRS00] D. Q. Mayne, J. B. Rawlings, C. V. Rao, and P. O. M. Scokaert. Constrained model predictive control: stability and optimality. *Automatica*, 36:789–814, 2000.

[MS97]   L. Magni and R. Sepulchre. Stability magins of nonlinear receding horizon control via inverse optimality. *System & Control Letters*, pages 241–245, 1997.

[MSR96]   Juan Martin-Sanchez and Jose Rodellar. *Adaptive Predictive Control: Industrial Plant Optimization*. Prentice Hall, 1996.

[NMS96a] G. D. Nicolao, L. Magni, and R. Scattolini. On the robustness of receding-horizon control with terminal constraints. *IEEE Trans. Automat. Contr.*, 41(3):451–453, 1996.

[NMS96b] G. De Nicolao, L. Magni, and R. Scattolini. Stabilizing nonlinear receding horizon control via a nonquadratic penalty. In *Proc. of the IMACS Multiconference CESA*, Lille, France, 1996.

[NP97]   V. Nevistic and J. Primbs. Finite receding horizon linear quadratic control: a unifying theory for stability and performance analysis. Technical Report CIT-CDS 97-001, Automatic Control Lab., Swiss Fedral Institute of Technology, 1997.

[NS97]   G. D. Nicolao and S. Strada. On the stability of receding-horizon LQ control with zero-state terminal constraint. *IEEE Trans. Automat. Contr.*, 22(2):257–260, 1997.

[OOH95]   M. Ohshima, H. Ohno, and I. Hashimoto. Model predictive control: experience in the university industry joint projects and statistics on mpc applications in japan. In *Inernational Workshop on Predictive and Receding Horizon Control*, Seoul, Korea, 1995.

[PBG88]   M. A. Poubelle, R. R. Bitmead, and M. R. Gevers. Fake algebraic Riccati techniques and stability. *IEEE Trans. Automat. Contr.*, 33:479–481, 1988.

[PBGM62] L. S. Pontryagin, V. G. Boltyanskii, R. V. Gamkrelidze, and E. F. Mischenko. *The Mathematical Theory of Optimal Processes*. Interscience Publishers, New York, 1962.

[PND00]   James A. Primbs, Vesna Nevistic, and John C. Doyle. A receding horizon generalization of pointwise min-norm controllers. *IEEE Trans. Automat. Contr.*, 45(5):898–909, 2000.

[Pop64]   V. M. Popov. Hyperstability and optimality of automatic systems with several control functions. *Rev. Roum. Sci. Tech. Ser Electrotech. Energ.*, 9:629–690, 1964.

[QB97]   S. J. Qin and T. A. Badgwell. An overview of industrial model predictive control technology. In *In Fifth Int. Conf. on Chemical Process Control*, volume 316, 232–256, 1997.

[QB00]   S. J. Qin and T. A. Badgwell. An overview of nonlinear model predictive control applications. In *Nonlinear Model Predictive Control, volume 26 of Progress in Systems and Control Theory*, Birkhauser, 2000.

[QB03]   S. Joe Qin and Thomas A. Badgwell. A survey of industrial model predictive control technology. *Control Engineering Practice*, 11(7):733–764, 2003.

[RM93]   J. B. Rawlings and K. R. Muske. The stability of constrained receding horizon control. *IEEE Trans. Automat. Contr.*, 38:1512–1516, 1993.

[RMM94]   J. B. Rawlings, E. S. Meadows, and K. R. Muske. Nonlinear model: a tutorial and survey. In *ADCHEM 94 Proceedings, Kyoto, Japan*, 185–197, 1994.

[Ros03]   J. A. Rossiter. *Model-Based Predictive Control: A Practical Approach.* CRC Press, 2003.

[RRTP78]   J. Richalet, A. Rault, J.L. Testud, and J. Papon. Model predictive heuristic control: applications to industrial processes. *Automatica*, 14:413–428, 1978.

[SC94]   P. O. M. Scokaert and D. W. Clarke. Stability and feasibility in constrained predictive control. In D. W. Clarke, editor, *Advances in model-based predictive control*, Oxford, England., 1994. Oxford University Press.

[Sch73]   F. C. Schweppe. *Uncertain Dynamic Systems.* Prentice-Hall, NJ, Englewood Cliffs, 1973.

[Soe92]   R. Soeterboek. *Predictive Control: A Unified Approach.* Prentice-Hall, 1992.

[SR98]   P. O. M. Scokaert and J. B. Rawlings. Constrained linear quadratic regulation. *IEEE Trans. Automat. Contr.*, 43(8):1163–1169, 1998.

[Str68]   Aaron Strauss. *An Introduction to Optimal Control Theory.* Springer-Verlag, 1968.

[VB96]   L. Vandenberghe and S. Boyd. Semidefinite programming. *SIAM Review*, 38:49–95, 1996.

[Won68]   W. M. Wonham. On the separation theorem of stochastic control. *SIAM Journal of Control and Optimization*, 6(0):312–326, 1968.

[Yak62]   V. A. Yakubovic. The solution of certain matrix inequalities in automatic control theory. *Dokl. Akad. Nauk*, 143:1304–1307, 1962.

[Yaz84]   E. Yaz. A suboptimal terminal controller for linear discrete-time systems. *Int. J. Contr.*, 40:271–289, 1984.

[ZDG96]   Kemin Zhou, John C. Doyle, and Keith Glover. *Robust and Optimal Control.* Prentice Hall, 1996.

[ZM95]   A. Zheng and M. Morari. Stability of model predictive control with mixed constraints. *IEEE Trans. Automat. Contr.*, 40(10):1818–1823, 1995.

Index

Printing: Mercedes-Druck, Berlin
Binding: Stein+Lehmann, Berlin